Handbook of Analytical Chemistry

Volume I

Handbook of Analytical Chemistry
Volume I

Edited by **Jessica Carol**

NY RESEARCH PRESS
New York

Published by NY Research Press,
23 West, 55th Street, Suite 816,
New York, NY 10019, USA
www.nyresearchpress.com

Handbook of Analytical Chemistry: Volume I
Edited by Jessica Carol

International Standard Book Number: 978-1-63238-220-7 (Hardback)

Printed in China.

Contents

Preface

Analytical chemistry is the branch of chemistry that studies about the quantification, separation and identification of the different chemical components of natural and artificial materials. This is a field that is also focused on improvements in chemo-metrics, experimental design as well as the conception of new measurement tools and methodologies to provide better chemical information. Analytical chemistry has applications in bio-analysis, materials analysis, clinical analysis, environmental analysis and forensics. This scientific discipline is highly interdisciplinary and spans almost all areas of chemistry. It also involves the development of methods and tools to compute physical properties of substances and apply such techniques to the identification and classification of their presence, called qualitative analysis as well as quantifying the amount present of various species in a wide variety of settings, called quantitative analysis. Qualitative analysis gives an indication of the identity of the chemical species in the sample, and quantitative analysis determines the amount of certain components in the substance. This is a field of study that has various applications in multiple industries and there are a variety of research projects that are looking into different aspects of this discipline. This also increases the demand for skilled researchers and scientists looking to work in this field.

This book is an attempt to compile and collate all available research on analytical chemistry under one umbrella. I am grateful to those who put their hard work, effort and expertise into these researches as well as those who were supportive in this endeavour. I would also like to thank my family for their continuous support.

Editor

Two Validated Spectrofluorometric Methods for Determination of Gemifloxacin Mesylate in Tablets and Human Plasma

Noha N. Atia,[1] Ashraf M. Mahmoud,[1,2] Salwa R. El-Shabouri,[1] and Wesam M. El-Koussi[1]

[1] *Department of Pharmaceutical Analytical Chemistry, Faculty of Pharmacy, Assiut University, Assiut 71526, Egypt*
[2] *Department of Pharmaceutical Chemistry, Faculty of Pharmacy, Najran University, P.O. Box 1988, Najran 61441, Saudi Arabia*

Correspondence should be addressed to Wesam M. El-Koussi; wesamelkosi@yahoo.com

Academic Editor: Hian Kee Lee

Two new, sensitive, and selective spectrofluorometric methods were developed for the determination of gemifloxacin mesylate (GFX) in tablets and spiked human plasma. Method A was based on measurement of the enhanced fluorescence spectral behaviour of GFX in a sodium dodecyl sulphate (SDS) micellar system. In aqueous solution of acetate buffer pH 5.5, the fluorescence intensity of GFX was greatly enhanced about tenfold in the presence of SDS. The fluorescence intensity was measured at 402 nm after excitation at 274 nm. Method B was based on Hantzsch condensation reaction between the primary amino group of GFX with acetylacetone and formaldehyde in acetate buffer of pH 3.5 yielding a highly yellow fluorescent derivative. The reaction of GFX with acetylacetone-formaldehyde system solution resulted in bathochromic shift of both emission (476 nm) and excitation (420 nm) wavelengths. The fluorescence intensity was directly proportional to the concentration over the range 10–1000 ng/ml and 100–2000 ng/ml for method A and B, respectively. The proposed methods were applied successfully for determination of GFX in its tablets and spiked plasma. Therefore, these methods can be considered of real interest for reliable and practical quality control analysis of GFX.

1. Introduction

Gemifloxacin mesylate (GFX) is a fluoronaphthyridone with a novel oxime functionalised pyrrolidine (Figure 1) [1]. It possesses a dual mechanism of action by inhibiting the bacterial topoisomerase IV and gyrase enzymes, resulting in interruption of bacterial DNA synthesis. Therefore, it has broad-spectrum activity against Gram-positive and Gram-negative organisms comparable to those of its quinolone counterparts which have the same pyrrolidine side chain [1, 2]. GFX was first approved by the FDA for clinical use in 2003 for the treatment of community acquired pneumonia and acute bacterial exacerbation of chronic bronchitis [3].

Several analytical methods were reported for determination of gemifloxacin in pharmaceutical preparations or human plasma by visible spectrophotometry [4, 5], capillary electrophoresis [6], HPLC-MS-MS [7], HPLC [8], and HPTLC [9]. However, these methods showed some drawbacks such as being time-consuming, tedious, or requiring expensive instruments that limit their use in quality control and routine clinical studies in developing countries where the cost is a main concern.

Spectrofluorimetric analysis constitutes a widespread, effective technique to improve analysis selectivity and sensitivity. Although, limited literatures have been reported for the determination of gemifloxacin spectrofluorometrically [10, 11]. Micelle-enhanced spectrofluorimetric method has been reported for determination of many drugs [12–17] due to the ability of micelle formation to increase the fluorescence intensity of the weakly fluorescent compounds. Moreover, these methods introduced sensitive and nonpollutant methodology, since no organic solvents were used. Hantzsch reaction is a well-known condensation reaction that was reported as a useful pathway for determination of many drugs

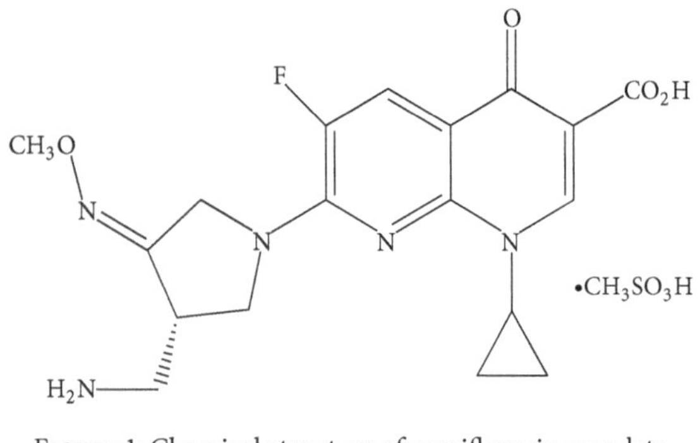

FIGURE 1: Chemical structure of gemifloxacin mesylate.

fluorometrically in biological fluids, air, or pharmaceutical preparations in trace amounts [18–22].

Therefore, the aim of this study was directed for using micelle-enhanced and Hantzsch reaction spectrofluorometric methods for developing new simple, selective, and sensitive methods for determination of GFX in its tablets and human plasma.

2. Experimental

2.1. Instrumentation. Spectrophotometric measurements were carried out using an LS 45 luminescence spectrometer (Perkin-Elmer, UK), equipped with a 150 W Xenon lamp. Slit widths for both monochromators set at 10 nm. Data acquisition was performed by FL WinLab software, version 4.00.03 (Perkin-Elmer, UK). A 1 cm quartz cell was used.

A solid phase extraction (SPE) vacuum manifold with 24-position configurations (Phenomenex, USA) was used for extraction of plasma samples. Vacuum was adjusted to 5 inches for proper flow through the SPE columns. The SPE cartridge used in this study was Strata C18-U (1 cm^3/100 mg; Phenomenex, USA).

2.2. Materials and Reagents. Gemifloxacin mesylate (99.8%, Tabuk pharmaceutical manufacturing Co., KSA) was used as received. The commercial formulation "Factive" (Oscient Pharmaceuticals, USA) was purchased from our local market, which was labeled to contain 320 mg of GFX per tablet. Simulated tablets were prepared in our laboratory according to similar dosage form in India "G-CIN-A" [23, 24] which are labeled to contain 320 mg of GFX and 75 mg of Ambroxol HCl per tablet. We used sodium dodecyl sulphate (SDS) (El-Nasr chemical Co., Egypt), cetrimide (Danochemo a subsidiary of Ferrosan, manufacturing chemists, Copenhagen, Denmark), beta-cyclodextrin (β-CD) (Winlab Laboratory chemicals reagents fine chemicals, UK). Acetylacetone was obtained from Tedia CO., USA. 34–38% Formaldehyde solution and other solvents and materials used throughout this study were of analytical grade. Double distilled water was obtained through WSC-4D water purification system (Hamilton Laboratory Milton Glass Ltd., Kent, USA) and used throughout the work. 0.1 M acetate buffer solution of pH 3.5 and 50 mM phosphate buffer pH 3 [25] were prepared and adjusted using 3505 pH-meter (Jenway, UK). Blank human plasma samples used herein were supplied from Assiut University Hospitals (Assiut, Egypt), and they were stored in deepfreezer at −80°C until analysis.

2.2.1. Reagent Preparation

SDS Solution. 0.1 M SDS solution was prepared by dissolving 0.29 g of SDS in sufficient distilled water and diluted up to 10 mL.

Acetylacetone-Formaldehyde Solution. Into 5 mL calibrated flask, the reagent was freshly prepared by mixing 0.5 mL of 0.1 M acetate buffer pH 3.5, 0.5 mL acetylacetone solution, and 1 mL formaldehyde solution and completed to the mark with double distilled water. The flask was protected from light with aluminum foil.

2.3. Standards and Stock Solutions

2.3.1. Gemifloxacin Mesylate Standard Solutions. An accurately weighed amount (25 mg) of GFX was transferred into 250 mL calibrated flask, dissolved in appropriate volume of double distilled water. Then, the void volume was completed with water to produce a stock solution of 100 μg/mL. The stock solution was further diluted with water to obtain working standard solution of 5 μg/mL.

2.3.2. Tablets Sample Solution. Twenty tablets were weighed and finely powdered. An accurately weighed quantity of the powder equivalent to 25 mg of GFX alone or with Ambroxol in combined dosage form was transferred into a 100 mL calibrated flask and dissolved in about 40 mL of distilled water. The contents of the flask were swirled, sonicated for 5 min, and then completed to volume with water. The contents were mixed well, filtered, and rejecting the first portion of the filtrate. The prepared solutions were diluted quantitatively with water to obtain stock solution of 5 μg/mL as a suitable concentration for the analysis.

2.3.3. Plasma Sample Processing

Sample Preparation. Plasma aliquot (0.25 mL) was transferred into a 2 mL Eppendorf tubes. Subsequently, 100 μL of GFX working solutions (2, 12.5, 25, 50, and 100 μg/mL) was added in each tube. After gentle mixing, the void volume was completed to 1.5 mL with 50 mM Na_2HPO_4 (pH 3). A blank plasma sample was treated similarly.

Solid-Phase Extraction. A polymeric sorbent (Strata C18-U) was used to prepare the samples. Before extraction, the cartridges were prewashed with 2 × 1 mL of methanol, followed by 2 × 1 mL of distilled water. After application of the samples, the cartridges were washed with 2 × 1 ml portions of distilled water and 0.5 mL of 50 mM Na_2HPO_4 (pH 3). Finally, GFX was eluted with 1 mL of methanol and 50 mM Na_2HPO_4 (pH 3, 90 : 10 v/v). Then 0.5 mL was taken from each eluent and the general procedure was followed to obtain

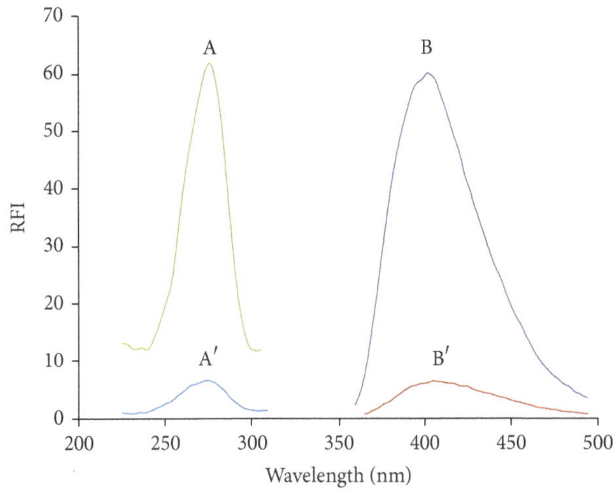

FIGURE 2: Fluorescence spectra of (A, B) GFX (500 ng/mL) in acetate buffer, pH 5.5/SDS system; (A′, B′) GFX (500 ng/mL) in acetate buffer, pH 3.5, where (A, A′) are the excitation spectra and (B, B′) are the emission spectra.

final concentrations of 20, 125 and 500 ng/mL for method A or 250, 500 and 1000 ng/mL for method B.

2.4. General Procedures

2.4.1. Micelle-Enhanced Spectrofluorometric Method (Method A). Aliquot of 10–1000 μL of GFX standard solution (5 μg/mL) was transferred into a series of 5 mL volumetric flasks to give final concentrations of 10–1000 ng/mL. 1 mL 0.2 M acetate buffer solution, pH 5.5, was added to each flask, followed by 100 μL of 0.1 M SDS solution. The volume was completed with distilled water, the contents of the flasks were mixed well, and the fluorescence intensity was measured at 402 nm after excitation at 274 nm, against a blank solution treated similarly.

2.4.2. Hantzsch Reaction Method (Method B). Aliquot of 0.1–2.0 mL of GFX standard solution (5 μg/mL) was mixed with 1 mL of acetylacetone-formaldehyde solution in a glass-stoppered tube that was protected from light with aluminum foil. The mixture was heated at 100°C for 20 min in a thermostatic water bath and after that it was cooled in an ice bath. The volume was adjusted to 5 mL with 2-propanol to provide a final concentration ranging from 100–2000 ng/mL. The fluorescence intensity was measured at 476 nm after excitation at 420 nm against a blank prepared similarly.

3. Results and Discussion

3.1. Spectral Characteristics. For method A, the fluorescence spectra of gemifloxacin in both aqueous and SDS systems were studied (Figure 2). GFX showed native fluorescence in aqueous solution measured at 406 nm after an excitation at 274 nm. In the presence of SDS, the fluorescence intensity of GFX was enhanced nearly tenfold in comparison with its native fluorescence intensity in aqueous medium. Moreover,

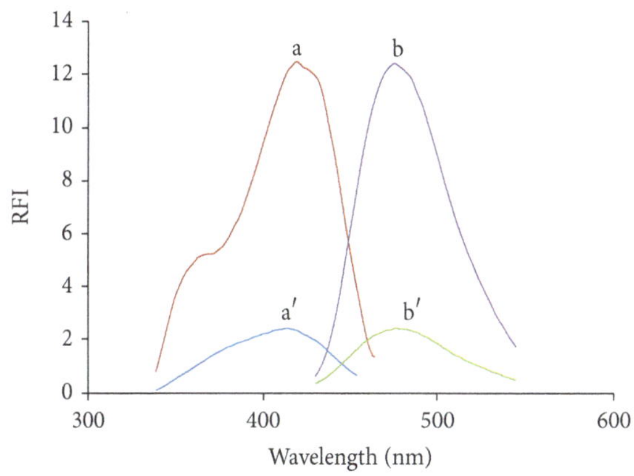

FIGURE 3: Fluorescence spectra of (a, b) gemifloxacin (1000 ng/mL) with acetylacetone-formaldehyde reagent solution, (a′, b′) acetylacetone-formaldehyde reagent, where (a, a′) are the excitation spectra and (b, b′) are the emission spectra. λex/λem = 420/476 nm.

the enhancement was associated with a slight blue shift (λem at 402 nm). It reflects that the microenvironment around GFX is quite different from that in aqueous solution. This can be attributed to restrictions imposed on the free rotational motions which are competitive with luminescent emission [26].

For method B, GFX reacts with acetylacetone and formaldehyde in an acidic-buffered medium yielding a highly yellow fluorescent product. The fluorescence intensity of the product was measured at 476 nm after excitation at 420 nm (Figure 3). The notable advantage of this reaction is the enhanced red shift in the excitation and emission wavelengths that improves the selectivity of GFX.

3.2. Optimization of Experimental Conditions. All different experimental factors influencing the development of the fluorescent product were carefully studied and optimized. Such factors were changed individually while others were kept constant. These factors were pH, volume of the reagent, temperature and diluting solvent.

3.2.1. Micelle-Enhanced Spectrofluorometric Method. The fluorescence properties of gemifloxacin in various micellar media were studied using anionic (SDS), cationic (cetrimide) and nonionic (β-CD) surfactants. It was observed that the fluorescence intensity of GFX showed decrease or no significant effect by using non-ionic or cationic surfactants. On the other hand, there was an obvious enhancement of the fluorescence intensity of GFX in the presence of SDS about tenfold in comparing with its aqueous solution (Figure 4(a)). Therefore, the influence of SDS on the RFI was studied using increasing volumes of 0.1 M SDS. It was found that increasing volumes of SDS solution resulted in a corresponding increase in RFI up to 100 μL, after which gradual decrease in RFI was attained. Therefore, 100 μL of 0.1 M SDS solution was selected as the optimum volume (Figure 4(b)). It was noted that, the SDS aggregation equilibrium in presence of GFX in our optimal experimental conditions showed a quite different

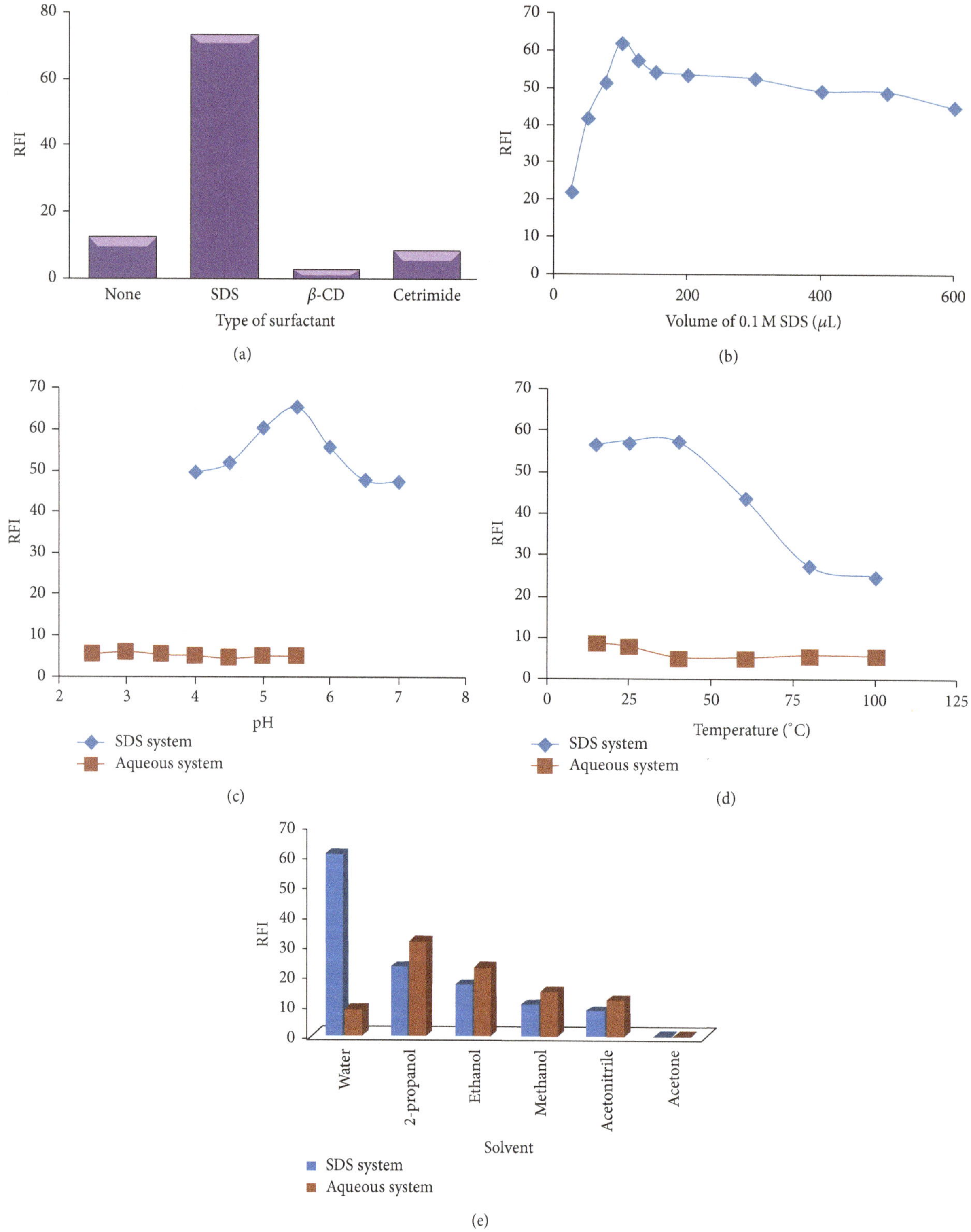

FIGURE 4: Effect of the different optimization factors. (a) Type of surfactant, (b) concentration of SDS surfactant, (c) pH, (d) temperature, and (e) diluting solvents on RFI using micelle-enhanced spectrofluorometric method.

TABLE 1: Analytical parameters for the determination of GFX using the two proposed methods.

Parameter	Micelle-enhanced spectrofluorometric method	Hantzsch reaction method
Range (ng/mL)	10–1000	100–2000
Intercept (a) ±SD[a]	−0.50 ± 0.10	1.28 ± 0.07
Slope (b) ± SD[a]	$0.13 \pm 2 \times 10^{-4}$	$0.01 \pm 4 \times 10^{-3}$
Correlation coefficient (r)	0.9999	0.9988
LOD (ng/mL)	2.32	20.45
LOQ (ng/mL)	7.74	68.18

[a]Mean of six determinations.

TABLE 2: Intraday and interday precision of the proposed methods.

Method	Conc. (ng/mL)	Recovery (%) ± SD[a]	Intraday precision		Interday precision	
			Mean ± SD[a]	%RSD	Mean ± SD[a]	%RSD
Micelle-enhanced spectrofluorometric	20	99.86 ± 1.17	100.63 ± 1.13	1.18	99.10 ± 1.50	1.51
	125	96.49 ± 0.40	99.82 ± 0.82	0.82	99.63 ± 1.85	1.86
	500	99.76 ± 1.66	100.34 ± 1.09	1.08	100.23 ± 1.43	1.42
Hantzsch reaction	250	99.24 ± 2.25	99.31 ± 0.58	0.58	98.36 ± 1.89	1.90
	500	100.71 ± 1.98	100.56 ± 0.70	0.70	100.71 ± 1.44	1.43
	1000	99.35 ± 1.44	99.86 ± 0.44	0.44	99.57 ± 0.82	0.82

[a]Mean of six determinations.

behavior than pure SDS aqueous solution. For GFX-SDS system, the fluorescent response showed maximum intensity at 2×10^{-3} M which is lower than the reported critical micellar concentration (cmc) of SDS (8×10^{-3} M) [27]. The change in the cmc of SDS in our results suggests the formation of mixed aggregates at concentrations below the reported cmc that was in concordance to other authors whose reported similar behavior for SDS systems [28–30].

The influence of pH on the micelle-enhanced fluorescence of GFX was studied carefully. The fluorescence intensity was maximal in the pH interval of 5.0–6.0. Thus, pH 5.5 was selected to be the most successful for further studies (Figure 4(c)). It may be suggested that at this pH value (5.5) GFX is present in its protonated form, because the fluorescence intensity of protonated species is always higher than that of neutral species [31, 32]. This can be inferred that protonated forms interact more strongly with the anionic micelles of SDS than the neutral forms of the drugs. Different buffer solutions (acetate, phosphate, and Torell and Stenhagen) were tested. The results revealed that the acetate buffer solution of pH 5.5 achieved the maximum fluorescence intensity, and the variation in the buffer concentration did not show any significant change in the fluorescence intensity. A 0.2 M acetate buffer was selected to obtain an adequate buffering capacity for further measurements.

The ionic strength can also influence significantly the solubilization of a drug in micellar solutions, especially in case of ionic surfactants [33]. Therefore, the effect of the addition of inert salt such as KCl on micellar solutions of GFX was tested. It was found that an increase of concentration above 5×10^{-2} M provoked a clouding phenomenon to the system, and below this value, no significant effect was observed.

Another factor that affects the fluorescence intensity of the micellar system of GFX is the temperature. The effect of temperature was studied in the range 25–100°C in a thermostatically controlled water bath. It was found that increasing the temperature resulted in a decrease in the RFI (Figure 4(d)). This effect can be explained by higher internal conversion as the temperature increases, facilitating nonradiative deactivation of the excited singlet state [34]. The results indicated that the fluorescence intensity was immediately developed at room temperature and remained stable for at least 2 hr.

Finally, the influence of different diluting solvents (water, methanol, ethanol, isopropanol, acetonitrile, or acetone) on the fluorescence intensity of GFX-SDS system was also investigated (Figure 4(e)). The results revealed that water was the best solvent for dilution in presence of SDS, as it gave the highest RFI and the lowest blank reading, while distinct and sharp decrease in the relative fluorescence intensities was observed in the SDS system using other solvents. This effect is attributed to their denaturating effect on the micelles, where short-chain alcohols (methanol, ethanol, and propanol) are solubilized mainly in the aqueous phase and affect the micellization process by modifying the solvent properties. Addition of these alcohols also results in a reduction of the size of the micelles but with a progressive breakdown of the surfactant aggregate at very high concentration [35].

3.2.2. Hantzsch Reaction Method. Effect of reagents composition; the effect of acetylacetone and formaldehyde reagents

Table 3: Robustness of the proposed methods.

Method	Experimental parameter variation	Recovery (%) ± SD[a]
Micelle-enhanced spectrofluorometric	No variation[b]	101.07 ± 0.34
	SDS volume (μL)	
	90	96.10 ± 1.13
	110	99.16 ± 0.44
	pH	
	5.3	101.23 ± 1.27
	5.7	98.52 ± 0.55
	Temperature (°C)	
	20	102.13 ± 0.91
	30	100.81 ± 1.92
Hantzsch reaction	No variation[b]	101.57 ± 0.53
	pH	
	3.3	100.05 ± 1.83
	3.7	101.02 ± 2.79
	Acetylacetone volume (mL)	
	0.4	99.16 ± 1.40
	0.6	99.86 ± 0.91
	Formaldehyde volume (mL)	
	0.8	97.60 ± 0.92
	1.2	100.90 ± 2.30
	Temperature (°C)	
	95	97.06 ± 1.40
	105	102.36 ± 1.90
	Heating time (min)	
	18	99.31 ± 2.74
	22	102.33 ± 1.83

[a]Mean of three determinations.
[b]Following the general assay procedure conditions.

volume in the final reagent solution were studied individually to show their influence on RFI (Figure 5). Different volumes ranging from 0.1 to 1.5 mL of either acetylacetone or formaldehyde solution (34–38%) were tested. A reagent composed of both 0.5 mL of acetylacetone and 1 mL of formaldehyde resulted in the maximum RFI and it was selected for further studies (Figures 5(a) and 5(b)).

In order to select the most appropriate pH, the reaction was carried out at different pHs. First, the pH of the reaction medium was changed over pH range 3–6 using 0.1 M acetate buffer to obtain the highest RFI of the resulted product. The maximum RFI was obtained at pH 3.5 (Figure 5(c)). Then, a series of different buffer systems (acetate, Mcllvaine, and Torell and Stenhagen buffers) of pH 3.5 was studied; the results indicated that acetate buffer was still the superior one. Also different volumes of the optimum buffer solution ranging from 0.1 to 2 mL were tested to obtain the maximum sensitivity. 0.5 mL of acetate buffer was selected for further investigations (Figure 5(d)).

After individual optimization of the reagent components, the effect of the final volume of the reagent system solution was studied over the range 0.1–1.5 mL. It was found that 1 mL of the reagent system was the optimum volume for further studies (Figure 5(e)).

The optimum temperature for the reaction was determined by investigating the RFI at different temperatures in the range 25–110°C in a thermostatically controlled water bath. The results revealed that the heating step is essential. The maximum RFI was reached after 20 min at 100°C (Figure 5(f)). Moreover, to select the most appropriate diluting solvent, the reaction mixture was diluted using different solvents (water, methanol, ethanol, 2-propanol, acetonitrile, or acetone). The results showed that 2-propanol was the best solvent for dilution as it achieves the highest RFI.

3.3. Method Validation. The method was validated according to ICH guidelines of the validation of analytical methods [36]. All results were expressed as percentages, with n representing the number of values. Microsoft office excel 2007 was used for statistical analysis. A 5% significance level was used for evaluation.

3.3.1. Linearity, Limits of Detection, and Quantitation. Under the optimum conditions, linear plots with good correlation coefficients (0.9999 and 0.9988) were obtained in the concentration ranges of 10–1000 and 100–2000 ng/mL for micelle-enhanced fluorescence and Hantzsch reaction methods, respectively. The limits of detection (LOD) and quantitation (LOQ) were determined using the formula: LOD or LOQ $= \kappa SD_a/b$, where $\kappa = 3.3$ for LOD and 10 for LOQ, SD_a is the standard deviation of the intercept, and b is the slope. The LOD values were 2.32 and 20.45 ng/mL for micelle-enhanced fluorescence and Hantzsch reaction methods, respectively. The parameters for the analytical performance of the proposed method are summarized in Table 1.

3.3.2. Precision and Accuracy. The precision of the proposed methods was determined by replicate analysis of six separate sample solutions at three concentration levels of GFX. The relative standard deviations (RSD) were 0.82–1.86 and 0.44–1.90% for micelle-enhanced fluorescence and Hantzsch reaction methods, respectively. Table 2 indicates the good reproducibility of the proposed methods. The accuracy of both methods was determined by investigating the recovery of GFX at three concentrations levels covering the specified range (six replicates of each concentration). The results shown in Table 2 depict good accuracy for the proposed methods.

3.3.3. Robustness. It was estimated by testing the susceptibility of measurements to deliberate variation of the analytical conditions. It was found that minor changes that may take place during the experimental operation did not affect the

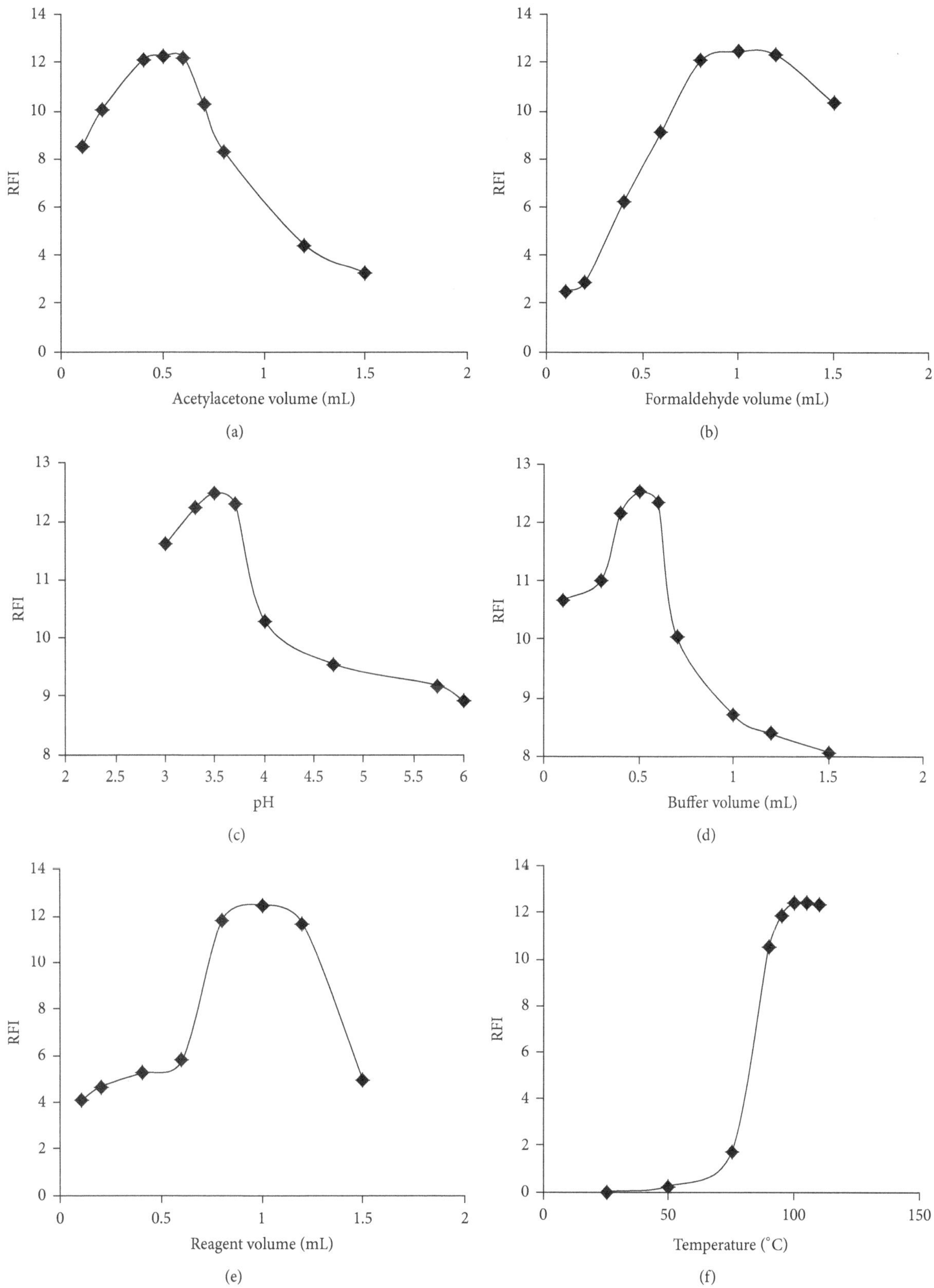

Figure 5: Effect of the different optimization factors. (a) Volume of acetylacetone, (b) volume of formaldehyde, (c) pH, (d) volume of buffer, (e) volume of reagent components, and (f) temperature on RFI using Hantzsch reaction.

Table 4: Determination of GFX in its pharmaceutical dosage forms using the proposed methods.

Dosage form	Recovery % ± SD (n = 4)		
	Micelle-enhanced spectrofluorometric method	Hantzsch reaction method	Reported method[a]
Factive	99.78 ± 0.88 $t = 0.61^b$ $F = 3.17^b$	99.44 ± 0.81 $t = 1.35^b$ $F = 2.69^b$	100.09 ± 0.50
Simulated tablets contain Ambroxol HCl	99.07 ± 0.66 $t = 1.72^b$ $F = 2.49^b$	100.85 ± 0.56 $t = 1.19^b$ $F = 3.49^b$	100.14 ± 1.05

[a]Reference [9].
[b]Theoretical value for t and F at 95% confidence limit, $t = 2.45$ and $F = 9.23$.

Table 5: Determination of GFX in spiked human plasma samples using the proposed methods.

Method	Spiked amount (ng/mL)	Found (ng/mL)	Recovery % ± SD[a]
Micelle-enhanced spectrofluorometric	20	19.52	97.60 ± 0.92
	125	123.98	99.18 ± 1.23
	500	495.50	99.10 ± 0.69
Hantzsch reaction	250	248.11	99.25 ± 0.97
	500	498.25	99.65 ± 1.42
	1000	1007.07	100.71 ± 0.88

[a]Mean of five determinations.

RFI of both methods. The results for the proposed methods are summarized in Table 3.

3.3.4. Specificity. The specificity of the proposed methods was investigated by considering the interference liabilities from Ambroxol HCl in the combined dosage form. Results present in Table 4 indicate that no interference from Ambroxol HCl was observed.

3.4. Applications of the Proposed Methods

3.4.1. Determination of GFX in Dosage Forms. The applicability of the proposed methods was tested by the determination of GFX in its marketed product (Factive Tablet) and simulated tablet with ambroxol HCl as a combined dosage form. The results obtained are accurate and precise as indicated by the excellent percentage recovery (Table 4).

Statistical analysis of the results obtained by the proposed methods and those given by reference method [9] was performed using the Student's t-test and the variance ratio F-test. The calculated values did not exceed the theoretical ones, indicating no significant difference between the compared methods regarding accuracy and precision, respectively. The recovery results of simulated tablets also indicated the selectivity of the proposed methods for GFX in presence of Ambroxol HCl (Table 4). Therefore, the proposed methods are recommended for the quality control analysis of GFX in its pharmaceutical preparations.

3.4.2. Determination of GFX in Plasma. The high sensitivity of the proposed methods allowed the determination of GFX in spiked human plasma. Allen et al. [37] studied the pharmacokinetic parameters of GFX in healthy volunteers after a single oral dose administration. This method indicated that the maximum concentration (C_{max}) of the drug was achieved approximately one hour after dosing, and the mean C_{max} value was found as 1.48 ± 0.39 μg/mL following a single oral dose of 320 mg GFX. Therefore, the drug level in plasma is within the working linearity range of the proposed method (Table 1).

Recently, solid phase extraction (SPE) becomes the most commonly used technique for sample extraction (especially of biological origins) due to its environmental safety. Compared to liquid-liquid extraction [11], the SPE method has several advantages such as making complete phase separations, high quantitative recoveries, no need for using expensive breakable specialty glassware, and disposal of large quantities of organic solvents. Hence, SPE represents efficient separation of interfering substances from analytes without tedious and time-consuming steps. Therefore, a polymeric SPE cartridge was used to prepare the samples in this study.

Hobara et al. introduced simple SPE method for determination of fleroxacin in rat plasma [38]. This method was used with slight modification to be more compatible with our method. This modification includes increasing the volume of spiked plasma to ensure the efficiency of this method (as the reported method was proposed for real plasma samples). This simple and efficient extraction procedure introduces successful method for GFX analysis in human plasma. Table 5 indicates that the obtained results are satisfactorily accurate and precise.

4. Suggested Mechanism of the Proposed Methods

4.1. Micelle-Enhanced Spectrofluorometric Method. This method has been reported for determination of many

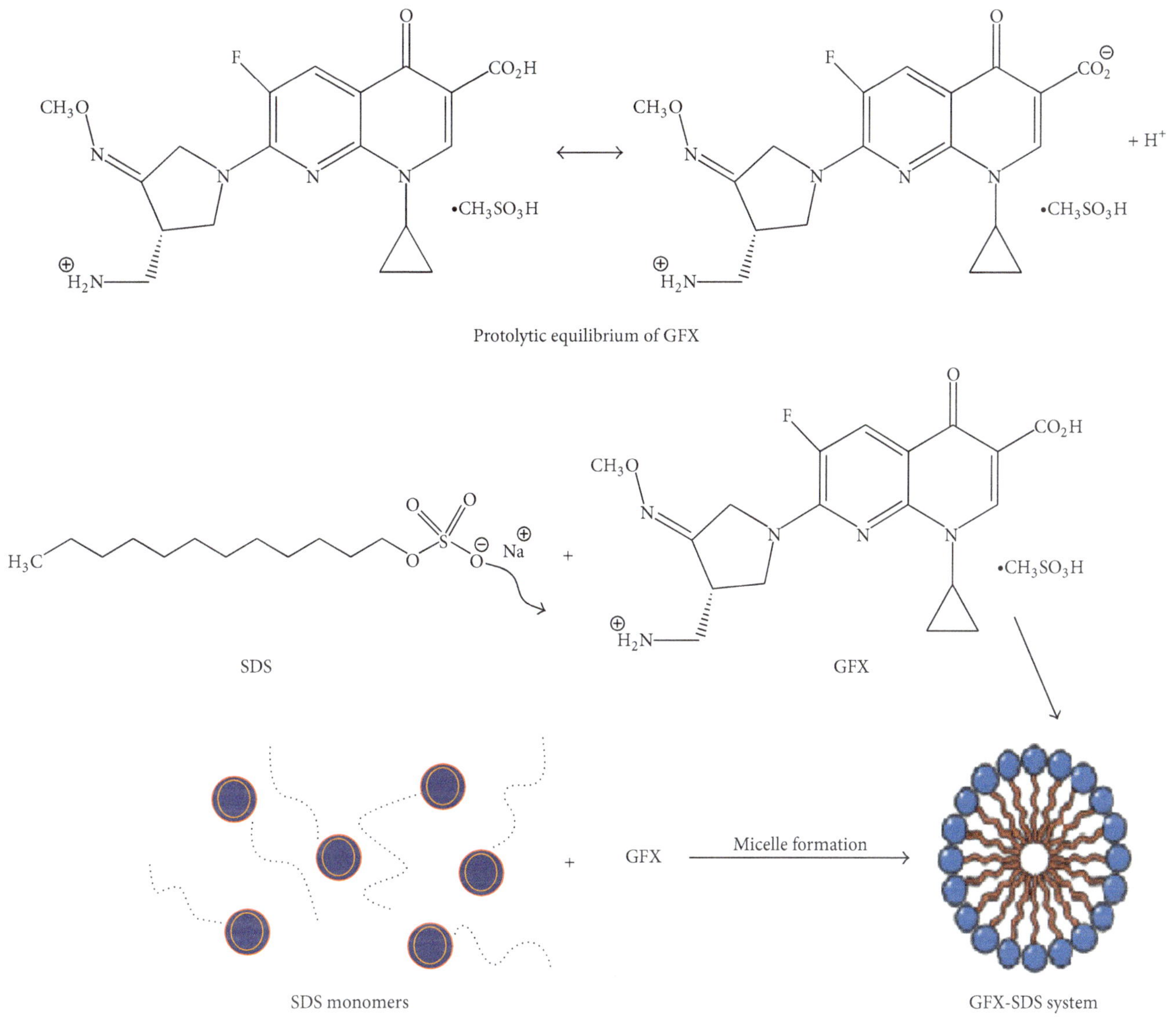

SCHEME 1: The suggested mechanism for GFX-SDS micelle formation.

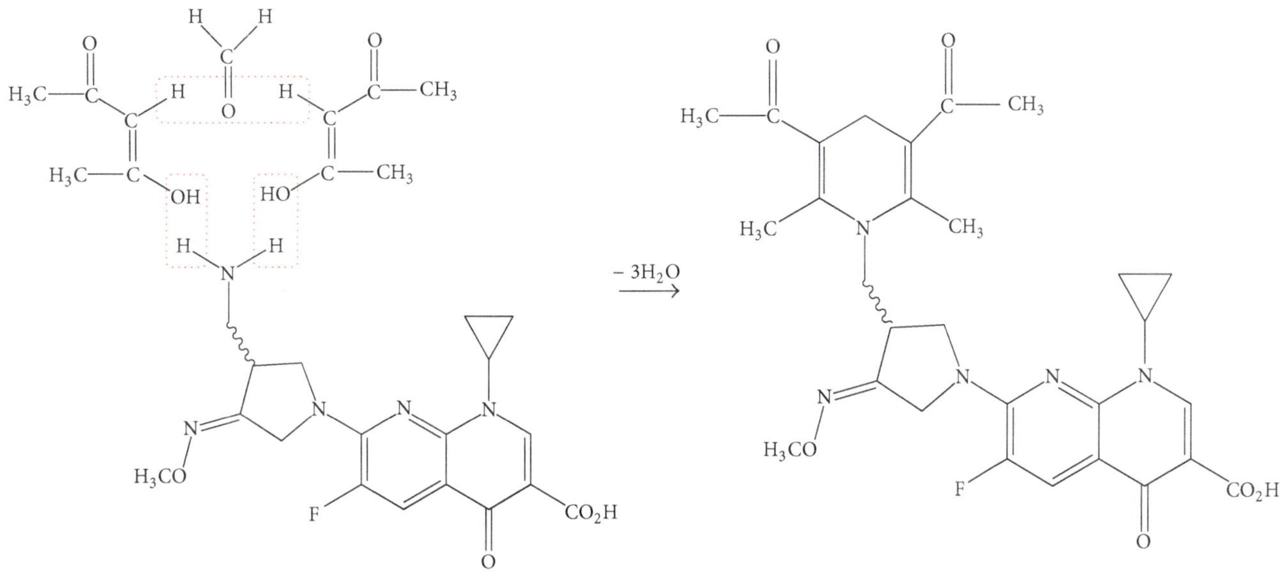

SCHEME 2: The suggested mechanism for the Hantzsch reaction between gemifloxacin and acetylacetone-formaldehyde reagent.

fluoroquinolones [12–15] and other compounds [16, 17] using different types of surfactants. In the presence of SDS anionic micelles, gemifloxacin is solubilized in a more favourable microenvironment that produces an important improvement in fluorescence intensity. GFX possesses two ionizable functional groups: a carboxylic group (pKa_1-6.5) and a basic amino group (pKa_2-8.9). It is thought that in the micellar phase, SDS could form an ion paired species with GFX through the interaction between the SDS negative sulphonyl (OSO_3^-) group and the protonated basic amino group of GFX (Scheme 1).

4.2. Hantzsch Reaction Mechanism. Hantzsch reaction occurs between acetylacetone (β-diketone derivative) in combination with formaldehyde and the aliphatic amino group containing compounds [18–22]. GFX reacts via its primary aliphatic amino group toward this combination in an acidic-buffered medium yielding a highly yellow fluorescent condensation product (Scheme 2).

5. Conclusion

The proposed methods are quite simple, accurate, precise, and do not require tedious extraction procedure. Considering the limits of detection and/or concentrations ranges, the developed methods are highly sensitive. Moreover, the micelle-enhanced spectrofluorometric method has additional advantages that it is rapid and nonpollutant (organic solvents free) that can be used as an alternative to the more time-consuming, expensive HPLC methods. The proposed methods were successfully applied for determining GFX in its tablets or plasma samples without any interference from the matrices. Thus, it can be effectively used for routine analysis of GFX in pharmaceutical industries, hospitals, and research laboratories.

Conflict of Interests

All authors declare that there is no conflict of interests in their submitted paper.

References

[1] J. I. Oh, K. S. Paek, M. J. Ahn et al., "In vitro and in vivo evaluations of LB20304, a new fluoronaphthyridone," *Antimicrobial Agents and Chemotherapy*, vol. 40, no. 6, pp. 1564–1568, 1996.

[2] A. Schulte and P. Heisig, "In vitro activity of gemifloxacin and five other fluoroquinolones against defined isogenic mutants of Escherichia coli, Pseudomonas aeruginosa and Staphylococcus aureus," *Journal of Antimicrobial Chemotherapy*, vol. 46, no. 6, pp. 1037–1038, 2000.

[3] L. D. Saravolatz and J. Leggett, "Gatifloxacin, gemifloxacin, and moxifloxacin: the role of 3 newer fluoroquinolones," *Clinical Infectious Diseases*, vol. 37, no. 9, pp. 1210–1215, 2003.

[4] M. V. Krishna and D. G. Sankar, "Utility of σ and π-acceptors for the spectrophotometric determination of gemifloxacin mesylate in pharmaceutical formulations," *E-Journal of Chemistry*, vol. 5, no. 3, pp. 493–498, 2008.

[5] M. V. Krishna and D. G. Sankar, "Spectrophotometric determination of gemifloxacin mesylate in pharmaceutical formulations through ion-pair complex formation," *E-Journal of Chemistry*, vol. 5, no. 3, pp. 515–520, 2008.

[6] A. A. Elbashir, B. Saad, A. S. M. Ali, K. M. M. Al-Azzam, and H. Y. Aboul-Enein, "Validated stability indicating assay of gemifloxacin and lomefloxacin in tablet formulations by capillary electrophoresis," *Journal of Liquid Chromatography and Related Technologies*, vol. 31, no. 10, pp. 1465–1477, 2008.

[7] E. Doyle, S. E. Fowles, D. F. McDonnell, R. McCarthy, and S. A. White, "Rapid determination of gemifloxacin in human plasma by high-performance liquid chromatography-tandem mass spectrometry," *Journal of Chromatography B*, vol. 746, no. 2, pp. 191–198, 2000.

[8] B. M. H. Al-Hadiya, A. A. Khady, and G. A. E. Mostafa, "Validated liquid chromatographic-fluorescence method for the quantitation of gemifloxacin in human plasma," *Talanta*, vol. 83, no. 1, pp. 110–116, 2010.

[9] A. R. Rote and S. P. Pingle, "Reverse phase-HPLC and HPTLC methods for determination of gemifloxacin mesylate in human plasma," *Journal of Chromatography B*, vol. 877, no. 29, pp. 3719–3723, 2009.

[10] N. F. Youssef and L. I. Bebawy, "Spectrofluorimetric methods for the determination of gemifloxacin mesylate and cefamandole nafate in bulk powder and pharmaceutical preparations," *Bulletin of the Faculty of Pharmacy*, vol. 44, no. 3, pp. 215–227, 2006.

[11] S. K. Tekkeli and A. Önal, "Spectrofluorimetric methods for the determination of gemifloxacin in tablets and spiked plasma samples," *Journal of Fluorescence*, vol. 21, no. 3, pp. 1001–1007, 2011.

[12] M. Rizk, F. Belal, F. Ibrahim, S. Ahmed, and N. El-Enany, "Spectrofluorimetric analysis of certain 4-quinolone in pharmaceuticals and biological fluids," *Pharmaceutica Acta Helvetiae*, vol. 74, no. 4, pp. 371–377, 2000.

[13] J. L. Vílchez, J. Taoufiki, O. Ballesteros, and A. Navalón, "Micelle-enhanced spectrofluorimetric method for the determination of antibacterial trovafloxacin in human urine and serum," *Microchimica Acta*, vol. 150, no. 3-4, pp. 247–252, 2005.

[14] J. A. Ocaña, F. J. Barragán, and M. Callejón, "Spectrofluorimetric and micelle-enhanced spectrofluorimetric determination of gatifloxacin in human urine and serum," *Journal of Pharmaceutical and Biomedical Analysis*, vol. 37, no. 2, pp. 327–332, 2005.

[15] J. A. O. González, M. Callejón Mochón, and F. J. B. De La Rosa, "Spectrofluorimetric determination of levofloxacin in tablets, human urine and serum," *Talanta*, vol. 52, no. 6, pp. 1149–1156, 2000.

[16] Q. Wang, Z. Liu, H. Yuan, and R. Cai, "Micelle enhanced catalytic-fluorimetric determination of hemoglobin," *Chinese Journal of Analytical Chemistry*, vol. 29, no. 4, pp. 423–424, 2001.

[17] J. L. Manzoori and M. Amjadi, "Spectrofluorimetric and micelle-enhanced spectrofluorimetric methods for the determination of gemfibrozil in pharmaceutical preparations," *Journal of Pharmaceutical and Biomedical Analysis*, vol. 31, no. 3, pp. 507–513, 2003.

[18] J. Al-Zehouri, S. Al-Madi, and F. Belal, "Determination of the antiepileptics vigabatrin and gabapentin in dosage forms and biological fluids using Hantzsch reaction," *Arzneimittel-Forschung*, vol. 51, no. 2, pp. 97–103, 2001.

[19] S. M. Sabry, M. Abdel-Hady, M. Elsayed, O. T. Fahmy, and H. M. Maher, "Study of stability of methotrexate in acidic solution

Spectrofluorimetric determination of methotrexate in pharmaceutical preparations through acid-catalyzed degradation reaction," *Journal of Pharmaceutical and Biomedical Analysis*, vol. 32, no. 3, pp. 409–423, 2003.

[20] C. Lee, X. Qiao, D. E. Goeger, and K. E. Anderson, "Fluorometric measurement of 5-aminolevulinic acid in serum," *Clinica Chimica Acta*, vol. 347, no. 1-2, pp. 183–188, 2004.

[21] Q. Peng and C. Jiang, "A new spectrofluorimetric method for determination of trace amounts 5-hydroxytryptamine in human urine and serum," *Journal of Fluorescence*, vol. 17, no. 3, pp. 339–343, 2007.

[22] Q. Peng, J. He, and C. Jiang, "A new spectrofluorimetric method for determination of trace amounts histamine in human urine and serum," *Luminescence*, vol. 24, no. 3, pp. 135–139, 2009.

[23] S. B. Wankhede, A. M. Mahajan, and S. S. Chitlange, "Simultaneous spectrophotometric estimation of gemifloxacin mesylate and ambroxol hydrochloride in tablets," *Der Pharma Chemica*, vol. 3, no. 1, pp. 269–273, 2011.

[24] D. Nagavalli, G. Abirami, and S. K. Kumar, "Validated HPLC method for the simultaneous estimation of gemifloxacin mesylate and ambroxol hydrochloride in bulk and tablet dosage form," *Journal of Pharmacy Research*, vol. 4, no. 6, pp. 1701–1703, 2011.

[25] A. I. Vogel, G. H. Jeffery, and A. Israel, *Vogel's Textbook of Quantitative Chemical Analysis*, 1989.

[26] C. C. Wang, A. N. Masi, and L. Fernández, "On-line micellar-enhanced spectrofluorimetric determination of rhodamine dye in cosmetics," *Talanta*, vol. 75, no. 1, pp. 135–140, 2008.

[27] E. Pramauro and E. Pelezetti, *Surfactants in Analytical Chemistry: Applications of Organized Amphiphilic Media*, 1996.

[28] P. Paton-Morales and F. I. Talens-Alesson, "Effect of ionic strength and competitive adsorption of Na^+ on the flocculation of lauryl sulfate micelles with Al^{3+}," *Langmuir*, vol. 17, no. 20, pp. 6059–6064, 2001.

[29] A. J. M. Valente, H. D. Burrows, R. F. Pereira, A. C. F. Ribeiro, J. L. G. C. Pereira, and V. M. M. Lobo, "Effect of europium(III) chloride on the aggregation behavior of sodium dodecyl sulfate," *Langmuir*, vol. 22, no. 13, pp. 5625–5629, 2006.

[30] A. C. S. Neves, A. J. M. Valente, H. D. Burrows, A. C. F. Ribeiro, and V. M. M. Lobo, "Effect of terbium(III) chloride on the micellization properties of sodium decyl- and dodecyl-sulfate solutions," *Journal of Colloid and Interface Science*, vol. 306, no. 1, pp. 166–174, 2007.

[31] L. Ohannesian and A. Streeter, *Handbook of Pharmaceutical Analysis*, New York, NY, USA, 2001.

[32] L. Ohannesian and A. J. Streeter, *Handbook of Pharmaceutical Analysis*, Edited by M. Dekker, 2002.

[33] M. J. Rosen, *Surfactants and Interfacial Phenomena*, New York, NY, USA, 1989.

[34] D. A. Skoog, D. M. west, F. J. Holler, and S. R. Crouch, *Fundamentals of Analytical Chemistry*, Brooks Cole, Belmont, Calif, USA, 2004.

[35] R. Leung and D. O. Shah, "Dynamic properties of micellar solutions. I. Effects of short-chain alcohols and polymers on micellar stability," *Journal of Colloid And Interface Science*, vol. 113, no. 2, pp. 484–499, 1986.

[36] I. C. H. Guideline, *Q2(R1) Validation of Analytical Procedures: Text and Methodology*, 2005.

[37] A. Allen, E. Bygate, S. Oliver et al., "Pharmacokinetics and tolerability of gemifloxacin (SB-265805) after administration of single oral doses to healthy volunteers," *Antimicrobial Agents and Chemotherapy*, vol. 44, no. 6, pp. 1604–1608, 2000.

[38] N. Hobara, H. Kameya, N. Hokama, S. Ohshiro, and M. Sakanashi, "Rapid and simple determination of fleroxacin in rat plasma using a solid-phase extraction column," *Journal of Chromatography B*, vol. 703, no. 1-2, pp. 279–283, 1997.

2

Sensitive Determination of Terazosin in Pharmaceutical Formulations and Biological Samples by Ionic-Liquid Microextraction Prior to Spectrofluorimetry

Mohsen Zeeb[1,2] and Mahdi Sadeghi[1]

[1] *Agricultural, Medical and Industrial Research School, Nuclear Science and Technology Research Institute, P.O. Box 31485/498, Karaj, Iran*
[2] *Department of Analytical Chemistry, Faculty of Chemistry, University College of Science, University of Tehran, P.O. Box 14155-6455, Tehran, Iran*

Correspondence should be addressed to Mahdi Sadeghi, mahdisadeghi2003@yahoo.com

Academic Editor: P. Haglund

An efficient and environmentally friendly sample preparation method based on the application of hydrophobic 1-Hexylpyridinium hexafluorophosphate [Hpy][PF_6] ionic liquid (IL) as a microextraction solvent was proposed to preconcentrate terazosin. The performance of the microextraction method was improved by introducing a common ion of pyridinium IL into the sample solution. Due to the presence of the common ion, the solubility of IL significantly decreased. As a result, the phase separation successfully occurred even at high ionic strength, and the volume of the settled IL-phase was not influenced by variations in the ionic strength (up to 30% w/v). After preconcentration step, the enriched phase was introduced to the spectrofluorimeter for the determination of terazosin. The obtained results revealed that this system did not suffer from the limitations of that in conventional ionic-liquid microextraction. Under optimum experimental conditions, the proposed method provided a limit of detection (LOD) of 0.027 $\mu g\,L^{-1}$ and a relative standard deviation (R.S.D.) of 2.4%. The present method was successfully applied to terazosin determination in actual pharmaceutical formulations and biological samples. Considering the large variety of ionic liquids, the proposed microextraction method earns many merits, and will present a wide application in the future.

1. Introduction

Terazosin is a selective potent α_1 adreno-receptor antagonist. It is an effective drug for hypertension and benign prostatic hyperplasia [1, 2]. The needs of monitoring trace levels of terazosin in treated patients with initiate or chronic treatment routine in chronic regimen, especially in renal impairment cases, are necessary. To prevent excess employed dosage, it is critical to develop the sensitive and accurate techniques for its quantitative analysis. Several analytical methods have been developed for the measurement of terazosin in bulk, pharmaceuticals, or biological fluids [3–8].

Sample preparation is a critical step to isolate the analytes of interest from the sample matrix as well as to increase the concentration of analytes prior to their determination, especially when their levels are lower than the detection limit of analytical techniques. Liquid-liquid extraction (LLE) and solid-phase extraction (SPE) have been widely used as basic, simple, and adaptable methods for sample preconcentration. However, some drawbacks such as the use of large amount of hazardous organic solvents, and the employment of multistep procedures associated with the high risk of analyte losses make LLE and SPE tedious, expensive, and environmentally unfriendly. In recent years, reduction or complete removal of hazardous organic solvents in sample pretreatment techniques has received great attention. For this goal, many microextraction methods such as dispersive liquid-liquid microextraction (DLLME) [9–12], single-drop microextraction (SDME) [13–15], stir bar sorptive extraction [16], and so forth, have been developed to reduce the amount of toxic solvents. In some of these procedures, however, the use of toxic materials like benzene, toluene, or chloroform as the microextraction solvent is a common practice.

Ionic liquids (ILs) are ionic media resulting from combinations of organic cations and different anions. ILs have a variety of chemical and physical properties such as low vapor pressure, excellent thermal stabilities, adjustable miscibility, moderate solubility of organic compounds and metal ions, and so forth, which make them highly attractive in microextraction methods [17]. ILs have been widely applied in many sample pretreatment procedures as green extraction solvent including ionic liquid-based liquid-liquid microextraction (IL-LLME) [18, 19], cold induced aggregation microextraction (CIAME) [20], ionic liquid-based single-drop microextraction (IL-SDME) [21, 22], ionic liquid-based dispersive liquid-liquid microextraction (IL-DLLME) [23, 24], in situ solvent formation microextraction (ISFME) [25], temperature-controlled ionic liquid dispersive liquid phase microextraction (TILDLME) [26], and so forth.

The use of separation and preconcentration steps based on IL-based microextraction methods offers several advantages such as high recovery, no need of hazardous extraction solvent, simplicity of operation, and compatibility with many analytical techniques. Despite the many advantages of using ionic liquid-based microextraction techniques, under high salt condition, the fine droplets of the extraction phase can not be settled due to increase in sample solution density. The solubility of ionic liquids increases as the salt amount of sample increases. As a result, the volume of the settled phase depends significantly on the ionic strength of the sample solution. On the other hand, accuracy and performance of the extraction method are significantly influenced by variations in the ionic strength. In this work, in order to solve these problems and obtain higher accuracy, a novel, simple, and powerful ionic-liquid microextraction method which put very few dangers to environment was developed. In the proposed approach, a hydrophobic pyridinium IL was added to the aqueous sample solution containing one of the IL's ions as a common ion to act as the extraction phase. Hydrophobic 1-hexylpyridinium hexafluorophosphate [Hpy][PF_6] ionic liquid and $NaPF_6$ were used as the extraction phase and common ion source, respectively. Due to the presence of the common ion in the sample, the solubility of IL decreases, even at high ionic strength. Therefore, the volume of the enriched phase is independent from the ionic strength. In this work, [Hpy][PF_6] was employed as the extractant solvent due to some physicochemical properties such as high hydrophobicity, water immiscibility, and suitable density. The present methodology does not need heating, antisticking agent, long equilibration time, and cooling before or after centrifugation prior to analysis. To the best of our knowledge, up until now, no method based on the microextraction with ILs and spectrofluorimetric detection has been proposed for preconcentration, extraction, and determination of drugs in their trace levels. Spectrofluorimetric method was applied due to ease, proper selectivity, sensitivity, wide dynamic range, and low cost of operation. The goal of this paper is to study the applicability of the proposed microextraction method followed by spectrofluorimetry to determine terazosin as a model compound in pharmaceutical and biological samples. Furthermore, the obtained results reveal that the proposed methodology is simple and practical for the routine determination of terazosin in aqueous media.

2. Experimental

2.1. Instrumentation. Fluorescence measurements were performed using a Perkin-Elmer LS 50 spectrofluorimeter equipped with xenon discharge lamp, and quartz microcell with a volume of 100 μL. A centrifuge from Hettich (Tuttlingen, Germany) was utilized to accelerate the phase separation process. An adjustable sampler (10–100 μL) was prepared from Eppendorf (Hamburg, Germany). The pH-meter model 692 (Herisau, Switzerland) supplied with a glass-combined electrode and universal pH indicator (pH 0–14) from Merck (Darmstadt, Germany) were used for the pH measurements.

2.2. Reagents and Materials. All chemicals used were of analytical reagent grade and all solutions were prepared with ultrapure water. 1-Hexylpyridinium hexafluorophosphate [Hpy][PF_6] (97%) (Acros organics, Belgium) was applied as an extractant solvent. Acetone, acetonitrile, methanol, ethanol and sodium hexafluorophosphate ($NaPF_6$) were purchased from Merck (Darmstadt, Germany). Stock solution of terazosin at concentration of 1000 mg L^{-1} was prepared by dissolving the required amount of pure drug in ultrapure water, and working standard solutions were obtained by serial dilutions of this stock solution. [HPy][PF_6] IL is not liquid at room temperature (melting point: 45°C). Therefore, this IL was dissolved in acetonitrile to obtain a working solution of 75 mg mL^{-1}. A solution of 250 mg mL^{-1} $NaPF_6$ was obtained by dissolving appropriate amount of this salt in ultrapure water. 0.5 mol L^{-1} of sodium hydroxide and concentrated hydrochloric acid were used for adjusting the pH value of sample solutions. Terazosin tablets (labeled as containing 2 and 5 mg terazosin per tablet) were obtained from Daroupakhsh (Iran, Tehtan).

2.3. General Analytical Procedure. Aliquots of 10.0 mL sample or standard solution (pH 10) containing terazosin in the range of 0.1 to 115 μg L^{-1} was placed into a glass test tube with conic bottom, 1.4 mL of $NaPF_6$ (250 mg mL^{-1}) was added. Then, 150 mg of [HPy][PF_6] (dissolved in acetonitrile) as the extraction solvent was added to the sample solution. After shaking, a cloudy condition was immediately formed at room temperature. Due to the formation a large contact area between extraction solvent and analyte, the uncharged form of terazosin was rapidly trapped into IL-phase. In order to accelerate phase separation, the resultant solution was centrifuged for 5 min at 4000 rpm. The upper aqueous solution was then removed by a pipette. Finally, in order to reduce the viscosity of IL-phase and facilitate sample handling prior to analysis, the settled phase (about 16 μL) in the test tube was diluted to 250 μL by adding appropriate amount of ethanol. 100 μL of the diluted settled phase was removed by a sampler and transferred to the microcell of the fluorimeter. The fluorescence intensity was measured

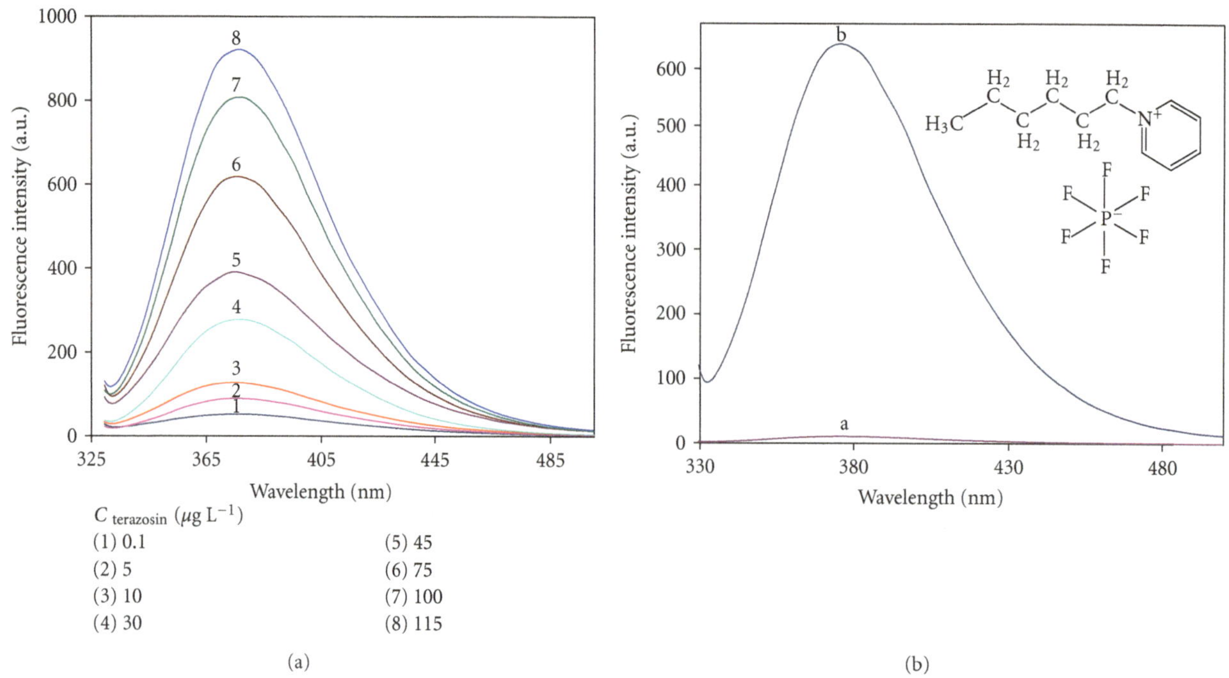

FIGURE 1: (a) Emission spectra obtained for 8 standard solutions of terazosin with different concentrations. (b) Emission spectrum of terazosin ($75\,\mu g\,L^{-1}$) treated the same as previously described in the general analytical procedure (a), and emission spectrum of reagent blank in $[Hpy][PF_6]$ ionic liquid (b). Inset: chemical structure of $[Hpy][PF_6]$ ionic liquid. Experimental conditions were as indicated in the Table 1.

at 376 nm with the excitation wavelength set at 330 nm. A reagent blank was prepared using a similar procedure but without adding terazosin.

2.4. Analysis of Tablets. Five terazosin tablets, labeled as containing 2 mg terazosin each, were weighed and the average mass per tablet was determined. An amount of the powder equivalent to 2 mg of terazosin was accurately weighed and dissolved in 50 mL ultrapure water. The solution was sonicated for 5 min and filtered into a 100 mL volumetric flask. Further dilutions were made up with ultrapure water to achieve a final concentration of $20\,\mu g\,mL^{-1}$. In addition, appropriate dilution was performed for terazosin analysis.

2.5. Analysis of Spiked Human Plasma. Human plasma samples (1.0 mL) were spiked with terazosin solutions, deproteinized by adding 4 mL acetonitrile, and centrifuged at 4000 rpm for 15 min. Then, 2.0 mL of the clear supernatant was diluted to 50 mL and subjected to the proposed method.

2.6. Analysis of Spiked Human Urine. Fresh matutinal human urine (10 mL) was transferred into graduated centrifuge tubes. These solutions were centrifuged for 5 min at 4000 rpm. Then, aliquots of 2 mL from clear supernatant were put in new centrifuge tubes and spiked with different amounts of terazosin (2 to $100\,\mu g\,L^{-1}$); the general analytical procedure was followed.

3. Results and Discussion

In this work, a practical and simple ionic-liquid microextraction method was combined with spectrofluorimetry for preconcentration and determination of terazosin. Hydrophobic $[Hpy][PF_6]$ and $NaPF_6$ were applied as the extraction phase and common ion source, respectively. In the proposed microextraction method, the volume of the enriched phase was independent form the ionic strength of the sample solution (up to 30% w/v) which as an advantage significantly improved the performance of the extraction method. The influence of different factors affecting extraction conditions were studied and optimized to obtain a compromise between sensitivity, simplicity, and reproducibility.

3.1. Spectral Characteristics of Terazosin and $[HPy][PF_6]$. The intense fluorescence of terazosin is due to its cyclic conjugated structure which benefits of having π-electron system. The emission spectra of terazosin were recorded as described in the general analytical procedure (see Figure 1(a)). As it can be ascertained, the emission peak of terazosin is at 376 nm, while its excitation peak is at 330 nm. In order to obtain accurate fluorescence intensity, it is important to study the effect of the extraction phase on fluorescence spectrum of the analyte of interest. For this goal, fluorescence spectrum of the reagent blank was evaluated. As it can be seen in Figure 1(b), the emission of $[HPy][PF_6]$ has no notable effect on the determination of terazosin. Thus, the aforementioned wavelengths were selected for quantitative analysis of terazosin.

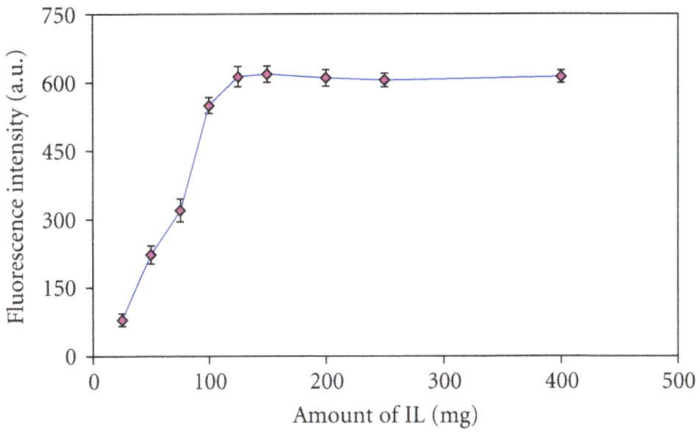

FIGURE 2: Influence of amount of [Hpy][PF_6] on the fluorescence intensities. Experimental conditions: terazosin concentration 75 $\mu g\,L^{-1}$; pH 10; $NaPF_6$ 350 mg.

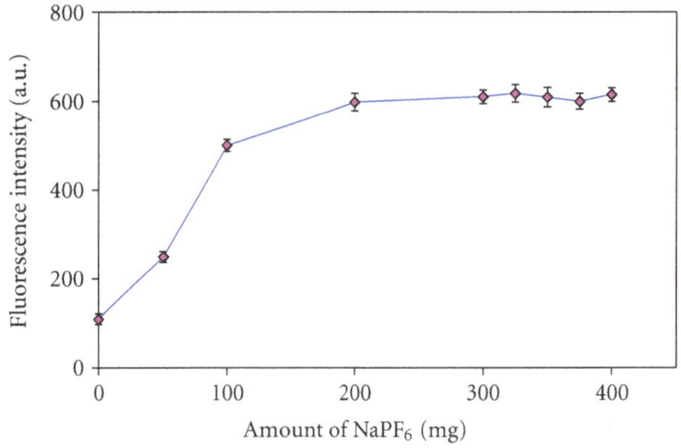

FIGURE 3: Influence of $NaPF_6$ on the analytical signals obtained for terazosin. Experimental conditions: terazosin concentration 75 $\mu g\,L^{-1}$; pH 10; [Hpy][PF_6] 150 mg.

3.2. Influence of IL Amount. The amount of [HPy][PF_6] is a critical parameter which can affect the analyte recoveries. Thus, the microextraction procedure was carefully evaluated in order to achieve a compromise between the amount of IL and the analytical responses. The Effect of [HPy][PF_6] amount on the analytical responses was studied within the range of 25–400 mg. Figure 2 shows the variation of the fluorescence signal versus the amount of [HPy][PF_6]. As it can be ascertained, the signal intensity increases as the amount of [HPy][PF_6] increases and remains nearly constant from 125 mg. No significant variations were obtained on the microextraction efficiency for higher amounts of IL. Thus, in order to achieve sensitive analytical response, 150 mg of IL was selected as an optimum value.

3.3. Influence of Common Ion. The effect of $NaPF_6$ amount as a common ion source on the fluorescence signal was studied within the range of 0–400 mg (see Figure 3). Due to the common ion influence, an increase in the amount of $NaPF_6$ causes a notable decrease in the solubility of [HPy][PF_6]. As a result, the extraction performance and the analytical intensity increase. As it can be seen in Figure 3, the fluorescence intensity increases with increasing the amount of $NaPF_6$ up to 325 mg and then remains nearly stable. However, 350 mg of $NaPF_6$ was chosen through the rest of the work.

3.4. Influence of Diluting Agent. Due to the high viscosity of the IL, it must be conditioned with a diluting agent prior to its introduction to the spectrofluorimeter. Several diluting agents including methanol, ethanol, acetone, and acetonitrile were evaluated in order to choice a diluting agent which can dissolve the IL-phase completely and provide the best sensitivity. Type of the organic solvent had no significant effect on the analytical response. Finally, due to less toxicity of ethanol, it was applied as diluting agent in all experiments.

3.5. Influence of pH. The pH of the sample is an important factor which affects the chemical structure of terazosin. The effect of pH on the extraction of terazosin from aqueous samples was studied over the range of 5–12 using HCl and NaOH. For ionizable organic molecules, maximum extraction efficiency is obtained at pH values where the uncharged form of the analyte overcomes, and therefore, the analyte of interest is favored to be partitioned into the hydrophobic IL-phase. The results illustrated in Figure 4 reveal that analytical signals, obtained for terazosin, depend on pH. The best conditions of terazosin microextraction were achieved at alkaline pH values due to dissociation of the aromatic amine group, which acts as a weak base (see Figure 4) (pK_b = 6.9). Based on the results obtained, pH 10 was selected as the optimum value.

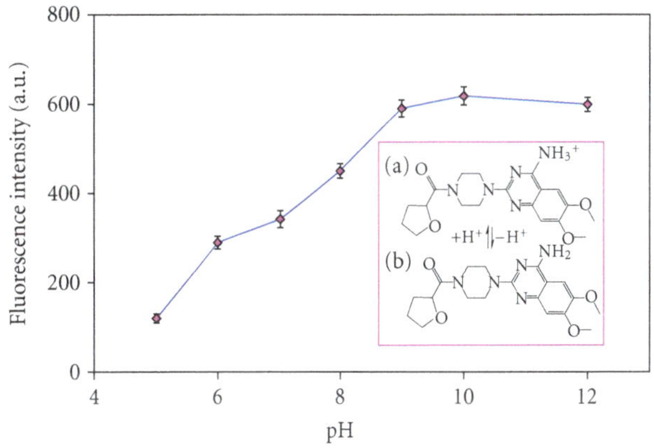

FIGURE 4: Influence of pH on the fluorescence intensities. Experimental conditions: terazosin concentration 75 $\mu g\,L^{-1}$; [Hpy][PF_6] 150 mg; $NaPF_6$ 350 mg. Inset: chemical structure of terazosin. Acid base equilibrium involved at working pH. (a) Protonated form of this terazosin; (b) dissociated form of terazosin.

3.6. Influence of Equilibration Temperature and Extraction Time. In order to obtain complete extraction of terazosin and easy phase separation, the effect of equilibration temperature was studied from 5 to 40°C. Based on the results obtained in this study, the temperature had no significant and benefic effect on the extraction efficiency. Therefore, room temperature as an equilibration temperature was used through the rest of the work. Microextraction time is one of the important factors affecting the extraction efficiency, especially in microextraction methods such as SPME and LPME. The dependence of extraction efficiency upon extraction time was studied from 5 sec to 30 min. The results obtained in this experiment revealed that signal variations versus extraction time were not significant. It was demonstrated that after formation of the cloudy condition, the surface area between extraction phase and aqueous media was infinitely large. Thus, the transfer of the target analyte from aqueous media to IL-phase was very fast. In order to keep extraction time as short as possible, the cloudy solution was centrifuged immediately after the preparation.

3.7. Influence of Centrifugation Condition. The effect of centrifugation rate on the fluorescence intensity was evaluated from 1000 to 6000 rpm. The obtained results showed that over 3500 rpm the IL-phase was completely transferred to the bottom of the test tube, and the analytical response remained constant. Thus, 4000 rpm was selected for all experiments. At the optimum rate, the effect of centrifugation time upon analytical responses was investigated within the range of 1–15 min. Over 4 min, the fluorescence intensity remained stable indicating entire transfer of IL-phase to the bottom of the centrifuge tube. Therefore, a centrifugation time of 5 min was chosen as the optimum value.

3.8. Influence of Salt Concentration. Influence of salt concentration on the extraction performance was studied by adding various amounts of $NaNO_3$ from 0 to 40% (w/v) while other experimental conditions were kept constant. In the proposed microextraction method, due to the presence of PF_6^- as a common ion in the sample solution, the volume of the enriched phase was independent from the ionic strength, and phase separation occurred up to 30% w/v of $NaNO_3$. Table 1 summarizes both instrumental and experimental conditions selected.

3.9. Influence of Interfering Substances. The effect of different chemical species on the determination of target analyte was studied using solutions containing 75 μg L^{-1}of terazosin, and adding 100 mg L^{-1} of the possible interfering species. The tolerance limit of each interfering substance was taken into account as the largest amount yields an error in the determination of the target analyte which does not exceed 5%. No interference was observed from commonly interfering species such as Na^+, NH_4^+, Ca^{2+}, Zn^{2+}, Mg^{2+}, Cl^-, PO_4^{3-}, SO_4^{2-}, starch, glucose, lactose, fructose, sucrose, ascorbic acid, citric acid, dye species (as yellow quinoline), urea, and saccharin. The obtained results revealed the excellent selectivity of the proposed methodology in detecting the studied drug in dosage form and biological samples.

4. Application

4.1. Analytical Performance. Under the optimum conditions, calibration graph was achieved by analyzing 10.0 mL of standard solutions containing known amounts of terazosin.

Table 1: The experimental conditions for terazosin determination.

Microextraction parameters	Studied range	Selected condition
Amount of $NaPF_6$ (mg)	0–400	350
Amount of [Hpy][BF_4] (mg)	25–400	150
pH	5–12	10
Equilibration temperature (°C)	5–40	Room temperature
Centrifugation time (min)	1–15	5
Ionic strength (% w/v)	0–40	NE[a]
Instrumental parameters		
Excitation wavelength (nm)	200–350	330
Emission wavelength (nm)	350–700	376
Excitation and emission slit widths (nm)	5–20	10

[a]No effect.

Table 2: Determination of terazosin in tablets by the proposed methodology and by a reported voltammetric method [8].

Claimed (mg/tablet)	Proposed methodology (mg)[a]	Reported method (mg)[a]	Error (%)[b]	Error (%)[c]
2	1.95 (±0.08)	2.07 (±0.10)	−2.5	−5.8
5	5.1 (±0.21)	5.27 (±0.37)	+2.0	−3.2

[a]Values in parenthesis show the standard deviation based on three replicates.
[b]Error against the declared value.
[c]Error against the reported method.

The settled ionic phase was diluted to 250 μL with ethanol and the fluorescence intensity was recorded. Therefore, a preconcentration factor of 40 was achieved. The enhancement factor calculated as the slope ratio of calibration graph after and before extraction was about 33. The linear concentration range was from 0.1 to 115 μg L^{-1} of terazosin with linear regression equation as

$$I = 51.33 + 7.5608\ C, \tag{1}$$

where I is the fluorescence intensity at 376 ± 3 nm and C is the terazosin concentration in μg L^{-1}. The relative standard deviation (R.S.D.) obtained for the determination of 75 μg L^{-1} of terazosin was 2.4% ($n = 5$). The limit of detection (LOD), calculated as three times the standard deviation of the measurement of blanks divided by the slope of the calibration curve, was found to be 0.027 μg L^{-1}.

4.2. Analysis of Terazosin in Pharmaceutical Formulations. In order to show the validity of the proposed method, it was applied for terazosin determination in commercial tablets. Three replicate determinations were performed, and satisfactory results were achieved. Table 2 shows the results obtained by applying the present method and those obtained by a reported voltammetric method [8]. Until date no official method has been reported for terazosin determination. The results show the applicability of the proposed methodology and its accuracy for quantitative analysis of terazosin in this type of samples.

Table 3: Determination of terazosin in spiked urine and spiked plasma by present work.

Drug	Spiked urine			Spiked plasma		
	Amount added ($\mu g\,L^{-1}$)	Amount found ($\mu g\,L^{-1}$) ± S.D.[a]	Recovery (%)	Amount added ($\mu g\,L^{-1}$)	Amount found ($\mu g\,L^{-1}$) ± S.D.[a]	Recovery (%)
Terazosin	2	1.83 ± 0.09	91.5	2	1.73 ± 0.11	86.5
	5	4.81 ± 0.28	96.2	5	4.82 ± 0.33	96.4
	10	9.40 ± 0.30	94.0	10	9.60 ± 0.51	96.0
	100	105.81 ± 4.11	105.8	100	93.10 ± 4.88	93.1

[a] Average of three independent measurements.

Table 4: Comparison of the proposed approach with other reported methods for determination of terazosin.

Method	Sample	LOD ($\mu g\,L^{-1}$)	R.S.D. (%)	LR ($\mu g\,L^{-1}$)	References
Spectrofluorimetry	Serum, urine	210	2.5	Up to 3486.9	[4]
X-ray fluorescence spectrometry	Drug	732	—	732–843000	[5]
HPLC with fluorescence detection	Plasma	0.25	<7	Up to 100	[6]
HPLC with electrospray ionization mass spectrometry detection	Plasma	0.0625	—	—	[7]
Square-wave adsorptive cathodic stripping voltammetry	Plasma, drug	0.0058	<1.1	0.39–11.62	[8]
ionic-liquid microextraction spectrofluorimetry	Urine, plasma, drug	0.027	2.4	0.1–115	This work

LOD: limit of detection, LR: linear range, R.S.D.: relative standard deviation.

4.3. Analysis of Terazosin in Spiked Human Plasma and Urine. The accuracy of the present method was evaluated by determination of terazosin in spiked human urine and spiked human plasma. The recovery of the studied drug was investigated at four concentration levels. The obtained results are shown in Table 3. As can be seen, calculated amounts of recoveries varied between 91.5–105.8% and 86.5–96.0% for human urine and human plasma, respectively, indicating both accuracy and precision.

4.4. Comparison with Other Reported Methods. Determination of terazosin by the proposed methodology was compared with other reported techniques. The results are shown in Table 4. As it can be seen, compared to the previously reported methods, the present method has a relatively low LOD, wide dynamic range, and excellent reproducibility. The method developed in this work is proposed as a suitable alternative to expensive instrumental methods for trace determination of terazosin in pharmaceutical and biological samples. These results reveal that the proposed approach is a very sensitive, low-cost, rapid, environmentally friendly and accurate technique that can be used for terazosin determination in routine analytical laboratories.

5. Conclusion

A novel and efficient mode of ionic-liquid microextraction was combined with spectrofluorimetric detection to preconcentrate and determine trace levels of terazosin. Hydrophobic [Hpy][PF_6] ionic liquid was applied as a green extraction solvent and an alternative to hazardous organic solvents. Due to the presence of one of the IL's ions as a common ion in the sample solution, the performance of the microextraction method was not influenced by variations in the ionic strength of the aqueous solution (up to 30%). The method developed in this work was demonstrated to be simple, rapid, practical, inexpensive, and environmentally friendly for preconcentration and separation of the trace analytes. In addition, the present method revealed to be a practical tool for routine quality control of drugs in pharmaceutical and biological samples with low operation cost and simplicity of instrumentation.

Acknowledgments

Support of this investigation by the Research Council of Agricultural, Medical and Industrial Research School through grant is gratefully acknowledged. The aythors also thank Miss Shakila Savadkouhi for help to check voltammetric procedure as a reference method.

References

[1] R. Mudiyala and A. Ahmed, "Effect of terazosin on clinical benign prostatic hyperplasia in older adults," *Journal of the American Geriatrics Society*, vol. 51, no. 3, pp. 424–426, 2003.

[2] H. Lepor, K. Jones, and W. Williford, "The mechanism of adverse events associated with terazosin: An analysis of the Veterans Affairs cooperative study," *Journal of Urology*, vol. 163, no. 4, pp. 1134–1137, 2000.

[3] H. H. Abdine, F. A. El-Yazbi, S. M. Blaih, and R. A. Shaalan, "Spectrophotometric and spectrofluorimetric methods for the determination of terazosin in dosage forms," *Spectroscopy Letters*, vol. 31, no. 5, pp. 969–980, 1998.

[4] C. Q. Jiang, M. X. Gao, and J. X. He, "Study of the interaction between terazosin and serum albumin -Synchronous fluorescence determination of terazosin," *Analytica Chimica Acta*, vol. 452, no. 2, pp. 185–189, 2002.

[5] R. A. Silva and A. N. Masi, "Sensitive determination of terazosin by x-ray fluorescence spectrometry based on the formation of ion-pair associates with zinc thiocyanate," *X-Ray Spectrometry*, vol. 36, no. 4, pp. 279–283, 2007.

[6] P. Y. Cheah, K. H. Yuen, and M. L. Liong, "Improved high-performance liquid chromatographic analysis of terazosin in human plasma," *Journal of Chromatography B*, vol. 745, no. 2, pp. 439–443, 2000.

[7] A. P. Zavitsanos and T. Alebic-Kolbah, "Enantioselective determination of terazosin in human plasma by normal phase high-performance liquid chromatography-electrospray mass spectrometry," *Journal of Chromatography A*, vol. 794, no. 1-2, pp. 45–56, 1998.

[8] M. M. Ghoneim, M. A. El Ries, E. Hammam, and A. M. Beltagi, "A validated stripping voltammetric procedure for quantification of the anti-hypertensive and benign prostatic hyperplasia drug terazosin in tablets and human serum," *Talanta*, vol. 64, no. 3, pp. 703–710, 2004.

[9] M. Zeeb, M. R. Ganjali, and P. Norouzi, "Dispersive liquid-liquid microextraction followed by spectrofluorimetry as a simple and accurate technique for determination of thiamine (vitamin B1)," *Microchimica Acta*, vol. 168, no. 3, pp. 317–324, 2010.

[10] P. Hemmatkhah, A. Bidari, S. Jafarvand, M. R. M. Hosseini, and Y. Assadi, "Speciation of chromium in water samples using dispersive liquid-liquid microextraction and flame atomic absorption spectrometry," *Microchimica Acta*, vol. 166, no. 1-2, pp. 69–75, 2009.

[11] S. Jafarvand, A. Bidari, P. Hemmatkhah, M. R. M. Hosseini, and Y. Assadi, "Dispersive liquid-liquid microextraction of silver prior to determination by microsample introduction-flame atomic absorption spectrometry," *Analytical Letters*, vol. 42, no. 14, pp. 2214–2231, 2009.

[12] F. Rezaei, A. Bidari, A. P. Birjandi, M. R. Milani Hosseini, and Y. Assadi, "Development of a dispersive liquid-liquid microextraction method for the determination of polychlorinated biphenyls in water," *Journal of Hazardous Materials*, vol. 158, no. 2-3, pp. 621–627, 2008.

[13] A. Jain, A. K. K. V. Pillai, N. Sharma, and K. K. Verma, "Headspace single-drop microextraction and cuvetteless microspectrophotometry for the selective determination of free and total cyanide involving reaction with ninhydrin," *Talanta*, vol. 82, no. 2, pp. 758–765, 2010.

[14] A. Garbi, V. Sakkas, Y. C. Fiamegos, C. D. Stalikas, and T. Albanis, "Sensitive determination of pesticides residues in wine samples with the aid of single-drop microextraction and response surface methodology," *Talanta*, vol. 82, no. 4, pp. 1286–1291, 2010.

[15] Y. He and H. K. Lee, "Liquid-phase microextraction in a single drop of organic solvent by using a conventional microsyringe," *Analytical Chemistry*, vol. 69, no. 22, pp. 4634–4640, 1997.

[16] A. Sanchez-Ortega, N. Unceta, A. Gómez-Caballero et al., "Sensitive determination of triazines in underground waters using stir bar sorptive extraction directly coupled to automated thermal desorption and gas chromatography-mass spectrometry," *Analytica Chimica Acta*, vol. 641, no. 1-2, pp. 110–116, 2009.

[17] S. Pandey, "Analytical applications of room-temperature ionic liquids: a review of recent efforts," *Analytica Chimica Acta*, vol. 556, no. 1, pp. 38–45, 2006.

[18] N. Hirayama, M. Deguchi, H. Kawasumi, and T. Honjo, "Use of 1-alkyl-3-methylimidazolium hexafluorophosphate room temperature ionic liquids as chelate extraction solvent with 4,4,4-trifluoro-1-(2-thienyl)- 1,3-butanedione," *Talanta*, vol. 65, no. 1, pp. 255–260, 2005.

[19] P. Berton, E. M. Martinis, L. D. Martinez, and R. G. Wuilloud, "Room temperature ionic liquid-based microextraction for vanadium species separation and determination in water samples by electrothermal atomic absorption spectrometry," *Analytica Chimica Acta*, vol. 640, no. 1-2, pp. 40–46, 2009.

[20] M. Gharehbaghi, F. Shemirani, and M. D. Farahani, "Cold-induced aggregation microextraction based on ionic liquids and fiber optic-linear array detection spectrophotometry of cobalt in water samples," *Journal of Hazardous Materials*, vol. 165, no. 1–3, pp. 1049–1055, 2009.

[21] L. Vidal, A. Chisvert, A. Canals, and A. Salvador, "Ionic liquid-based single-drop microextraction followed by liquid chromatography-ultraviolet spectrophotometry detection to determine typical UV filters in surface water samples," *Talanta*, vol. 81, no. 1-2, pp. 549–555, 2010.

[22] E. Aguilera-Herrador, R. Lucena, S. Cárdenas, and M. Valcárcel, "Determination of trihalomethanes in waters by ionic liquid-based single drop microextraction/gas chromatographic/mass spectrometry," *Journal of Chromatography A*, vol. 1209, no. 1-2, pp. 76–82, 2008.

[23] M. Gharehbaghi, F. Shemirani, and M. Baghdadi, "Dispersive liquid-liquid microextraction based on ionic liquid and spectrophotometric determination of mercury in water samples," *International Journal of Environmental Analytical Chemistry*, vol. 89, no. 1, pp. 21–33, 2009.

[24] H. Abdolmohammad-Zadeh and G. H. Sadeghi, "Combination of ionic liquid-based dispersive liquid-liquid microextraction with stopped-flow spectrofluorometry for the preconcentration and determination of aluminum in natural waters, fruit juice and food samples," *Talanta*, vol. 81, no. 3, pp. 778–785, 2010.

[25] M. Zeeb, M. R. Ganjali, P. Norouzi, and M. R. Kalaee, "Separation and preconcentration system based on microextraction with ionic liquid for determination of copper in water and food samples by stopped-flow injection spectrofluorimetry," *Food and Chemical Toxicology*, vol. 49, no. 5, pp. 1086–1091, 2011.

[26] Q. Zhou, H. Bai, G. Xie, and J. Xiao, "Temperature-controlled ionic liquid dispersive liquid phase micro-extraction," *Journal of Chromatography A*, vol. 1177, no. 1, pp. 43–49, 2008.

3

Application of CZE Method in Routine Analysis for Determination of B-Complex Vitamins in Pharmaceutical and Veterinary Preparations

Marina Franco, Renata Jasionowska, and Elisa Salvatore

Dipartimento di Chimica, Sapienza Università, 00185 Roma, Italy

Correspondence should be addressed to Marina Franco, marina.franco@uniroma1.it

Academic Editor: Mohamed Mohamed

A competitive CZE method for quality control analysis of multivitamin preparations and veterinary products containing B-group vitamins was developed. Vitamins of interest are thiamine hydrochloride (B_1), thiamine monophosphate chloride (B_{1a}), riboflavine (B_2), riboflavine-5′monophosphate (B_{2a}), nicotinamide (B_3), d-pantothenic acid calcium salt (B_5), pyridoxine hydrochloride (B_6), folic acid (B_9), and 4-aminobenzoic acid (B_{10}). These analytes were separated optimizing the experimental conditions in 20 mM tetraborate buffer pH = 9.2 as a BGE (background electrolyte), on a Beckman P/ACE System MDQ instrument, using uncoated fused silica capillary. The effective capillary length was of 49.5 cm, I.D. = 50 μm, the applied voltage 20 kV and the temperature 25°C. Detection was performed by a diode array detector at 214 nm for all vitamins except B_5 (190 nm) and B_{2a} (260 nm). Separation time was about 9 min. After experimental conditions optimization, the proposed method was validated. Precision of migration time and corrected peak area, linearity range, LOD and LOQ, accuracy (recovery), robustness, and ruggedness were evaluated for each analyte demonstrating the good reliability of the method. Analyses of the pharmaceutical real samples were performed and confirmed the versatility of this method.

1. Introduction

Quality control (QC) plays an essential role in the pharmaceutical industry. In fact many analytical methods, employing different techniques, are being developed for evaluation of pharmaceutical preparations. Vitamins of B group were well separated by RP-HPLC. A method for the simultaneous determination of taurine and 10 water-soluble vitamins including vitamin B_1 (thiamine), B_2 (riboflavine), B_5 (pantothenic acid), B_6 (pyridoxine and pyridoxal), B_8 (biotin), B_9 (folic acid), C (ascorbic acid), and B_3 (nicotinamide and nicotinic acid) in multivitamin tablets was developed and validated. Detection of components was by ESI-MS [1]. Some water-soluble vitamins of our interest (ascorbic acid, thiamine hydrochloride, riboflavine-5′-phosphate sodium, pyridoxine hydrochloride, nicotinamide, and (+)-panthenol) and two preservatives (methylparaben and sodium benzoate) in multivitamin syrup were well-separated on Zorbax SB-Aq (C18) column [2]. A HPLC-UV method was established for the simultaneous determination of eight vitamins, including B_1, B_2, B_3, B_6, B_9, cyanocobalamin, ascorbic acid in baby milk powder [3]. A validated HPLC-UV method for the determination of seven B-complex vitamins (B_1, B_2, B_3, B_6, B_9, and cyanocobalamin) in pharmaceuticals and biological fluids after solid-phase extraction was performed [4]. Simultaneous determination of water- and fat-soluble vitamins in pharmaceutical preparations by HPLC was performed in a single run using combined isocratic and linear gradient elution with a mobile phase consisting of trifluoroacetic acid and methanol. The method was applied for real sample. The results were in good agreement with the declared values. Vitamins analyzed were B_1, B_2, B_5, B_6, B_9, B_3, and B_{12} (cyanocobalamin) [5].

The RP ion-pair HPLC method was applied for determination of some vitamins of our interest in multivitamin with minerals from different authors. For nicotinamide, pyridoxine hydrochloride, thiamine mononitrate and riboflavine a HPLC method was validated using methanol 0.5% acetic acid as the mobile phase [6].

A sensitive RP-HPLC method was developed and validated for the simultaneous determination of B_1, B_3, B_6, and B_9 in Pentovit coated tablets. The procedures were carried out on a Supelcosil ABZ column with methanol, heptanesulfonic acid sodium salt, and triethylamine as the mobile phase [7]. For the determination of thiamine and riboflavine in Duoweiyuansu Tablet an HPLC-UV method was established. The mobile phase consisted of ion-pairing reagent (containing 1-hexanesulfonate, glacial acetic acid, and triethylamine)/methanol (80 : 20). The method can be used for the quality control of thiamine and riboflavine in Duoweiyuansu Tablet [8]. There are many papers about vitamins of B group analysed by capillary electrophoresis (CE) and its modifications. Some authors studied determination of vitamins in food based on supercritical fluid extraction prior to micellar electrokinetic capillary chromatographic (MEKC) analyses of individual components. The method was optimized using sodium cholate as the micellar phase for the separation of 11 B-vitamins, ascorbic acid, and 4 impurities in about 25 min [9]. Other authors investigated in capillary zone electrophoresis (CZE) analysis the effect of lower organic alcohols as cosurfactants (methanol, ethanol, n-propanol, isopropanol, propanediol, n-butanol and isoamyl alc.) and n-hexane as an organic modifier in phosphate buffer with varying SDS concentration using a set of vitamins and p-hydroxybenzoic acid as the test mixture. Optimum separations were achieved particularly at high concentrations of the surfactant [10]. The CZE analyses of three vitamins (B_1, B_2, and B_6 in tablets achieved using phosphate-borate buffer pH 9.0 short time of analysis (5 min) [11]. One year later the same authors studied another specific, precise, sensitive, and accurate method for separation of the same vitamins obtaining high separation efficiency and shorter analysis time (3 min) [12]. Four B vitamins, as B_1, B_2, pyridoxal, pyridoxine, and pyridoxamine (B_6) in a pharmaceutical product were determined simultaneously using CZE. An HCl soln. was used for the extraction of the vitamins from a multivitamin-multimineral tablet. BGE employed was sodium phosphate buffer (pH 9.0) [13].

Mixture of four water-soluble B-group vitamins were analysed both by CZE and MECK. The quantitative analyses of different pharmaceutical formulations were compared with the LC method of the US Pharmacopeia obtaining a good correlation [14].

Rapid methods were developed for simultaneously separation of five vitamins: B_1, B_2, B_6, nicotinamide, nicotinic, and ascorbic acid. They were tested on 15 real samples obtaining good resolution by (CZE) and (MECC). CZE was performed with 0.02 M borate buffer, while MECC in 0.02 M borate/phosphate buffer with 4% acetonitrile containing. 0.1 M sodium dodecyl sulfate [15].

Some authors studied a method for the analysis of six water-soluble vitamins (thiamine, nicotinamide, riboflavin, pyridoxine, pantothenic, and ascorbic acid) in a pharmaceutical formulation, by CZE. A good compromise between resolution, analysis time, and analyte stability was obtained by use of a 50 mM borax buffer of pH 8.5. This CZE method was very useful for the separation of more complex samples, but cyanocobalamin could not be separated from nicotinamide in this CZE system. In fact the 2 compounds were in uncharged form the pH used. Instead, they were resolved by MECC using SDS in BGE. Good results with respect to linearity, precision, and accuracy were obtained in the concentration range studied for the 6 vitamins [16].

There are some studies about B-complex by spectrophotometric methods.

Spectrophotometric determination of ternary mixtures of thiamine, riboflavin, and pyridoxal in pharmaceutical and human plasma by least-squares support vector machines were performed. The partial least squares (PLS) modelling and least-squares support vector machines were used for the multivariate calibration of the spectrophotometric data [17]. The same procedure was applied successfully for determination of B_1, B_2, B_3, and B_6 in pharmaceuticals by other authors [18].

Simultaneous determination of fourteen water-soluble vitamins (13 of B-group and vitamin C) in selected food matrices by LC/MS/MS technique is described in a recent paper [19]. Analytes were separated in less than 10 min with recoveries between 30% and 70%.

The QC process in pharmaceutical industry needs methods that are able to determine, in a single run, the majority of components. In many cases, to determine all components of a multivitamin pharmaceutical preparation, it is necessary to perform different analyses applying different techniques.

Capillary electrophoresis (CE) is a powerful analytical technique that is widely used in research, and in quality control of pharmaceuticals. As demonstrated exploring the literature data, the MEKC was the only method [9] able to separate up to 11 B-group vitamins. Other CE methods offer the possibility to separate only a few (max 5) molecules of this class. There are no publications about simultaneous determination of nine B-group vitamins by CZE, the simplest in CE system, which offers several advantages over HPLC such as rapid analysis, lower solvent consumption, then lower costs and minor environmental impact, and higher efficiency.

For this purpose, the aim of this work is to develop a validated CZE method for assay of a multicomponent pharmaceutical and veterinary formulations in routine analysis.

2. Experimental

2.1. Apparatus. The analyses were carried out on a P/ACE system from Beckman Instrument Fullerton, CA (USA), with a UV-DAD detector. For ruggedness evaluation, electrophoretic separations were carried out on the SpectraPhoresis 1000 instrument from Thermo-Quest Corporation, CA (USA).

The uncoated fused silica capillary ID = 50 μm, 59.5 cm total length and 49.5 cm effective length was supplied by SGE (Melbourne, Australia). Detection wavelength was set at 214 nm. The sample injections were performed in a hydrodynamic mode (5 s under 0.5 psi).

2.2. Chemicals and Materials. All reagents were of analytical grade purity.

Borax, disodium hydrogen phosphate dihydrate, sodium dihydrogen phosphate monohydrate, potassium chloride, and CTAB (Cetyl trimethylammonium bromide) were from FLUKA (Buchs, Switzerland). Phosphoric acid, sodium hydroxide, chloric acid, and acetonitrile were supplied by Carlo Erba (Milan, Italy). Trizma base (Tris [hydroxymethyl]aminomethane), EDTA (ethylenediaminetetraacetic acid). Sodium bicarbonate, citric acid, oxalic acid, and sodium bitartrate monohydrate were from Sigma-Aldrich (Steinheim, Germany).

The real samples were commercially available pharmaceutical products: Berocca Plus (Bayer S.p.A), Gabbrovital B forte (CEVA VETEM SpA), Trinidex (LDB LAB.DIACO BIOMEDICALI SpA), and Biochetasi (Sigma-Tau SpA). For solution filtering the syringe filters 0.45 μm (Millex HV, Millipore, MA, USA) were used.

2.2.1. The Standards of Vitamins. thiamine hydrochloride (B_1), thiamine pyrophosphate chloride (B_{1a}), riboflavine (B_2), riboflavine-5′monophosphate (B_{2a}), nicotinamide (B_3), d-pantothenic acid calcium salt (B_5), pyridoxine hydrochloride (B_6), biotin (B_8), folic acid (B_9), B_{10} (4-aminobenzoic acid), cyanocobalamin (B_{12}), hydroxocobalamin (B_{12a}), and 2-3 dihydroxybenzoic acid were purchased from Fluka (Buchs, Switzerland).

2.3. Standard, Buffer, and Real Sample Preparation. Standard stock solutions of studied analytes (at 1 mg/mL concentration) were prepared in distilled water, riboflavine and folic acid were dissolved in HCl, respectively, 3 M and 2 M.

The calibration solutions were obtained by dilution of the stock solution with distilled water to give a desired analyte concentration. The real samples preparation was different for each product.

For Berocca Plus preparation, twenty tablets were weighted and grounded in a mortar. The quantity of powder equivalent at one tablet was weighted and dissolved in a 50 mL volumetric flask with distilled water. After sonication for 10 min in an ultrasonic bath, the supernatant solution was centrifuged for 5 min and then filtered through the 0.45 μm syringe filter. Finally the solution was diluted with distilled water to reach a concentration order inside the calibration range.

For Trinidex sample, the intravenous injection solution was filtered through the 0.45 μm syringe filter without dilution and injected.

For Gabbrovital preparation, the veterinary intramuscular injection solution was diluted 1 : 1000 because of its high concentration, then filtered and analysed.

For Biochetasi sample, the powder contained in the vial for the intramuscular injection, was diluted with distilled water to reach a desired concentration and then filtered through the 0.45 μm syringe filter.

3. Results and Discussion

3.1. Method Development and Optimization. In order to propose a suitable method for routine analysis, it was necessary to evaluate the experimental conditions for the best resolution of studied analytes.

First, because of different solubility and stability for each vitamin, many conditions for their dissolution were tested. Several parameters for the CZE analysis, such as BGE composition and concentration, pH, injection conditions, wavelengths, applied voltage, and temperature were explored.

3.1.1. Effect of BGE Composition, Concentration, and pH. The BGE buffer composition is one of the most important parameters in capillary electrophoresis.

In order to obtain maximum precision level, it is necessary to choose the buffer very carefully.

Most of vitamins are nitrogen-containing, protonable molecules, so starting point of this work was to explore the low range of pH values for their separation.

In the preliminary step of study, the BGEs tested pH ranging between 1.5–4.2 and concentration ranging between 20 mM to 50 mM, are HCl/KCl, phosphate, phosphate-oxalate, phosphate with 5% acetonitrile, phosphate with CTAB, phosphate-tartrate, and tartrate.

Only in 20 mM phosphate buffer at pH = 2.3 good separation of eleven analytes was realised in about 23 minutes. In Figure 1, analytes from 1 to 7 are cations, the others (8–10) are in anionic form. In these operating conditions, riboflavine (B_2) and cyanocobalamin (B_{12}) are overlaid. Since these compounds are often contained in the same pharmaceutical preparation, it is impossible to determinate both of them. Long analysis time, high intensity of current, and low repeatability of migration time (RSD% $\sim$ 10%) did not allow to propose this procedure for routinely analyses.

Consequently the possibility of separation of studied vitamins in alkaline buffers were evaluated.

BGE buffers at pH between 8.5 and 9.3 into the range of concentration 20 mM to 80 mM were explored: tris/HCl, bicarbonate/carbonate, tris/borate, borate, borate with EDTA, and borate with CTAB.

Best separation of nine vitamins was realized in 20 mM borate buffer at pH 9.2 in about 9 minutes (Figure 2).

The peaks are symmetric and baseline resolved for individual B vitamins with exception of d-pantothenic acid calcium salt (B_5) and riboflavine-5′monophosphate (B_{2a}), that are coeluted. Since a good compromise between peak symmetry, resolution, and analysis time resulted in 20 mM borate buffer at pH = 9.2, these experimental conditions were chosen as optimal.

3.1.2. Effect of Wavelength. Primarily it was necessary to select the most suitable wavelength for the simultaneous detection of vitamins, since spectral properties of each one differs enough. The UV spectra obtained by UV-DAD show three suitable wavelengths for quantitative analysis. 214 nm for all the vitamins, at exception of d-pantothenic acid calcium salt (B_5) and riboflavine-5′monophosphate (B_{2a}), that are overlaid. Then the quantitative analysis of these two analytes was realized at the wavelengths of 190 and 260 nm, because the considered compounds do not interfere themselves. In fact vitamin B_5 has only maximum of absorbance at 190 nm,

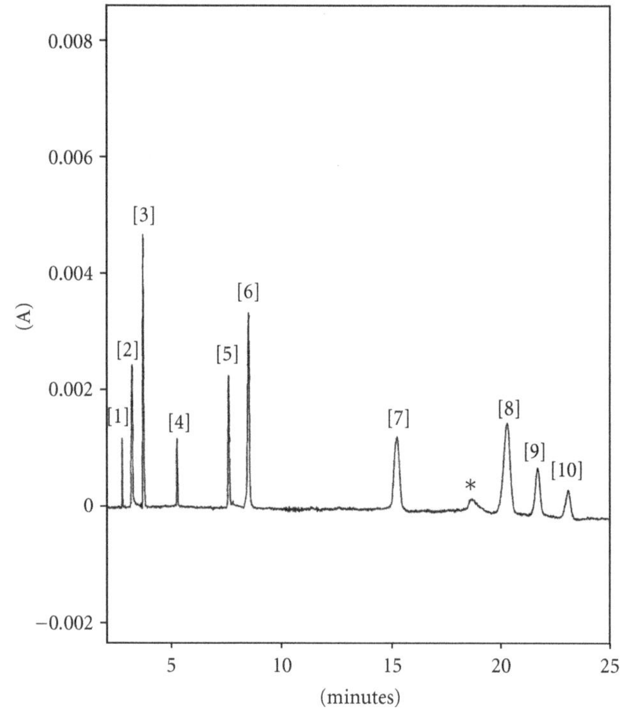

FIGURE 1: Electropherogram of standard mixture in acidic BGE. Peak identification: (1) B_1 20 μg/mL; (2) B_3 10 μg/mL; (3) B_6 5 μg/mL; (4) B_{12a} 20 μg/mL; (5) B_{1a} 10 μg/mL; (6) B_{10} 5 μg/mL; (7) B_9 15 μg/mL; (8) $B_2 + B_{12}$ (10 + 20) μg/mL; (9) B_7 10 μg/mL; (10) B_5 40 μg/mL. Experimental conditions: phosphate buffer 20 mM; pH = 2.3; V = 30 kV; i = 80 μA; T = 25°C; injection at 0.5 psi for 5 s, fused silica capillary i.d. = 50 μm, L = 59.5 cm, l = 49.5 cm instrument P/ACE MDQ System (Beckman).

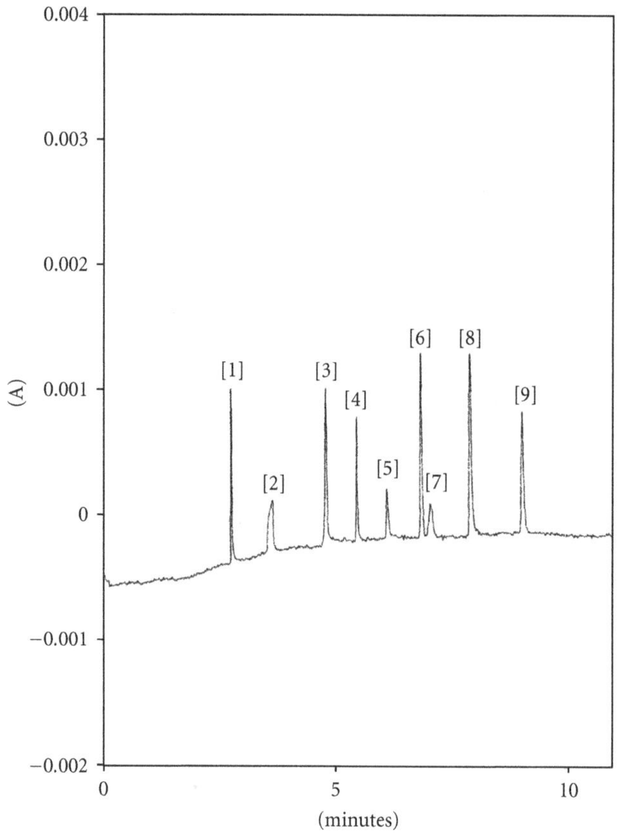

FIGURE 2: Electropherogram of standard mixture in basic BGE. Peak identification: (1) B_1 20 μg/mL; (2) B_3 10 μg/mL; (3) B_2 10 μg/mL; (4) B_6 5 μg/mL; (5) B_{1a} 10 μg/mL; (6) B_{10} 5 μg/mL; (7) $B_5 + B_{2a}$ (40 + 10) μg/mL; (8) S.I. 3 μg/mL; (9) B_9 20 μg/mL. Experimental conditions: borate buffer 20 mM; pH = 9.2; V = 20 kV; i = 25 μA; T = 25°C; t_{inj} = 5 s p_{inj} = 0.5 psi fused silica capillary ID = 50 μm; L = 59.5 cm; l = 49.5 cm, instrument P/ACE MDQ System (Beckman).

while vitamin B_{2a} presents two maximums at 214 and 260 nm.

3.1.3. Effect of Applied Voltage. For the optimization of separation voltage the analyses ranging the applied voltage between 10 and 30 kV were carried out. At low voltage the separation of all the analytes was reached but the analysis time increased to 20 min with the broadening of peaks. Increasing the voltage (max 30 kV) the optimal value results in 20 kV producing the current intensity of 25 uA and good repeatability. At 30 kV the separation of analytes was incomplete.

3.1.4. Effect of Temperature. The increase of temperature (30°C) produced the shortening of analysis time (the viscosity of the BGE decrease). It should be advantageous setting high temperature of cartridge but B-complex is thermolabile. So the suitable temperature for analyses of these compounds resulted in 25°C.

3.2. Method Validation. The method was validated following the ICH guideline [20].

3.2.1. Precision. The precision was evaluated in terms of RSD% of migration time (t_m) and corrected peak area (A_c) for intraday and interday analyses. For the intra-day precision evaluation the standard solutions at three concentrations levels (low, intermediate, and high) were injected in the same day. To estimate the interday precision, the standard solutions were analysed for five consecutive days performing five consecutive injections every day. The calculated RSD% values are reported in Table 1.

3.2.2. Calibration Range. The linearity of detector response was tested in different ranges as reported in Table 2. For the quantitative analysis the Internal Standard method was applied. For its property, was choosen Internal Standard 3,4 dihydroxy-benzoic acid.

Six standard solutions, containing nine analytes of interest and the Internal Standard, were injected in triplicate. The calibration lines were obtained plotting R (corrected areas ratio) versus the standard solutions concentration using Microsoft EXCEL.

TABLE 1: Intraday and interday precision of migration time (t_m) and corrected peak area (A_c) in optimal condition at low-, intermediate-, and high-concentration levels.

	RSD% t_m		RSD% A_c			
	Intraday	Interday	Intraday Low	Intraday Intermediate	Intraday High	Interday Intermediate
B_1	0.20	0.31	5.51	2.89	1.76	4.37
B_3	0.17	0.57	3.16	2.54	2.30	4.31
B_2	0.24	0.99	4.89	1.89	1.41	4.15
B_6	0.32	0.77	5.48	2.39	1.91	3.22
B_{1a}	0.34	0.81	6.29	1.95	1.26	3.87
B_{10}	0.32	0.95	3.83	1.91	2.14	4.28
B_5	0.27	0.87	4.91	1.04	1.56	3.17
B_{2a}	0.34	0.57	5.00	2.59	1.51	4.91
SI	0.34	0.82	4.51	2.15	1.99	2.61
B_9	0.38	1.01	5.74	1.16	2.84	4.33

TABLE 2: Method linearity data.

Analyte	Calibration range μg/mL	Equation	r^2
B_1	5–200	$y = 0.0235x - 0.0111$	0.9952
B_3	10–120	$y = 0.0722x - 0.2297$	0.9931
B_2	5–40	$y = 0.0742x - 0.0104$	0.9989
B_6	3–40	$y = 0.0597x - 0.0024$	0.9993
B_{1a}	5–40	$y = 0.0183x - 0.0194$	0.9956
B_{10}	3–25	$y = 0.0887x - 0.0240$	0.9973
B_5	30–120	$y = 0.0077x - 0.0666$	0.9970
B_{2a}	5–40	$y = 0.0224x + 0.0053$	0.9972
B_9	5–30	$y = 0.0557x - 0.0488$	0.9912

TABLE 3: Commercial product analysis.

Commercial product	Berocca plus		Gabbrovital B		Trinidex		Biochetasi	
Analyte	mg/tab		mg/100 mL		mg/100 mL		mg/vial	
	Found	Declared	Found	Declared	Found	Declared	Found	Declared
B_1	1.55 ± 0.06	1.54	15.24 ± 0.38	15.00	0.98 ± 0.05	1.00	—	—
B_{1a}	—	—	—	—	—	—	40.41 ± 1.20	47.00
B_2	1.77 ± 0.36	1.76	—	—	—	—	—	—
B_{2a}	—	—	—	—	—	—	31.10 ± 1.19	28.30
B_3	17.93 ± 0.97	18.00	3.03 ± 0.17	3.00	11.03 ± 0.91	10.00	—	—
B_5	6.48 ± 0.50	6.60	—	—	—	—	—	—
B_6	2.18 ± 0.11	2.20	1.50 ± 0.03	1.50	20.18 ± 0.24	20.00	12.35 ± 0.23	15.00
B_9	0.24 ± 0.03	0.25	—	—	—	—	—	—

3.2.3. LOD and LOQ. LOD, the lowest concentration of analytes that can be distinguished from the noise, defined as signal to noise ratio S/N of 3 : 1 was ranging between 0.9 μg/mL to 9.0 μg/mL.

LOQ, the lowest concentration of analytes that can be quantified with good precision, defined as signal to noise ratio S/N of 10 : 1 was ranging between 3.0 μg/mL to 30.0 μg/mL.

3.2.4. Analysis of Real Samples. Four pharmaceutical formulation containing B-complex vitamins were analyzed. The experimental results are given in Table 3. No interference was observed in pharmaceutical solutions. RSD% of migration time and corrected peak area values were lower, respectively, than 0.5% and 3.0%. A typical electropherogram of commercial pharmaceutical preparation (Berocca Plus) is shown in Figure 3.

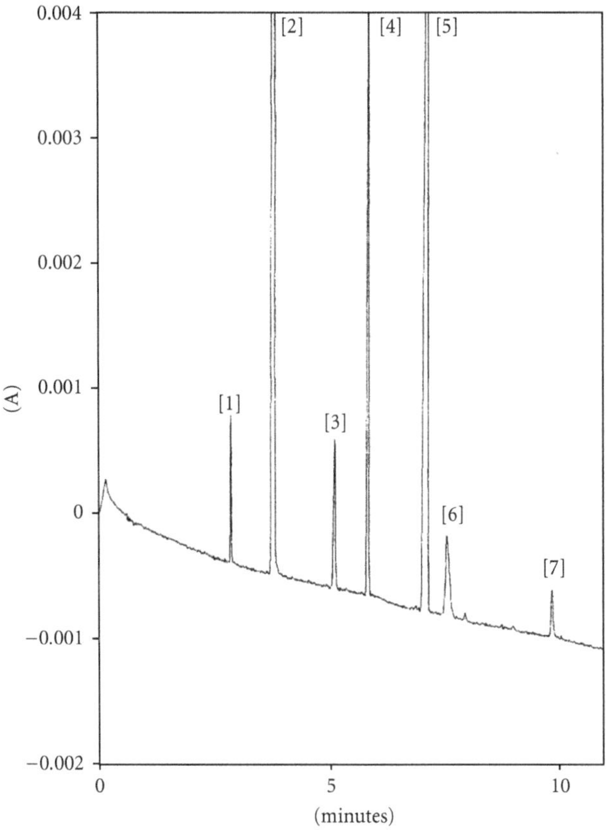

FIGURE 3: Electropherogram of real sample Berocca Plus. Experimental conditions: see Figure 2. Peak identification: (1) B_1 31 μg/mL; (2) B_3 36 μg/mL; (3) B_2 35 μg/mL; (4) B_6 40 μg/mL; (5) ascorbic acid (not determined); (6) B_9 5 μg/mL; (7) SI 3 μg/mL. The concentrations are the nominal values.

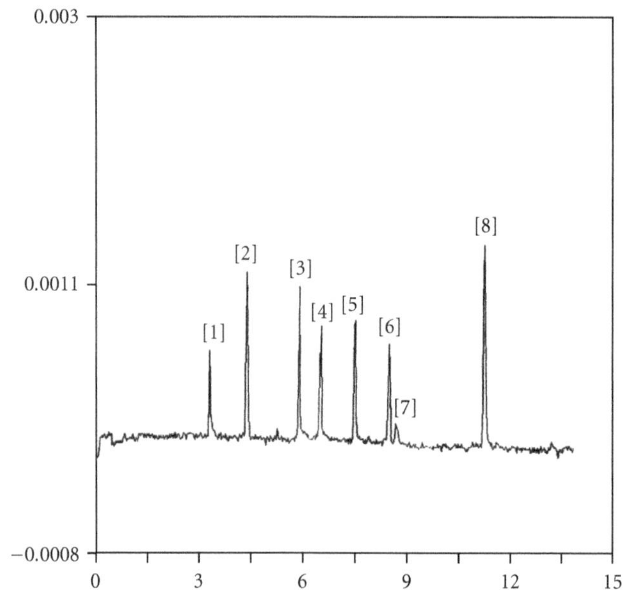

FIGURE 4: Electropherogram of standard mixture. Peak identification: (1) B_1, (2) B_3, (3) B_2, (4) B_6, (5) B_{1a}, (6) B_{10}, (7) B_5 + B_{2a}, (8) B_9. The vitamin concentrations are the same as in Figure 2. Experimental conditions: borate buffer 20 mM; pH = 9,2; V = 10 kV; i = 17 μA; T = 25°C; hydrodynamic injection t_{inj} = 1 s; fused silica capillary ID = 50 μm; L = 44 cm; l = 36 cm; instrument Spectra Phoresis 1000.

3.2.5. Accuracy. Accuracy was evaluated by recovery assays on the commercial product. After fortification, the real sample undergo the analysis and the resulted concentration were compared with the data obtained without fortification. The resulting recoveries data are ranging from 97.0 to 101.4%.

3.2.6. Ruggedness. For this purpose, measurements on the SpectraPhoresis 1000 apparatus in the optimal experimental conditions were performed. In this instrument the cartridge form is different and the capillary total length is 44 cm, the effective length 36 cm (shorter than optimal length of capillary utilized on the Beckman instrument). Consequently, on this instrument, it was necessary to apply lower voltage (10 kV). The electropherogram, obtained on this apparatus, injecting the standard mixture in hydrodynamic mode for 1 sec, is shown in Figure 4. Repeatability of migration time is ranging from 0.32% to 0.52%.

4. Conclusion

Proposed method was developed for the simultaneous determination of nine water-soluble vitamins (B_1, B_{1a}, B_2, B_{2a}, B_3, B_5, B_6, B_9, and B_{10}) in the multivitamin pharmaceutical formulations by CZE. The best resolution was obtained in about 9 min using a simple tetraborate buffer 20 mM, pH = 9.2 at 25°C, and constant voltage of 20 kV in uncoated fused silica capillary. The method is suitable for the vitamins B analysis as the validation confirmed: it is precise, accurate, and rugged. The analyses of real samples proved its applicability. Since no expensive reagent and no pretreatment of samples are required in this procedure, it offers a valid alternative for quality control analysis in the pharmaceutical industry.

References

[1] Z. Chen, B. Chen, and S. Yao, "High-performance liquid chromatography/electrospray ionization-mass spectrometry for simultaneous determination of taurine and 10 water-soluble vitamins in multivitamin tablets," *Analytica Chimica Acta*, vol. 569, no. 1-2, pp. 169–175, 2006.

[2] S. Vidović, B. Stojanović, J. Veljković, L. Pražić-Arsić, G. Roglić, and D. Manojlović, "Simultaneous determination of some water-soluble vitamins and preservatives in multivitamin syrup by validated stability-indicating high-performance liquid chromatography method," *Journal of Chromatography A*, vol. 1202, no. 2, pp. 155–162, 2008.

[3] N. Liu, D. Chen, H. Tang, H. Yu, and L. Li, "Simultaneous determination by high performance liquid chromatography of water-soluble vitamins in powdered milk for infants of eight species," *Fenxi Ceshi Xuebao*, vol. 27, no. 4, pp. 408–411, 2008.

[4] P. F. Chatzimichalakis, V. F. Samanidou, R. Verpoorte, and I. N. Papadoyannis, "Development of a validated HPLC method for the determination of B-complex vitamins in pharmaceuticals and biological fluids after solid phase extraction," *Journal of Separation Science*, vol. 27, no. 14, pp. 1181–1188, 2004.

[5] B. Klejdus, J. Petrlová, D. Potěšil et al., "Simultaneous determination of water- and fat-soluble vitamins in pharmaceutical preparations by high-performance liquid chromatography coupled with diode array detection," *Analytica Chimica Acta*, vol. 520, no. 1-2, pp. 57–67, 2004.

[6] K. Li, "Simultaneous determination of nicotinamide, pyridoxine hydrochloride, thiamine mononitrate and riboflavin in multivitamin with minerals tablets by reversed-phase ion-pair high performance liquid chromatography," *Biomedical Chromatography*, vol. 16, no. 8, pp. 504–507, 2002.

[7] R. Amidžić, J. Brboric, O. Čudina, and S. Vladimirov, "RP-HPLC determination of vitamins B_1, B_3, B_6, folic acid and B_{12} in multivitamin tablets," *Journal of the Serbian Chemical Society*, vol. 70, no. 10, pp. 1229–1235, 2005.

[8] W. Chen, J. Ye, and W. Tian, "Determination of thiamine and riboflavin in "Duoweiyansu" tablet by HPLC," *Zhongguo Yaoye*, vol. 17, no. 5, pp. 18–19, 2008.

[9] S. Buskov, P. Møller, H. Sørensen, J. C. Sørensen, and S. Sørensen, "Determination of vitamins in food based on supercritical fluid extraction prior to micellar electrokinetic capillary chromatographic analyses of individual vitamins," *Journal of Chromatography A*, vol. 802, no. 1, pp. 233–241, 1998.

[10] M. S. Bellini, G. Manetto, Z. Deyl, F. Tagliaro, and I. Mikšík, "Capillary electrophoretic separation of vitamins in sodium dodecyl sulfate containing buffers with lower aliphatic alcohols and n-hexane as organic modifiers," *Journal of Chromatography B*, vol. 741, no. 1, pp. 67–75, 2000.

[11] D. Rajzer, E. Sell, and A. Kraczkowska, "Determination of vitamins of the B group in "Vitaminum B Compositum" by capillary zone electrophoresis," *Annales Academiae Medicae Gedanensis*, vol. 31, pp. 89–94, 2001.

[12] D. Rajzer, E. Sell, A. Kraczkowska, and K. Sawicz, "Identification and determination of vitamins of the B group in "extract of Equisetum arvense with vitamins" by capillary zone electrophoresis," *Annales Academiae Medicae Gedanensis*, vol. 32, pp. 293–300, 2002.

[13] R. Huopalahti and J. Sunell, "Use of capillary zone electrophoresis in the determination of B vitamins in pharmaceutical products," *Journal of Chromatography*, vol. 636, no. 1, pp. 133–135, 1993.

[14] S. Boonkerd, R. Detaevernier, and Y. Michotte, "Use of capillary electrophoresis for the determination of vitamins of the B group in pharmaceutical preparations," *Journal of Chromatography A*, vol. 670, no. 1-2, pp. 209–214, 1994.

[15] S.-C. Su, S.-S. Chou, D.-F. Hwang, P.-C. Chang, and C.-H. Liu, "Capillary zone electrophoresis and micellar electrokinetic capillary chromatography for determining water-soluble vitamins in commercial capsules and tablets," *Journal of Food Science*, vol. 66, no. 1, pp. 10–14, 2001.

[16] L. Fotsing, M. Fillet, I. Bechet, P. Hubert, and J. Crommen, "Determination of six water-soluble vitamins in a pharmaceutical formulation by capillary electrophoresis," *Journal of Pharmaceutical and Biomedical Analysis*, vol. 15, no. 8, pp. 1113–1123, 1997.

[17] A. Niazi, J. Zolgharnein, and S. Afiuni-Zadeh, "Spectrophotometric determination of ternary mixtures of thiamin, riboflavin and pyridoxal in pharmaceutical and human plasma by least-squares support vector machines," *Analytical Sciences*, vol. 23, no. 11, pp. 1311–1316, 2007.

[18] R. C. Barthus, L. H. Mazo, and R. J. Poppi, "UV spectrophotometry and chemometrics methods for simultaneous determinations of riboflavin (VB_2), thiamine (VB_1), pyridoxine (VB_6) and nicotinamide (VPP)," *Quimica Nova*, vol. 30, no. 7, pp. 1638–1643, 2007.

[19] A. Gentili, F. Caretti, G. D'Ascenzo et al., "Simultaneous determination of water-soluble vitamins in selected food matrices by liquid chromatography/electrospray ionization tandem mass spectrometry," *Rapid Communications in Mass Spectrometry*, vol. 22, no. 13, pp. 2029–2043, 2008.

[20] ICH, *Harmonised Tripartite Guideline*, IFPMA Ed., Geneva, Switzerland, 1994.

4

Rapid Screening for Flavone *C*-Glycosides in the Leaves of Different Species of Bamboo and Simultaneous Quantitation of Four Marker Compounds by HPLC-UV/DAD

Jin Wang, Yong-de Yue, Hao Jiang, and Feng Tang

SFA Key Laboratory of Bamboo and Rattan Science and Technology, International Centre for Bamboo and Rattan, No. 8 Futong Dongdajie, Wangjing, Chaoyang District, Beijing 100102, China

Correspondence should be addressed to Yong-de Yue, yueyd@icbr.ac.cn

Academic Editor: M. A. Raggi

A strategy for analyzing flavone *C*-glucosides in the leaves of different species of bamboo was developed. Firstly, the flavone *C*-glycosides were extracted from the bamboo leaves (51 species in 17 genera) with methanol and chromatographed on silica gel 60 plates in automatic developing chamber (ADC2), and a qualitative survey using simple derivatization steps with the NP reagent was carried out. The flavone *C*-glycosides were found in 40 of 51 species of bamboo examined. Secondly, an HPLC method with photodiode array and multiple wavelength detector was optimized and validated for the simultaneous determination of flavone *C*-glycosides, including isoorientin, isovitexin, orientin, and vitexin in the leaves of three species of bamboo and the flavone *C*-glycosides were confirmed by LC/MS. The optimized HPLC method proved to be linear in the concentration range tested (0.2–100 μg/mL, $r^2 \geq 0.9997$), precise (RSD $\leq$ 1.56%), and accurate (88–106%). The concentration ranges of isoorientin, isovitexin, orientin, and vitexin in three bamboo leaves samples were 1.00–2.78, 0–0.31, 0–0.07, and 0.20–0.68 mg/g, respectively. The proposed method was validated to be simple and reliable and can be a tool for quality control of bamboo leaf extract or its commercial products.

1. Introduction

Bamboo leaves have been used in traditional Chinese medicine for treating fever and detoxification for over 1000 years. It was found that extract of bamboo leaves has multiple biological activities, such as cancer prevention [1, 2], anti-free radical, and antioxidation [3, 4], and can be used as a pharmaceutical intermediate and food additive. A new product named bamboo-leaf flavonoid has appeared on the market. The main functional components of the extract from bamboo leaf are flavone *C*-glycosides including isoorientin, isovitexin, orientin, and vitexin (Figure 1), which have been used as marker compounds for the determination of commercial products of bamboo-leaf flavonoids.

Bamboo is an important renewable resource and more than 1250 bamboo species, belonging to 75 genera, are distributed all over the world [5]. The quality of different batches of the same herb could vary significantly [6]. Hence, to investigate and determine these flavone *C*-glycosides in the leaves of different species of bamboo is needed.

Thin-layer chromatography (TLC) chemical screening is simple, fast, and of low cost in identification of chemical ingredient from plant extracts [7, 8]. Many papers about TLC chemical screening have been published in recent years [9, 10]. Analytical methods for determination of the *C*-glycosylflavone isomer pairs orientin/isoorientin and vitexin/isovitexin, including HPLC [11], HPCE [12], HPTLC [13, 14], and HPLC-MS [15], have been reported. However, the reported HPLC chromatographic conditions need to be optimized to obtain a good resolution of adjacent peaks. Furthermore, HPLC analysis for selected bamboo species in this study was not reported.

The work aimed to develop a two-step procedure to quickly survey a large number of bamboo leave samples (51 species in 17 genera) for the presence of flavone *C*-glycosides and to optimize and validate an HPLC method

Table 1: Results of TLC screening for isoorientin, isovitexin, orientin, and vitexin in the leaves of 51 different species of bamboo.

No.	Family	Bamboo species	Bands intensity[a]				Plant sources[b]
			Isoorientin	Isovitexin	Orientin	Vitexin	
1	*Acidosasa*	*A. gigantea*	−	−	−	−	A
2	*Bambusa*	*B. multiplex* cv. *Alphonse-Karr*	+	+	−	−	B
3		*B. textilis* McClure	+	−	+	−	A
4		*B. multiplex* cv. *Silverstripe*	+	+	−	−	A
5		*B. ventricosa* McClure	+	−	−	−	A
6		*B. vulgaris* cv. *Wamin*	−	−	−	−	A
7		*B. multiplex* var. *multiplex*	+	+	−	−	B
8		*B. multiplex* cv. *Fernleaf*	+	+	−	−	A
9		*B. multiplex* var. *riviereorum* R. Maire	+	+	−	−	A
10	*Bashania*	*B. fargesii* (E. G. Camus) Keng f. et Yi	+	−	+	+	B
11	*Brachystachyum*	*B. densiflorum* (Rendle) Keng	+	−	+	+	A
12	*Chimonobambusa*	*Ch. quadrangularis* (Fenzi) Makino	+	−	−	−	A
13	*Chimonocalamus*	*Chi. Delicates*	+	−	+	−	C
14	*Dendrocalamus*	*D. oldhamic* McClure	+	−	−	−	A
15		*D. latiflorus* Munro	−	−	−	−	D
16		*D. minor* var. *amoenus* (Q. H. Dai et C. F. Huang) Hsueh et D. Z. Li	+	+	−	+	E
17		*D. hamiltonii* Nees et Arn. ex Munro	−	−	−	−	D
18		*D. yunnanicus* Hsuch et	−	−	−	−	E
19	*Indocalamus*	*I. Tessellatus* (Munro) Keng f.	+	−	−	−	E
20		*I. latifolius* (Keng) McClure	+	−	−	−	E
21		*I. longiauritus* Hand.-Mazz	−	−	−	−	E
22	*Indosasa*	*I. sinica* C. D. Chu et C. S. Chao	−	−	−	−	A
23	*Menstruocalamus*	*M. sichuanensis*	+	−	+	−	B
24	*Neosinocalamus*	*N. affinis* (Rendle) Keng f.	+	−	−	−	E
25	*Phyllostachys*	*P. meyeri* McClure	−	−	−	−	E
26		*P. nidularia* Munro	+	−	+	−	A
27		*P. aureosulcata* cv. *spectabilis*	+	−	+	−	A
28		*P. glauca* McClure	+	−	+	−	B
29		*P. dulcis* McClure	+	−	−	−	A
30		*P. acuta* C. D. Chu et C. S. Chao	+	+	+	−	A
31		*P. stimulosa* H. R. Zhao et A. T. Liu	−	−	+	−	A
32		*P. heteroclada* Oliver	−	−	−	−	A
33		*P. sulphurea* cv. *Houzeau*	+	+	−	−	A
34		*P. heterocycla* cv. *pubescens*	+	+	+	+	A
35		*P. vivax* McClure	+	−	+	−	A
36		*P. mannii* Gamble	+	−	+	−	A
37		*P. bambusoides* f. *shouzhu*	+	−	+	−	A
38		*P. aurea* Carr.ex A. et C. Riviere	+	−	+	−	A
39	*Pleioblastus*	*P. maculatus* (McClure) C. D. Chu et C. S. Chao	+	−	+	−	B
40		*P. amarus* (Keng) Keng f.	−	−	+	+	B
41		*P. gramineus* (Bcan) Nakai	+	−	+	−	B
42		*P. gozadakensis* Nakai	+	−	+	−	B
43		*P. yixingensis*	−	−	—	−	B
44	*Pseudosasa*	*P. cantori* (Munro) Keng f.	−	−	—	−	A
45		*P. japonica* (Sieb. et Zucc.) Makino	−	−	+	−	B
46		*P. amabilis* (McClure) Keng f.	+	−	+	−	A

Table 1: Continued.

No.	Family	Bamboo species	Bands intensity[a]				Plant sources[b]
			Isoorientin	Isovitexin	Orientin	Vitexin	
47		*S. fortunei* (Van Houtte) Fiori	+	−	+	−	B
48	*Sasa*	*S. pygmaea* (Miq.) E. G. Camus	+	−	+	−	B
49		*S. argenteastriatus* E. G. Camus	+	−	+	−	B
50	*Shibataea*	*Sh. chinensis* Nakai	+	−	+	−	B
51	*Sinobambusa*	*Si. tootsik* (Sieb.) Makino	−	−	+	−	B

[a] + Presence. − not detected or absence;
[b] A: Bamboo Garden of Jiangxi Academy of Forestry (Nanchang, China);
B: Bamboo Garden of Nanjing Forestry University (Nanjing, China);
C: Forestry Bureau of Jinping county (Yunnan, China);
D: Xishuangbanna Tropical Bamboo Garden (Yunnan, China);
E: Changning Century Bamboo Garden (Sichuan, China).

Table 2: Regression equations, linear range, LOD, and LOQ for quantitative analysis by HPLC ($n = 3$).

Compound	Linear range (μg/mL)	Regression equation[a]	Correlation coefficient (r^2)	LOD (μg/mL)	LOQ (μg/mL)
Isoorientin	0.2–100	$Y = 2240X - 4620$	0.9997	0.02	0.07
Orientin	0.2–100	$Y = 2590X - 6080$	0.9997	0.02	0.07
Vitexin	0.2–100	$Y = 2600X + 516$	0.9998	0.03	0.10
Isovitexin	0.2–100	$Y = 2830X + 3430$	0.9998	0.02	0.07

[a] Y is the peak area; X is the amount (ng).

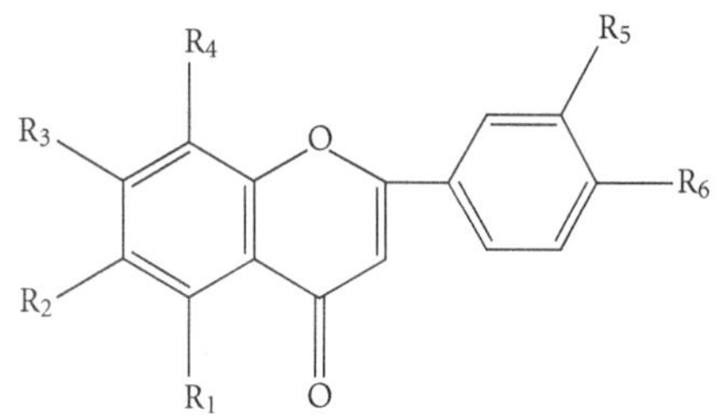

Figure 1: Chemical structures of isoorientin, isovitexin, orientin, and vitexin. Isoorientin: $R_1 = R_3 = R_5 = R_6 = OH$, $R_4 = H$, $R_2 =$ glucose; isovitexin: $R_1 = R_3 = R_6 = OH$, $R_4 = R_5 = H$, $R_2 =$ glucose; orientin: $R_1 = R_3 = R_5 = R_6 = OH$, $R_2 = H$, $R_4 =$ glucose; vitexin: $R_1 = R_3 = R_6 = OH$, $R_2 = R_5 = H$, $R_4 =$ glucose.

for the simultaneous quantitation of marker compounds in different bamboo leaves and its commercial products.

2. Experimental

2.1. Plant Material and Reagents. The leaves of fifty-one bamboo species were collected from five regions in China in 2007 and authenticated by Professor Jiusheng Peng from Jiangxi Academy of Forestry, Professor Yulong Ding from Nanjing Forestry University, and Professor Qirong Guo from International Centre for Bamboo and Rattan, respectively. Their family, species, and plant sources are shown in Table 1. Bamboo leaves were dried in the shade, ground to powder, and stored below −20°C.

The HPLC-grade acetonitrile and methanol were purchased from fisher scientific (USA). Glacial acetic acid (≥99.7%) was purchased from Sigma-Aldrich (USA). Water was purified with an ultrapure water system (Purelab Plus, Pall, USA). Analytical-grade chemical was obtained from Beijing Chemical Works (China). Commercial products of bamboo-leaf flavonoids labeled with total content of flavonoids (30%) and antioxidant potency were obtained from Jinhua Beikang Biochemical Development Co., Ltd. (Zhejiang, China). The standard compounds of isoorientin, orientin, isovitexin and vitexin were purchased from Shanghai Winherb Medical S & T Development Co., Ltd. (Shanghai, China).

2.2. Preparation of Standard Solutions. Isoorientin and isovitexin stock solutions were prepared in methanol at the concentration of 370 μg/mL and 480 μg/mL. Orientin and vitexin stock solutions were prepared in methanol/water (90/10, v/v) at the concentration of 480 μg/mL and 440 μg/mL. The standard solutions were prepared by transferring appropriate volumes of stock solutions of each compound to a 10 mL volumetric flask and diluting to volume with methanol/water (60/40, v/v). A series of standard solutions with concentrations 0.01–100 μg/mL were prepared.

2.3. TLC Screening of Flavone C-Glycosides. The dried bamboo leaf powder (1.0 g) was extracted by percolation with 20 mL of methanol and then sonicated for 30 min at room temperature. Thin-layer chromatography was performed on the plates precoated with silica gel 60 F254 (Merck, Germany). Samples (4 μL) were applied to the plate as bands of 8 mm wide by means of a Camag (Switzerland) Linomat V sample applicator. The plate was developed in

Table 3: Intra- and interday precision by the HPLC quantitation method.

Compound	Concentration (μg/mL)	Intraday (n = 4) RSD (%)			Interday (n = 12) RSD (%)
		Day 1	Day 2	Day 3	
Isoorientin	5	0.77	0.66	0.74	1.12
	15	0.62	0.08	0.15	0.90
	50	0.13	0.08	0.14	0.73
Orientin	5	0.64	0.27	0.18	1.18
	15	0.30	0.08	0.17	0.77
	50	0.21	0.08	0.08	0.73
Vitexin	5	0.16	0.15	0.22	0.64
	15	0.09	0.07	0.05	0.57
	50	0.25	0.06	0.08	0.57
Isovitexin	5	0.29	0.20	0.27	0.66
	15	1.56	0.11	0.16	1.33
	50	0.31	0.07	0.12	0.64

Table 4: Recoveries of isoorientin, isovitexin, orientin, and vitexin by the HPLC method.

Bamboo species	Compound	Original contents (μg/mL)	Spiked (μg/mL)	Amount found (μg/mL)	Recovery[a] (%)	RSD (%)
Bambusa multiplex cv. *Silverstripe*	Isoorientin	39.79	21.26	59.38	92.16	2.33
		39.79	42.52	81.80	98.80	1.11
		39.79	59.53	96.74	95.76	2.18
	Isovitexin	9.46	4.00	13.16	92.42	3.25
		9.46	8.00	17.45	99.92	0.81
		9.46	11.20	20.54	98.96	4.66
Phyllostachys heterocycla cv. *Pubesceus*	Isoorientin	15.18	7.59	2.81	100.53	3.02
		15.18	15.18	29.72	95.76	5.01
		15.18	22.77	37.78	99.25	1.53
	Orientin	3.89	2.00	6.01	105.83	3.22
		3.89	4.00	7.90	100.17	5.25
		3.89	6.00	10.23	105.61	4.03
	Vitexin	0.96	0.50	1.41	89.33	7.20
		0.96	1.00	1.93	96.67	7.48
		0.96	1.50	2.29	88.44	4.90
	Isovitexin	4.02	2.00	6.12	105.00	4.23
		4.02	4.00	8.20	104.42	2.88
		4.02	6.00	10.24	103.61	4.00
Phyllostachys acuta	Isoorientin	20.77	11.00	31.07	93.61	1.10
		20.77	22.00	41.72	95.21	3.92
		20.77	33.00	52.71	96.79	3.98
	Orientin	4.30	2.00	6.33	101.50	5.55
		4.30	4.00	7.89	89.75	4.69
		4.30	6.00	10.11	96.83	3.48
	Isovitexin	2.84	2.50	5.07	89.07	6.13
		2.84	5.00	7.89	101.00	4.98
		2.84	7.50	9.73	91.87	3.53

[a]mean ± SD (n = 3).

Table 5: Identification of flavone *C*-glycosides from bamboo leaves.

No.	RT (min)	$[M-H]^-$ (*m/z*)	$[2M-H]^-$ (*m/z*)	UV λ_{max} (nm)	Identification
1	14.09	447.1	895.2	270, 349	Isoorientin
2	15.89	447.1	895.2	256, 349	Orientin
3	23.55	431.0	863.2	269, 338	Vitexin
4	25.93	431.0	863.2	270, 338	Isovitexin

an automatic developing chamber (ADC2) (Camag, Switzerland) using ethyl acetate : formic acid : water (82 : 9 : 9, v/v/v) as the mobile phase. The developing distance was 80 mm. After development, plates were dried and derivatisation was carried out by immersion of the plates in a solution of NP reagent (diphenylborinic acid 2-aminoethyl ester (1.0 g) in ethyl acetate (200 mL) and PEG 400 (20 g) in dichloromethane (400 mL) [16].

2.4. Sample Preparation for HPLC and LC/MS Analysis. Dried bamboo leaves powder (1.0 g) was accurately weighed and transferred in a 150 mL conical flask with glass stopper, and 60% aqueous methanol (15 mL) was added [3, 17]. The solution was stirred for 12 h and sonicated for 30 min at room temperature and then centrifuged at 5000 rpm for 5 min. The residue was further extracted with 60% methanol (2×15 mL) as mentioned above. All the extracted solutions were combined in a 50 mL volumetric flask and diluted to the mark. All solutions were filtered through 0.45 μm membrane filter prior to analysis. An aliquot of 20 μL of solution was injected for HPLC and LC/MS analysis.

2.5. HPLC Analysis. The HPLC-UV/DAD analyses were carried out on a Waters Alliance 2695 (Milford, MA, USA) system connected to a model 2996 (DAD). The separation was performed using a C_{18} column (4.6×250 mm, 5 μm, YMC pack R&D ODS-A, YMC Japan) at 30°C. The solvent system consisted of a mixture of water with 0.5% (v/v) glacial acetic acid and acetonitrile (85/15, v/v) at a flow rate of 1.0 mL/min.

2.6. LC/MS Analysis. The LC-MS analysis was conducted using an Agilent 1100 system equipped with a photodiode array and multiple wavelength detector. The column and the chromatographic condition were the same as those used for the HPLC analysis. The LC effluent was split using a T-splitter to produce a flow of 0.25 mL/min. The identification of isoorientin, isovitexin, orientin, and vitexin was performed on the quadrupole mass spectrometer equipped with electrospray ionization (ESI-MS) in the negative mode. The collision energy was 60 eV. Capillary voltage was 3.5 kV, and the drying gas was set at 350°C. The scan range was set at *m/z* 100 to 1000.

2.7. Validation of HPLC Method. The HPLC method was validated in terms of linearity, limit of detection (LOD), limit of quantitation (LOQ) precision, specificity, and accuracy according to the International Conference on Harmonization (ICH) guidelines [18].

The calibration curves were constructed by plotting peak area versus concentrations with the help of the Empower software. The standard solutions of the analytes were diluted with 60% methanol to yield a series of appropriate concentrations. LOD and LOQ were estimated experimentally by injecting a series of dilute solutions with known concentrations until the signal-to-noise ratio for the standards reached a 3 : 1 ratio for LOD and 10 : 1 for LOQ.

Intraday and interday variations were chosen to determine the precision of the HPLC method. For intraday precision, three different concentrations of standard solutions (5, 15, and 50 μg/mL) were determined in one day and expressed as relative standard deviation (RSD). For interday precision, the standards were determined for three consecutive days.

The specificity of the method was ascertained by analyzing standard compounds and samples. The peaks for the marker compounds from sample solutions were confirmed by comparing the retention time and UV spectra with the reference standards. The peak purity was checked using DAD (λ = 210–400 nm). The peaks were considered pure when there was a coincidence between the two spectral sections.

Recovery experiments were performed to evaluate the accuracy of the method. The extract of bamboo leaves used for recovery studies was preanalyzed and spiked with the mixed standards at high, middle, and low concentration levels. Three replicates were performed for the test and variations were expressed by standard deviation (SD).

3. Results and Discussion

3.1. Result of Preliminary TLC Screening. To optimize the TLC chromatographic conditions, different mobile phases and derivatisation regents were evaluated. The mobile phase ethyl acetate : formicacid : water (82 : 9 : 9, v/v/v) gave a good seperation of the markers. Satisfactory derivatisation efficiency was obtained using diphenylborinic acid 2-aminoethyl ester and PEG 400 as derivatisation regents. After derivatisation, flavone *C*-glycosides including isoorientin, isovitexin, orientin, and vitexin showed the four characteristic fluorescent bands on the TLC plates with the colour of yellow, green, yellow and green, respectively. A typical TLC profile of extracts from bamboo leaves is shown in Figure 2.

The flavone *C*-glycosides, especially isoorientin, were found in most of these species of bamboo tested. The different species of bamboo showed different major constituents. Identification of isoorientin, isovitexin, orientin, and vitexin in the different species of bamboo was carried out by the same retention factors (R_f) and similar colour to the standards compounds. The results of the flavone *C*-glycosides survey are presented in Table 1. The flavone *C*-glycosides were found in 40 of 51 species of bamboo examined. TLC screening method proved to be an efficient strategy for the rapid screening of flavone *C*-glucosides. Among the fifty-one species of bamboo, *B. multiplex* cv. *Silverstripe*, *P. heterocycla* cv. *pubescens*, and *P. acuta* showed

Table 6: Content of isoorientin, isovitexin, orientin, and vitexin in different samples by HPLC method.

Compound	Content of marker compounds[a] (% w/w)			
	Bambusa multiplex cv. *Silverstripe*	*Phyllostachys heterocycla* cv. *pubescens*	*Phyllostachys acuta* C. D. Chu et C. S. Chao	Commercial product of bamboo-leaf flavonoids
Isoorientin	0.278 ± 0.0021	0.100 ± 0.0037	0.138 ± 0.0096	2.235 ± 0.0100
Orientin	ND	0.026 ± 0.0032	0.031 ± 0.0008	0.762 ± 0.0076
Vitexin	ND	0.007 ± 0.0004	ND	0.183 ± 0.0014
Isovitexin	0.068 ± 0.0006	0.028 ± 0.0012	0.020 ± 0.0011	0.547 ± 0.0029

[a]mean ± SD ($n = 3$).

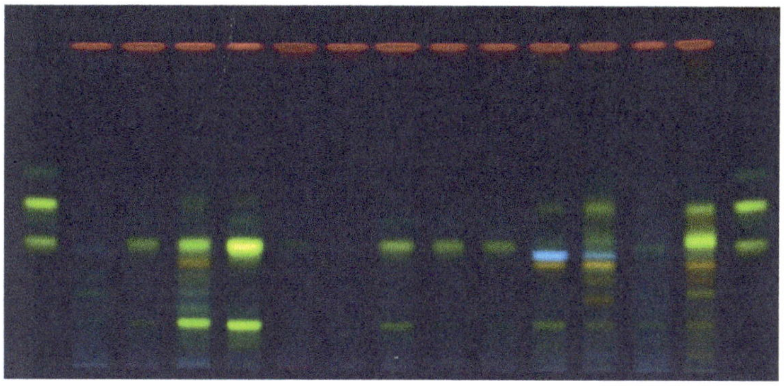

Figure 2: TLC profile of standards and extracts from the different species of bamboo. S: standard mixture: isoorientin ($R_f = 0.39$), isovitexin ($R_f = 0.47$), orientin ($R_f = 0.51$), and vitexin ($R_f = 0.61$). Different bamboo species are listed in Table 1.

strong band intensity as to flavone *C*-glycosides on the TLC plates, so they were examined further by HPLC.

3.2. Optimization of HPLC Conditions. To achieve a good chromatographic resolution, different types of chromatographic columns such as Xterra RP18 (4.6 × 250 mm, 5 μm; Waters USA), Zorbax SB-C18 (4.6 × 250 mm, 5 μm; Agilent USA), and YMC-Pack R&D ODS-A (4.6 × 250 mm, 5 μm; YMC Japan) were tested. To enhance the resolution and eliminate the peak tailing of the marker compounds, glacial acetic acid was added to the mobile phase. Different mobile phase compositions and different column temperatures were also optimized. The best results were obtained using a YMC-Pack R&D ODS-A column at 30°C, with acetonitrile and water containing 0.5% (v/v) glacial acetic acid as the mobile phase and isocratic elution (1.0 mL/min) was applied (Figure 3(a)).

3.3. Validation of HPLC Quantitation Method. Standard solutions of isoorientin, orientin, vitexin, and isovitexin were prepared at a concentration range of 0.2–100 μg/mL. The nine-point calibration curves of the analytes showed good linearity ($r^2 \geq 0.9997$) with given concentration ranges. The parameters of linearity and range were summarized in Table 2, and LOD and LOQ values showed the adequate sensitivity of the proposed method.

The RSD values for both parameters ranged from 0.05 to 1.56% (Table 3). Therefore, the method was considered repeatable.

The recovery values of the four flavone *C*-glycosides are given in Table 4. Average recoveries ranged from 88.44% to 105.83%, confirming the accuracy of the method.

The four flavone *C*-glycosides in bamboo leaves were identified by comparing their UV/DAD and MS spectra with those of the standards (Figures 3 and 4). The result of peak purity test showed that there was no coelution to interfere with the quantitation of marker compounds.

The four flavone *C*-glycosides from bamboo leaves were detected. The retention time, maximum wavelength, and characteristic fragment ions for individual peaks are presented in Table 5.

3.4. Sample Analysis. The optimized method was applied to determinate the four marker compounds in different samples. These HPLC chromatograms are depicted in Figure 3. The results are shown in Table 6. The concentration ranges of isoorientin, isovitexin, orientin, and vitexin in three bamboo leaves samples were 1.00–2.78, 0–0.31, 0–0.07, and 0.20–0.68 mg/g, respectively. A total of four flavone *C*-glycosides were found in the leaf of *P. heterocycla* cv. *Pubesceus*, which has been used as a raw material to produce commercial products of bamboo-leaf flavonoids. The quantitative data also showed that the content of isoorientin in *B. multiplex* cv. *Silverstripe* was higher than that in the two other bamboo

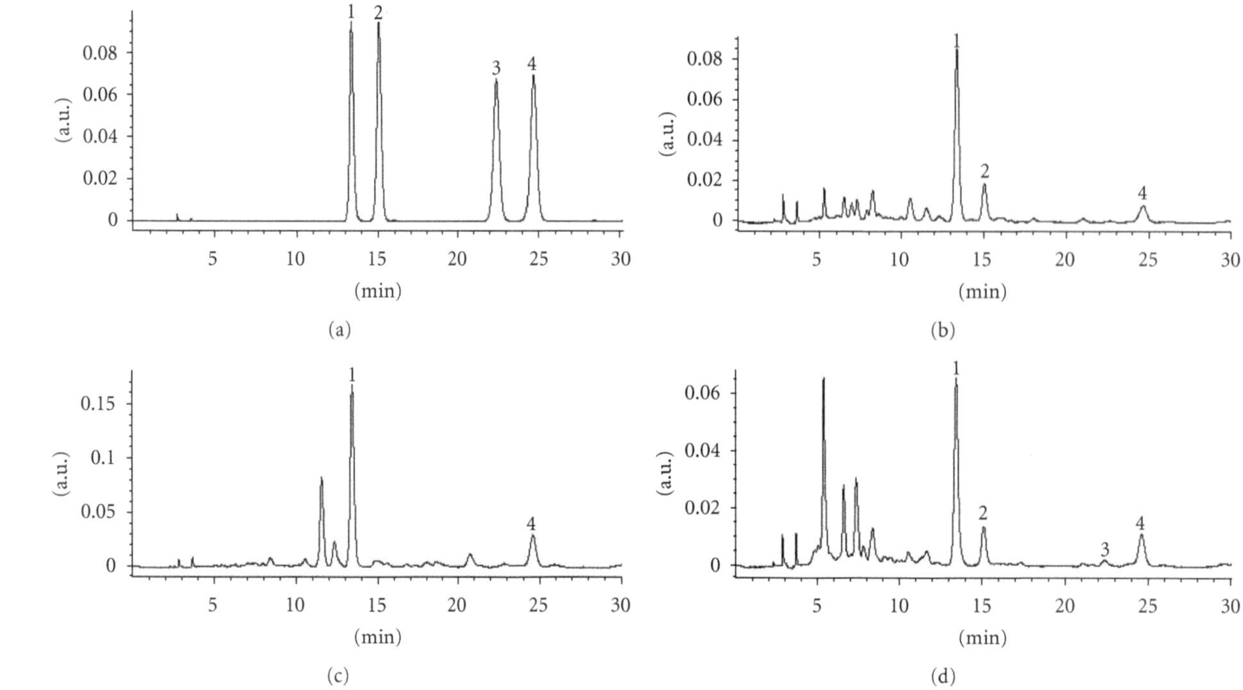

FIGURE 3: HPLC chromatograms of the standard mixture (a), 60% methanol extract of *P. acuta* (b), *B. multiplex* cv. *Silverstripe* (c), and *P. heterocycla* cv. *pubescens* (d) using diode array detection at 345 nm. (1) Isoorientin, (2) orientin, (3) vitexin, and (4) isovitexin.

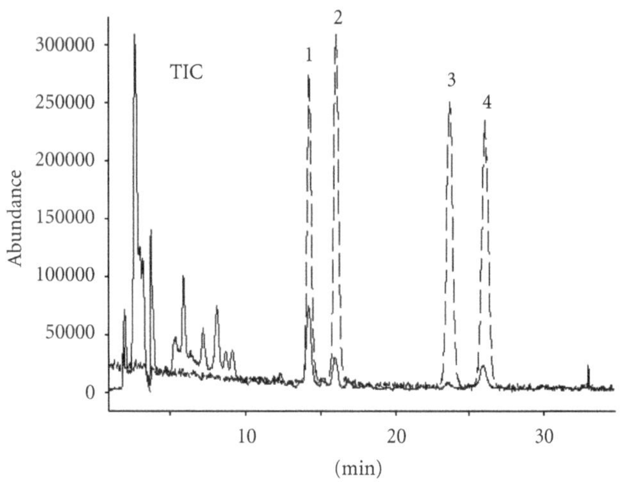

FIGURE 4: Overlap of TIC profiles (LC/MS ESI, negative mode) of mixed standards (dashed line), and the 60% methanol extract of *Phyllostachys heterocycla* cv. *pubescens* (solid line). For the identification of 1–4, see Table 5.

species. It could be considered a promising plant source for preparation of isoorientin.

4. Conclusion

In summary, this paper describes a two-step procedure for the rapid screening of flavone *C*-glucosides in a large number of samples and quantitative analysis of four marker compounds. The proposed procedure proved to be specific and sensitive and can be regarded as an alternative method to detect and quantify flavone *C*-glucosides in bamboo leaves and its commercial product.

Acknowledgment

This study was financially supported by the Central Public-Interest Scientific Institution Basal Research Fund, China (no. 1632010003).

References

[1] T. Seki, K. Kida, and H. Maeda, "Immunostimulation-mediated anti-tumor activity of bamboo (*Sasa senanensis*) leaf extracts obtained under "Vigorous" condition," *Evidence-Based Complementary and Alternative Medicine*, vol. 7, no. 4, pp. 447–457, 2010.

[2] T. Seki and H. Maeda, "Cancer preventive effect of Kumaizasa bamboo leaf extracts administered prior to carcinogenesis or cancer inoculation," *Anticancer Research*, vol. 30, no. 1, pp. 111–118, 2010.

[3] H. S. Park, H. L. Ju, J. K. Hyun, J. C. Hyun, and I. S. Lee, "Antioxidant flavone glycosides from the leaves of Sasa borealis," *Archives of Pharmacal Research*, vol. 30, no. 2, pp. 161–166, 2007.

[4] J. Mu, T. Uehara, J. Li, and T. Furuno, "Identification and evaluation of antioxidant activities of bamboo extracts," *Forestry Studies in China*, vol. 6, no. 2, pp. 1–5, 2004.

[5] Z. H. Jiang, *Bamboo and Rattan in the World*, China Forestry Publishing House, Beijing, China, 2007.

[6] X. Qiao, W. N. He, C. Xiang et al., "Qualitative and quantitative analyses of flavonoids in spirodela polyrrhiza by high

performance liquid chromatography coupled with mass spectrometry," *Phytochemical Analysis*, vol. 22, no. 6, pp. 475–483, 2011.

[7] R. C. Gueldner, M. E. Snook, N. W. Widstrom, and B. R. Wiseman, "TLC screen for maysin, chlorogenic acid, and other possible resistance factors to the fall armyworm and the corn earworm in Zea mays," *Journal Of Agricultural And Food Chemistry*, vol. 40, no. 7, pp. 1211–1213, 1992.

[8] M. E. Pascual, M. E. Carretero, K. V. Slowing, and A. Villar, "Simplified screening by TLC of plant drugs," *Pharmaceutical Biology*, vol. 40, no. 2, pp. 139–143, 2002.

[9] A. D. Talukdar, M. D. Choudhury, M. Chakraborty, and B. K. Dutta, "Phytochemical screening and TLC profiling of plant extracts of Cyathea gigantea (Wall. Ex. Hook.) Haltt. and Cyathea brunoniana. Wall. ex. Hook (Cl. & Bak.)," *Assam University Journal of Science and Technology*, vol. 5, no. 1, pp. 70–74, 2010.

[10] O. M. Wahab, A. E. Ayodele, and J. O. Moody, "TLC phytochemical screening in some Nigerian Loranthaceae," *Journal of Pharmacognosy and Phytotherapy*, vol. 2, no. 5, pp. 64–70, 2010.

[11] Y. Zhang, B. Bao, B. Lu, Y. Ren, X. Tie, and Y. Zhang, "Determination of flavone C-glucosides in antioxidant of bamboo leaves (AOB) fortified foods by reversed-phase high-performance liquid chromatography with ultraviolet diode array detection," *Journal of Chromatography A*, vol. 1065, no. 2, pp. 177–185, 2005.

[12] B. Lu, X. Wu, X. Tie, Y. Zhang, and Y. Zhang, "Toxicology and safety of anti-oxidant of bamboo leaves. Part 1: Acute and subchronic toxicity studies on anti-oxidant of bamboo leaves," *Food and Chemical Toxicology*, vol. 43, no. 5, pp. 783–792, 2005.

[13] J. Sun, Y. Yue, F. Tang, and X. Guo, "Simultaneous HPTLC analysis of flavonoids in the leaves of three different species of bamboo," *Journal of Planar Chromatography*, vol. 23, no. 1, pp. 40–45, 2010.

[14] J. Wang, F. Tang, Y. Yue, X. Guo, and X. Yao, "Development and validation of an HPTLC method for simultaneous quantitation of isoorientin, isovitexin, orientin, and vitexin in bamboo-leaf flavonoids," *Journal of AOAC International*, vol. 93, no. 5, pp. 1376–1383, 2010.

[15] C. A. M. Pereira, J. H. Yariwake, and M. McCullagh, "Distinction of the C-glycosylfavone isomer pairs orientin/isoorientin and vitexin/isovitexin using HPLC-MS exact mass measurement and in-source CID," *Phytochemical Analysis*, vol. 16, no. 5, pp. 295–301, 2005.

[16] C. A. M. Pereira, J. H. Yariwake, F. M. Lanças, J. N. Wauters, M. Tits, and L. Angenot, "A HPTLC densitometric determination of flavonoids from *Passiflora alata, P. edulis, P. incarnata* and *P. caerulea* and comparison with HPLC method," *Phytochemical Analysis*, vol. 15, no. 4, pp. 241–248, 2004.

[17] M. L. Zeraik and J. H. Yariwake, "Quantification of isoorientin and total flavonoids in *Passiflora edulis* fruit pulp by HPLC-UV/DAD," *Microchemical Journal*, vol. 96, no. 1, pp. 86–91, 2010.

[18] I. C. H. *International Conference on Harmonization of Technical Requirements for Registration of Pharmaceuticals for Human Use*: Validation of Analytical Procedures: Text and MethodologyQ2(R1), 2005, http://www.ich.org/fileadmin/Public_Web_Site/ICH_Products/Guidelines/Quality/Q2_R1/Step4/Q2_R1__Guideline.pdf.

5

Characterisation of Flavonoid Aglycones by Negative Ion Chip-Based Nanospray Tandem Mass Spectrometry

Paul J. Gates[1] and Norberto P. Lopes[2]

[1] *School of Chemistry, University of Bristol, Cantock's Close, Bristol BS8 1TS, UK*
[2] *Faculdade de Ciências Farmacêuticas de Ribeirão Preto, Universidade de São Paulo, Via do Café S/N, 14040-903 Ribeirão Preto, SP, Brazil*

Correspondence should be addressed to Paul J. Gates, paul.gates@bristol.ac.uk

Academic Editor: Michael Niehues

Flavonoids are one of the most important classes of natural products having a wide variety of biological activities. There is wide interest in a range of medical and dietary applications, and having a rapid, reliable method for structural elucidation is essential. In this study a range of flavonoid standards are investigated by chip-based negative ion nanospray mass spectrometry. It was found that the different classes of flavonoid studied have a combination of distinct neutral losses from the precursor ion $[M\text{-}H]^-$ along with characteristic low-mass ions. By looking only for this distinct pattern of product ions, it is possible to determine the class of flavonoid directly. This methodology is tested here by the analysis of a green tea extract, where the expected flavonoids were readily identified, along with quercetin, which is shown to be present at only about 2% of the most intense ion in the spectrum.

1. Introduction

Flavonoids are an important class of dietary natural products with a range of biological activities, such as antioxidant, UV-protection, antiparasitic, anti-inflammatory, and antifungal [1–6]. The flavonoids are subcategorised into eight different classes with some of the compounds also exhibiting possible beneficial properties such as health-promoting and anticancer activities [7]. The common C6-C3-C6 structural core for all flavonoids arises from the shikimate (C6-C3) and acetate (C6) biosynthetic pathways. In their review, Williams and Grayer pointed out that the theoretical number of possible flavonoid structures (with hydroxyl, methoxyl, methyl, isoprenyl benzyl, and sugar substituents) is enormous, and many new natural flavonoids are still to be isolated [8]. Until now, more than 9000 different flavonoids have been isolated. The majority were isolated and identified employing classical phytochemical procedures, and there is no doubt that many more new flavonoids remain to be discovered [8].

Many analytical methodologies have been developed to detect and quantify flavonoids, mostly using high-performance liquid chromatography with UV-VIS spectral detection. However, identification of flavonoids, as well as other natural products, through hyphenated systems (LC-UV) is limited since a complete chromatographic resolution for all chromophores is required to be sure that the correct conclusion is reached [9, 10]. Mass spectrometry (MS) with electrospray ionisation (ESI) has emerged as a complementary method for high sensitivity, selectivity, and fast analysis of natural products [11], such as sesquiterpene lactones [12] and alkaloids [13]. Among all mass spectrometry techniques, electrospray ionisation tandem mass spectrometry (ESI-MS/MS) using low-energy collision-induced dissociation (CID) has been the technique of choice for such studies through the technique's ability to analyse natural products with medium to high polarities [14].

Nanospray ionisation is an improvement over traditional ESI for the analysis of low volume low concentration samples [15]. With nanospray, it is possible to obtain mass spectra from picogram quantities of material with little sample clean-up being required. Standard nanospray uses disposable tips and as a result has problems with signal reproducibility between tips and difficulties with coupling to HPLC. With the development of automated "chip-based" nanospray

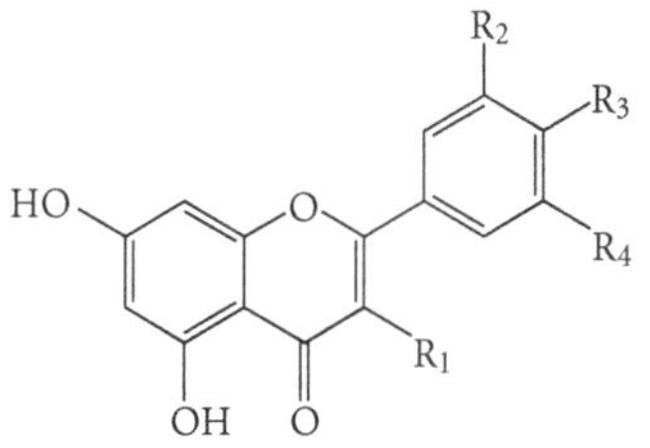

	R_1	R_2	R_3	R_4
(**1**)	OH	OH	OH	H
(**2**)	OH	OH	OH	OH
(**3**)	H	H	OH	H
(**4**)	H	OH	OH	H

(a)

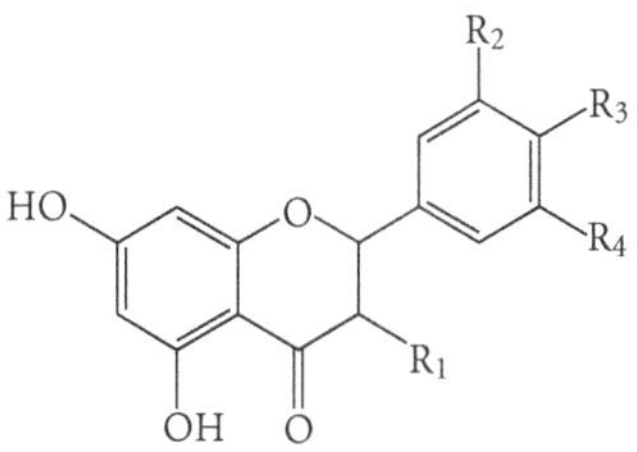

	R_1	R_2	R_3	R_4
(**5**)	H	H	OH	H
(**6**)	H	OH	OMe	H

(b)

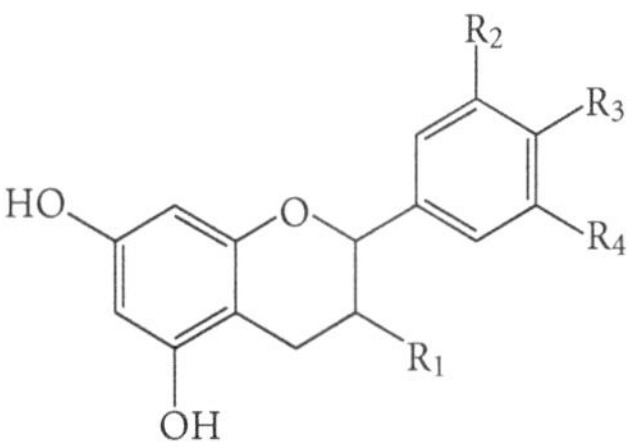

	R_1	R_2	R_3	R_4
(**7**)	OH	OH	OH	H
(**8**)	OH	OH	OH	OH

(c)

Figure 1: The structures of the flavonoids analysed. (**1**) Quercetin (molecular weight = 302); (**2**) Myricetin (molecular weight = 318); (**3**) Apigenin (molecular weight = 270); (**4**) Luteolin (molecular weight = 286); (**5**) Naringenin (molecular weight = 272); (**6**) Hesperetin (molecular weight = 302); (**7**) Catechin (molecular weight = 290) and (**8**) Epigallocatechin (molecular weight = 306).

systems, using arrays of uniform nanospray needles, the technique is becoming much more important [16]. In "chip-based" nanospray, the analyte solution is sprayed from a conductive pipette tip pressed against the rear of the chip using a small gas pressure and low voltage to create the spray. Each nanospray needle in the array is used only once to avoid contamination.

In recent years, nanospray ionisation has been applied to the analysis of natural products, but there are still some doubts about the applicability of the technique for the analysis of small molecules. Analysis of retinal, carotenoids, and xanthophylls showed some significant differences between the ions observed between nanospray and electrospray ionisation [14, 17, 18]. These results could be correlated to differences in the source design and ionisation conditions for nanospray and open up a new area of research in natural product chemistry. Based upon these previous studies and the increasingly recognised importance of flavonoids in the human diet along with the increase in metabolomic studies, the purpose of this study is to establish a sound basis for the ionisation and fragmentation of four aglycone flavonoid classes (Figure 1) in negative ion nanospray ionisation. The application and power of the technique to "real world" samples is exemplified with the identification of medium-polarity flavonoids from a simple extract of green tea without employing any prior sample preparation, clean-up, or chromatography.

2. Experimental

2.1. Materials. The flavonoid standards (Figure 1) were isolated as previously described [19] or obtained from Sigma-Aldrich (United Kingdom). Solutions of the analytes (approximetely 0.1 mg/mL) in 100% HPLC-grade methanol (Fisher Scientific) were prepared immediately prior to the analysis. The green tea sample was obtained from a local supermarket. A few grains were dissolved in 100% methanol with the sample centrifuged (13,000 rpm, 5 mins) prior to the analysis.

2.2. Instrumentation. Nanospray ionisation analyses were performed on a QStar-XL quadrupole-time-of-flight hybrid instrument (Applied Biosystems, Warrington, UK) using a NanoMate HD automatic chip-based nanospray system (Advion Biosciences, Norwich, UK). Instrument control, data acquisition, and data processing were performed through the Analyst QS version 1.1 software (Applied Biosystems, Warrington, UK). NanoMate control was through the ChipSoft software (Advion Biosciences, Norwich, UK). The NanoMate was set for 5 μL of solution to be aspirated and sprayed through a NanoMate 400 chip at 1.45 kV with a nitrogen back pressure of 0.4 psi. QStar acquisition parameters were ion source gas flow rate, 50; curtain gas flow rate, 20; ion spray voltage, 2700 V; declustering potential, 75 V; focusing potential, 280 V; declustering potential 2, 15 V. CID-MS/MS was performed at a collision energy in the range from −20 to −40 eV. The ion source gas, curtain gas, and collision gas were all nitrogen.

3. Results and Discussion

The compounds quercetin (flavonol, **1**), apigenin (flavone, **3**), naringenin (flavanone, **5**), and hesperetin (flavanone, **6**) (Figure 1) were used as standards to study their ability to produce high-intensity, stable-deprotonated molecule signals in negative ion mode nanospray ionisation. 100% HPLC methanol proved to be an excellent solvent for these studies with stable ion signals being produced for up to 20 minutes (Figure 2). This is essential as it allows for a number of tandem mass spectrometry (MS/MS) experiments to be performed on the same sample without any adjustments or tuning of the nanospray source. Use of methanol resulted in no observed methylation reactions as has previously been described for other natural products [20]. Over the range of source conditions used, all the aglycone flavonoids produced an intense and stable spray for at least 15 minutes from single 5 μL analyte solution aspirations. This demonstrates the possibility to work with more complex flavonoid samples and allows for setting up automatic MS/MS acquisitions from a batch analysis.

Following on from the ion formation studies, the systematic investigation was continued to determine the best

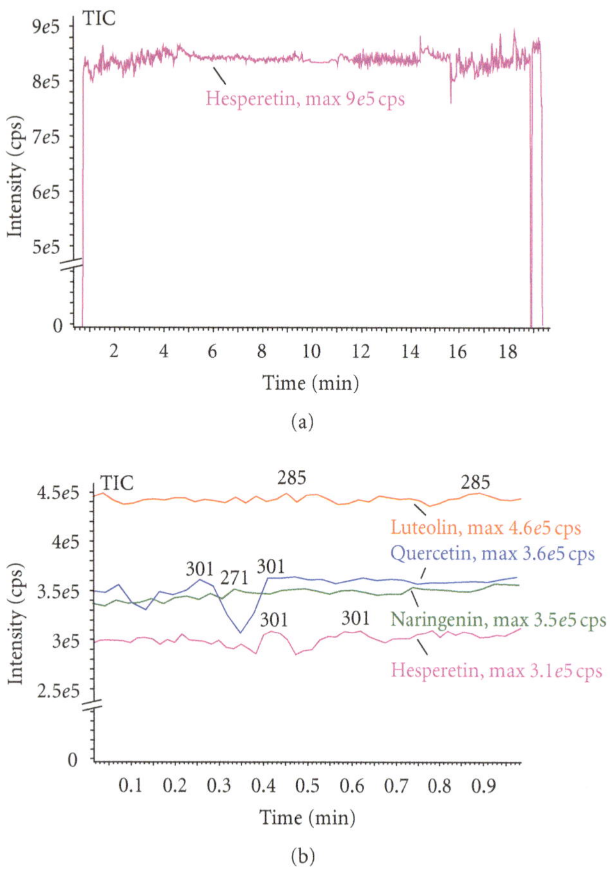

Figure 2: Demonstration of the stability of the chip-based nanospray infusion. The data shows plots of total ion count (from 5 μL aspirations) versus time for $[M\text{-}H]^-$ ions. Plot (a) is for hesperetin over a 20 minute run. The onset of nanospray is at about 30 s into the run with about 15 minutes of highly stable spray. After 15 minutes, the spray is less stable until the spray breaks down at about 19.5 minutes. Plot (b) is of luteolin, quercetin, naringenin, and hesperetin over a 1-minute run demonstrating intersample reproducibility.

CID collision energies required for effective product ion formation whilst eliminating unwanted gas-phase interactions. Collision energies from −20 to −40 eV resulted in good product ion spectra, with, as expected, more product ions being observed at higher voltages (more negative). A collision energy of around −35 eV was determined to result in the "best" product ion spectra (Figure 3). Examination of the spectra revealed high levels of complexity with many competing fragmentation routes. The main neutral molecules lost from the $[M\text{-}H]^-$ ions consisted of a combination of H_2O, CO, CO_2, and/or H_2CCO (Figure 3). A detailed analysis of all the spectra indicates that a combination of a specific order of neutral eliminations occurs along with the presence of a series of diagnostic low-mass product ions for each of the flavonoid classes analysed (Table 1 and Figure 3) resulting in the quick and reliable method for the identification of the flavonoid class. The diagnostic low-mass product ions result from ring contraction reactions which follow the same mechanisms as previously reported

Table 1: Table of the characteristic sequences of neutral losses, from their corresponding $[M\text{-}H]^-$ precursor ions and characteristic low-mass product ions, for the four flavonoid classes analysed in this study.

Flavonoid class	Characteristic neutral losses	Characteristic productions
Flavonols (**1** and **2**)	−28, −44, −18	151, 125, 107
Flavones (**3** and **4**)	−28, −44, −44, −28, −42	151, 121, 107
Flavanones (**5** and **6**)	−18, −44, −44, −18, −42	151, 125, 107
Flavanols (**7** and **8**)	−18, −44, −44, −18, −42	137, 125, 109

for flavonoids in negative mode ESI [21]. All of the flavonoids (except the flavanols) have the previously described ions at *m/z* 151 and 107 [21], whereas the flavanols catechin and epigallocatechin (with no oxidation at carbon 3, but following a similar ring contraction mechanism) result in the product ions at *m/z* 137 and 109. Also, all of the flavonoids except the flavones have an ion at *m/z* 125, and the flavones have an ion at *m/z* 121.

The flavanone hesperetin has a methoxyl substitution at the aromatic ring and showed elimination of a methyl radical ($^\bullet CH_3$) similar to that previously reported for mycosporine-like amino acids [22] and some other flavonoids [23]. Observation of this behaviour in nanospray allows the easy distinguishing of methoxylated flavonoids with identical molecular mass, for example, when screening plant extracts for flavonoid composition as previously report in ESI [23]. Increasing the collision energy for hesperetin results in an almost complete fragmentation of the radical ion, but allows for the observation of a loss of 16 mass units. An unusual CH_4 elimination has been previously described for heterocyclic aromatic amines which is proposed to be due to a gas-phase ion-molecule aromatic-nucleophilic substitution between β-carbolines and water vapour [24]. With hesperetin, the loss of 16 is suggested to be due to CH_4 elimination involving the methoxyl group and the *ortho*-hydroxyl group. Figure 4 shows the expansion of two product ion spectra of hesperetin at different collisional energies, clearly showing the competing losses of $^\bullet CH_3$ and CH_4. The mechanism for loss of $^\bullet CH_3$ proceeds through homolytic cleavage as previously described [22, 23]. The mechanism for water elimination from ortho-substituted aromatic esters is well known in electron ionisation. In this case we suggest that a similar cyclic rearrangement through homolytic cleavage is occurring, but involving the hydroxyl substitution, resulting in a stable quinonic ion (Figure 4). Both of these mechanisms, when taken together, are very useful for the structure elucidation of disubstituted flavonoids.

The analysis of a green tea extract in methanol was performed to demonstrate the utility of the technique. The analysis was performed without any chromatography or sample cleanup. The negative ion nanospray spectrum (Figure 5) is very complicated with a considerable number

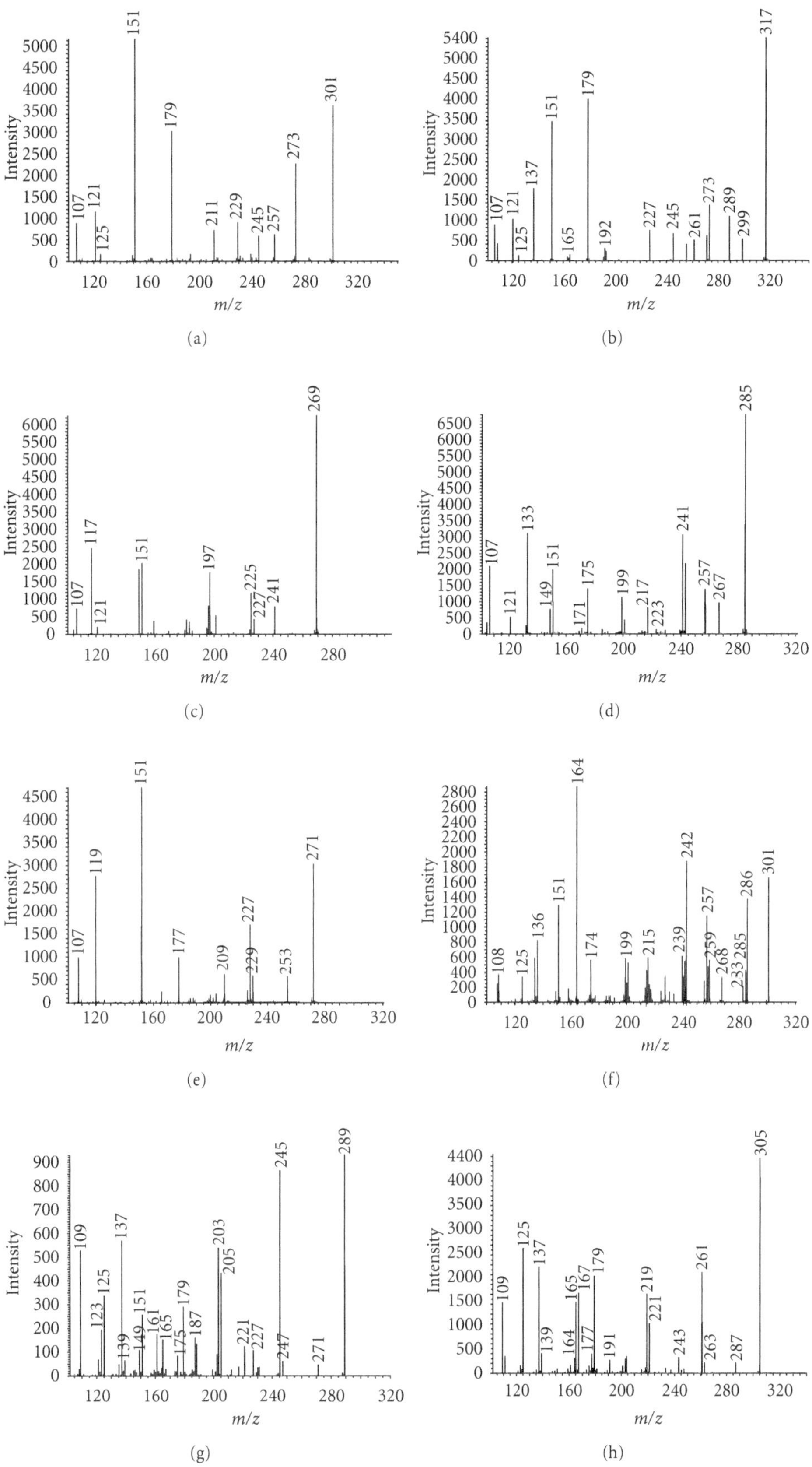

FIGURE 3: The negative ion nanospray product ion spectra of the eight flavonoids studied. Spectrum (a) is of quercetin, 1: (precursor ion (PI) *m/z* 301), (b) myricetin, 2: (PI *m/z* 317), (c) apigenin, 3: (PI *m/z* 269), (d) luteolin, 4: (PI *m/z* 285), (e) naringenin, 5: (PI *m/z* 271), (f) hesperetin, 6: (PI *m/z* 301), (g) catechin, 7: (PI *m/z* 289), and (h) epigallocatechin 8: (PI *m/z* 305).

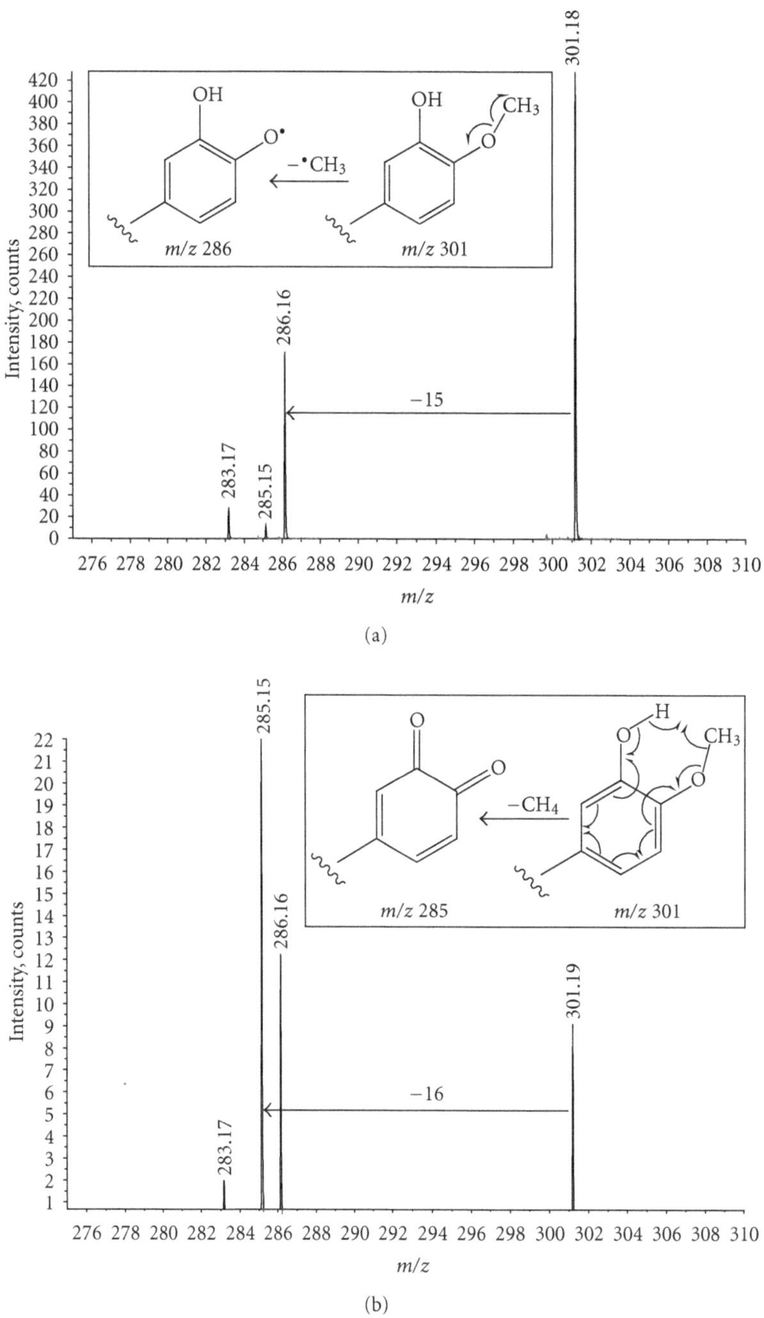

FIGURE 4: Enlargements of negative ion nanospray product ion spectra of hesperetin at low (a) and high (b) collision energies. The competition between losses of $^{\bullet}CH_3$ and CH_4 is clearly observed. At higher collision energy, the radical ion (*m/z* 286) has fragmented further to leave the quinonic ion (*m/z* 285) intact. The mechanism of formation of the two ions is shown in the inserts.

of ions over a wide mass range. Some of the observed masses (*m/z* 289, 305 and 317) match to the flavonoid standards already analysed in this study, and analysis of the MS/MS spectra (data not shown) of these proved them to be the expected flavonoids present in green tea: catechin, **7**, (flavanol), epigallocatechin, **8**, (flavanol) and myricetin, **2**, (flavonol). Other intense peaks (*m/z* 441 and 457) are gallate flavonoids not considered in this initial study. To test the detection limit of the technique, the peak at *m/z* 301 was studied further (see Figure 5). This peak occurs at approximately 2% of the most abundant ion in the spectrum, but performing MS/MS for about 1 minute still produced a good intensity product ion spectrum (Figure 5). A thorough study of this spectrum reveals an almost identical series of peaks to that of the flavonol quercetin, **1** (Figure 2). The differences between the two spectra are probably down to the different collision energies used. Quercetin is one of the most biologically active flavonoids and is more normally found in citrus fruits. The confirmation of the presence of quercetin in green tea (even at the low levels in this particular sample)

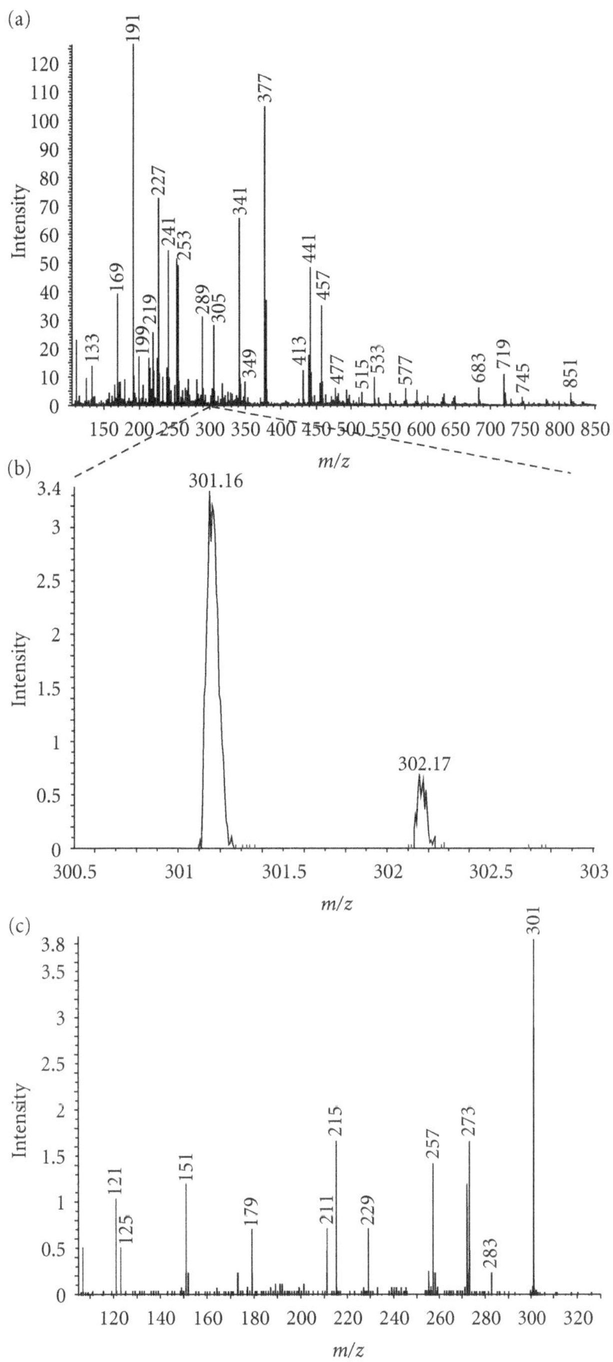

FIGURE 5: Negative ion nanospray spectra of the green tea extract. Spectrum (a) is the total extract recorded over a wide *m/z* range. Spectrum (b) is an enlargement of (a) to show the peak at *m/z* 301 at approximately 2% of the intensity of the most intense ion in spectrum (a). Spectrum (c) is the product ion spectrum of *m/z* 301 which clearly demonstrates the sensitivity of the technique.

is a highly significant result and a powerful demonstration of the sensitivity and application of this methodology.

4. Conclusions

In this initial study, the application of chip-based negative ion nanospray is demonstrated for the analysis of a series of flavonoid standards. The best spectra where produced from 100% HPLC methanol. MS/MS analysis of four of the classes of flavonoids have shown that they have a different, characteristic sequences of neutral losses from their corresponding $[M-H]^-$ precursor ions in combination with distinctive lower mass product ions. The application of this methodology is demonstrated for the analysis of a green tea extract where the expected flavonoids (catechin, epigallocatechin, and myricetin) were easily identified, along with the unexpected presence of quercetin (at approximately 2% of the most intense ion).

Acknowledgments

The authors thank Mark Allen of Advion BioSciences Ltd. (Norwich, UK) for helpful advice and discussions throughout this project. N. P. Lopes acknowledges FAPESP (São Paulo, Brazil) for financial support.

References

[1] A. Kanashiro, L. M. Kabeya, A. C. M. Polizello, N. P. Lopes, J. L. C. Lopes, and Y. M. Lucisano-Valim, "Inhibitory activity of flavonoids from *Lychnophora* sp. on generation of reactive oxygen species by neutrophils upon stimulation by immune complexes," *Phytotherapy Research*, vol. 18, no. 1, pp. 61–65, 2004.

[2] P. Chicaro, E. Pinto, P. Colepicolo, J. L. C. Lopes, and N. P. Lopes, "Flavonoids from *Lychnophora passerina* (Asteraceae): potential antioxidants and UV-protectants," *Biochemical Systematics and Ecology*, vol. 32, no. 3, pp. 239–243, 2004.

[3] R. Takeara, S. Albuquerque, N. P. Lopes, and J. L. C. Lopes, "Trypanocidal activity of *Lychnophora staavioides* Mart. (Vernonieae, Asteraceae)," *Phytomedicine*, vol. 10, no. 6-7, pp. 490–493, 2003.

[4] N. P. Lopes, P. Chicaro, M. J. Kato, S. Albuquerque, and M. Yoshida, "Flavonoids and lignans from *Virola surinamensis* twigs and their in vitro activity against *Trypanosoma cruzi*," *Planta Medica*, vol. 64, no. 7, pp. 667–669, 1998.

[5] L. Gobbo-Neto, M. D. Santos, A. Kanashiro et al., "Evaluation of the anti-inflammatory and antioxidant activities of di-*C*-glucosylflavones from *Lychnophora ericoides* (Asteraceae)," *Planta Medica*, vol. 71, no. 1, pp. 3–6, 2005.

[6] N. P. Lopes, M. J. Kato, and M. Yoshida, "Antifungal constituents from roots of Virolasurinamensis," *Phytochemistry*, vol. 51, no. 1, pp. 29–33, 1999.

[7] J. B. Harborne and C. A. Williams, "Anthocyanins and other flavonoids," *Natural Product Reports*, vol. 18, no. 3, pp. 310–333, 2001.

[8] C. A. Williams and R. J. Grayer, "Anthocyanins and other flavonoids," *Natural Product Reports*, vol. 21, no. 4, pp. 539–573, 2004.

[9] T. Guaratini, R. L. Vessecchi, F. C. Lavarda et al., "New chemical evidence for the ability to generate radical molecular ions of polyenes from ESI and HR-MALDI mass spectrometry," *Analyst*, vol. 129, no. 12, pp. 1223–1226, 2004.

[10] T. Guaratini, R. Vessecchi, E. Pinto, P. Colepicolo, and N. P. Lopes, "Balance of xanthophylls molecular and protonated molecular ions in electrospray ionization," *Journal of Mass Spectrometry*, vol. 40, no. 7, pp. 963–968, 2005.

[11] A. E. M. Crotti, R. Vessecchi, J. L. C. Lopes, and N. P. Lopes, "Electrospray ionization mass spectrometry: chemical processes invoeved in the ion formation from low molecular weight organic compounds," *Química Nova*, vol. 29, no. 2, pp. 287–292, 2006.

[12] A. E. M. Crotti, J. L. C. Lopes, and N. P. Lopes, "Triple quadrupole tandem mass spectrometry of sesquiterpene lactones: a study of goyazensolide and its congeners," *Journal of Mass Spectrometry*, vol. 40, no. 8, pp. 1030–1034, 2005.

[13] M. Pivatto, A. E. M. Crotti, N. P. Lopes et al., "Electrospray ionization mass spectrometry screening of piperidine alkaloids from *Senna spectabilis* (Fabaceae) extracts: fast identification of new constituents and co-metabolites," *Journal of the Brazilian Chemical Society*, vol. 16, no. 6, pp. 1431–1438, 2005.

[14] A. Fredenhagen, C. Derrien, and E. Gassmann, "An MS/MS library on an ion-trap instrument for efficient dereplication of natural products. Different fragmentation patterns for $[M+H]^+$ and $[M+Na]^+$ ions," *Journal of Natural Products*, vol. 68, no. 3, pp. 385–391, 2005.

[15] M. Wilm and M. Mann, "Analytical properties of the nanoelectrospray ion source," *Analytical Chemistry*, vol. 68, no. 1, pp. 1–8, 1996.

[16] G. A. Schultz, T. N. Corso, S. J. Prosser, and S. Zhang, "A fully integrated monolithic microchip electrospray device for mass spectrometry," *Analytical Chemistry*, vol. 72, no. 17, pp. 4058–4063, 2000.

[17] T. Guaratini, P. J. Gates, K. H. M. Cardozo, P. M. B. G. M. Campos, P. Colepicolo, and N. P. Lopes, "Letter: radical ion and protonated molecule formation with retinal in electrospray and nanospray," *European Journal of Mass Spectrometry*, vol. 12, no. 1, pp. 71–74, 2006.

[18] T. Guaratini, P. J. Gates, E. Pinto, P. Colepicolo, and N. P. Lopes, "Differential ionisation of natural antioxidant polyenes in electrospray and nanospray mass spectrometry," *Rapid Communications in Mass Spectrometry*, vol. 21, no. 23, pp. 3842–3848, 2007.

[19] P. A. dos Santos, J. L. C. Lopes, and N. P. Lopes, "Triterpenoids and flavonoids from *Lychnophoriopsis candelabrum* (Asteraceae)," *Biochemical Systematics and Ecology*, vol. 32, no. 5, pp. 509–512, 2004.

[20] N. P. Lopes, C. B. W. Stark, H. Hong, P. J. Gates, and J. Staunton, "A study of the effect of pH, solvent system, cone potential and the addition of crown ethers on the formation of the monensin protonated parent ion in electrospray mass spectrometry," *Analyst*, vol. 126, no. 10, pp. 1630–1632, 2001.

[21] N. Fabre, I. Rustan, E. de Hoffmann, and J. Quetin-Leclercq, "Determination of flavone, flavonol, and flavanone aglycones by negative ion liquid chromatography electrospray ion trap mass spectrometry," *Journal of the American Society for Mass Spectrometry*, vol. 12, no. 6, pp. 707–715, 2001.

[22] K. H. M. Cardozo, R. Vessecchi, V. M. Carvalho et al., "A theoretical and mass spectrometry study of the fragmentation of mycosporine-like amino acids," *International Journal of Mass Spectrometry*, vol. 21, pp. 3842–3848, 2008.

[23] U. Justesen, "Collision-induced fragmentation of deprotonated methoxylated flavonoids, obtained by electrospray ionization mass spectrometry," *Journal of Mass Spectrometry*, vol. 36, no. 2, pp. 169–178, 2001.

[24] N. P. Lopes, T. Fonseca, J. P. G. Wilkins, J. Staunton, and P. J. Gates, "Novel gas-phase ion-molecule aromatic nucleophilic substitution in *β*-carbolines," *Chemical Communications*, vol. 9, no. 1, pp. 72–73, 2003.

6

Application of Monoclonal Antibodies against Bioactive Natural Products: Eastern Blotting and Preparation of Knockout Extract

Hiroyuki Tanaka,[1] Osamu Morinaga,[2] Takuhiro Uto,[2] Shunsuke Fuji,[3] Frederick Asare Aboagye,[4] Nguyen Huu Tung,[2] Xiao Wei Li,[5] Waraporn Putalun,[6] and Yukihiro Shoyama[2]

[1] *Graduate School of Pharmaceutical Science, Kyushu University, 3-1-1 Maidashi, Higashi-ku, Fukuoka 812-0855, Japan*
[2] *Faculty of Pharmaceutical Science, Nagasaki International University, 2827-7 Huis Ten Bosch, Sasebo 859-3298, Japan*
[3] *Faculty of Health Management, Nagasaki International University, 2827-7 Huis Ten Bosch, Sasebo 859-3298, Japan*
[4] *Phytochemistry Department, Centre for Scientific Research into Plant Medicine, University of Ghana, P.O. Box 73, Mampong-Akuapem, Ghana*
[5] *State Key Laboratory of Natural and Biomimetic Drugs, School of Pharmaceutical Sciences, Peking University, No.38 Xue-yuan Road, Haidian District, Beijing 100191, China*
[6] *Faculty of Pharmaceutical Science, Khon Kaen University, Khon Kaen 40002, Thailand*

Correspondence should be addressed to Yukihiro Shoyama, shoyama@niu.ac.jp

Academic Editor: Norberto Peporine Lopes

Matrix-assisted laser desorption/ionization (MALDI) tof of mass spectrometry was used for the confirmation of hapten number in synthesized antigen. As application of MAb, the MAbs against ginsenosides and glycyrrhizin have been prepared resulting in the development of two new techniques that we named the eastern blotting method and the knockout extract preparation. In eastern blotting technique, glycosides like ginsenosides and glycyrrhizin separated by silica gel TLC were blotted to PVDF membrane that was treated with a $NaIO_4$ solution followed by BSA resulted in glycoside-BSA conjugate on a PVDF membrane. The blotted spots were stained by MAb. Double staining of eastern blotting for ginsenosides using antiginsenoside Rb_1 and Rg_1 MAbs promoted complete identification of ginsenosides in *Panax* species. The immunoaffinity concentration of glycyrrhizin was determined by immunoaffinity column conjugated with antiglycyrrhizin MAb resulting in the glycyrrhizin-knockout extract, which was determined by the synergic effect with glycyrrhizin on NO production using the cell line.

1. Introduction

Immunoassay systems using monoclonal antibody (MAb) against drugs and small molecular weight bioactive compounds have become an important tool for studies on receptor binding analysis, enzyme assay, and quantitative and/or qualitative analytical techniques in animals or plants. The immunoblotting method is based on the western blotting technique that utilizes the antigen-antibody binding properties and provides a specific and sensitive detection of higher molecule analytes like peptides and proteins. In our ongoing study on MAb, previously we prepared various kinds of MAb against natural products like forskolin [1], solamargine [2], crocin [3], marihuana compounds [4], opium alkaloids [5], ginsenosides [6, 7], berberine [8], sennosides [9], paeoniflorin [10], glycyrrhizin [11, 12], ginkgolic acid [13], aconitine alkaloid [14], baicalin [15], and so on and developed individual competitive enzyme-linked immunosorbent assay (ELISA) as a highly sensitive, specific, and simple methodology.

The confirmation of hapten number in synthesized antigens is most important in the first stage of MAb preparation. Therefore, its determination method will be discussed first of all. As an application of MAb, the MAb against ginsenosides and glycyrrhizin has been prepared resulting in the development of two new techniques that we have named the eastern blotting method [12] and the knockout extract preparation [16]. They will be introduced in this paper.

2. Preparation of MAb against Natural Products

Various methods have been employed for the determination of natural products. They include spectral methods such as infrared (IR), nuclear magnetic resonance (NMR), and circular dichroism (CD) and other chromatographic methods such as ion chromatography (IC), capillary electrophoresis (CE), and high-speed counter current chromatography (HSCCC), and so on. Compared to TLC, GLC and HPLC methods, the ELISA method was more sensitive and selective. Moreover, it is possible to study a large number of natural products. Since natural product extracts consist of various chemical constituents (e.g., licorice contains 470 components or more), in general, some pretreatment is necessary for HPLC and other chromatographic analysis methods. ELISA, however, can determine the concentration of components directly without any pretreatment. Therefore, ELISA was used to measure the concentration of ginsenoside Rb_1 in ginseng and traditional Chinese medicines (TCMs).

2.1. Analytical Method for Determination of Hapten Number in Antigen, Hapten-Carrier Protein Conjugate. For production of MAb, synthesis of hapten, which is derived from an immune antigen and linker bridge, and the carrier protein conjugate is necessary. There had been no direct and appropriate methods for the determination of haptens conjugated with carrier proteins without differential UV analysis, radiochemical, or chemical methods. Therefore, immunization by the injection of hapten-carrier protein conjugate was unreliable. Wengatz et al. [17] determined the hapten density of immunoconjugates by matrix-assisted UV laser desorption/ionization mass spectrometry. We also reported the direct analytical method of hapten and carrier protein conjugates by a matrix-assisted laser desorption/ionization mass spectrometry (MALDI) tof mass spectrometry using an internal standard [18–20].

Figure 1 shows the MALDI tof mass spectrum of the most pharmacologically active marihuana compound, tetrahydrocannabinolic acid (THCA)-bovine serum albumin (BSA) conjugate, and BSA used as an internal standard [20]. This shows only the singly, doubly, and triply ionized molecule ions of the intact conjugate. The sharp peak at m/z 66,465 is the $[M + H]^+$ peak of BSA. A small $[M + H]^+$ peak of the THCA-BSA conjugate is at m/z 70,792, indicating that the calculated molecular mass of the THCA-BSA conjugate is 70,581 using a calculated molecular mass of 66,267 for BSA. The calculated molecule mass of the THCA moiety is 4,314. From this result, 12.7 molecules of THCA are combined with BSA [20]. Since this method is suitable for small molecule natural products, we had been analyzing the hapten number of all natural products for MAbs including glycosides like ginsenosides and glycyrrhizin.

2.2. Preparation of MAb against Ginsenosides and ELISA as an Assay System. Ginseng, the crude drug of *Panax ginseng*, is one of the most important natural medicines in many countries. It has been used to enhance stamina and capacity to cope with fatigue and physical stress and as a tonic against cancers, disturbances of the central nervous system (memory, learning, and behavior), hypothermia, carbohydrate and lipid metabolism, immune function, the cardiovascular system, and radioprotection [21]. It contains more than 30 kinds of dammarane and oleanane saponins considered to be pharmacologically active components. Ginsenoside Rb_1 is the main saponin in ginseng. However, since the concentration in the ginseng root or the root extract varies depending on the method of extraction, subsequent treatment, or even the season of its collection [22], standardization of quality is required. For this purpose, we have prepared anti-ginsenoside Rb_1 [6] and Rg_1 MAbs [7]. The immunoassay system using MAb is not frequently used for naturally occurring smaller molecular weight bioactive compounds. Preparation of MAbs is difficult, but is one of the most important steps for the analysis of natural products. As a typical natural product, the preparation of MAb against the ginseng saponin ginsenoside Rb_1 will be discussed.

A hybridoma-producing MAb reactive to ginsenoside Rb_1 was obtained by the general procedure and classified into IgG2b which had κ light chains. The reactivity of IgG type MAb, 9G7 was tested by varying antibody concentration and by performing a dilution curve. The antibody concentration was selected for competitive ELISA. The free MAb following competition is bound to polystyrene microtiter plates precoated with ginsenoside Rb_1-HSA. Under these conditions, the full measurement range of the assay extends from 20 to 400 ng/mL [6].

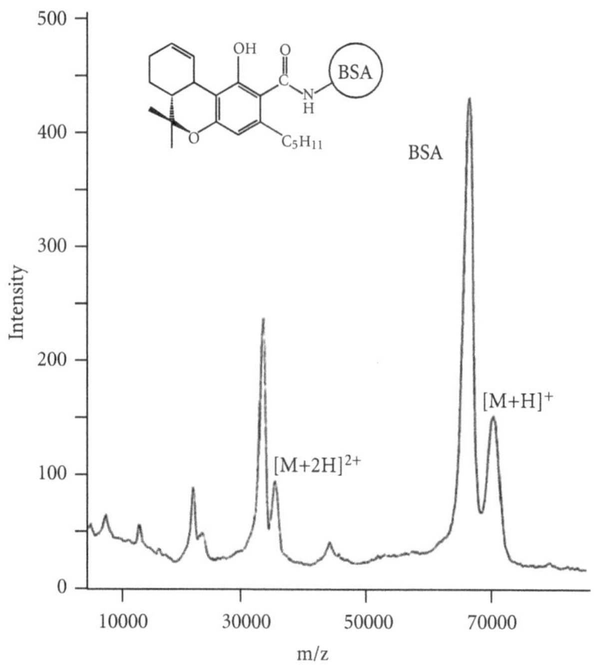

FIGURE 1: Matrix-assisted laser desorption/ionization tof mass spectrometry of tetrahydrocannabinolic-acid-BSA conjugate. [M + H] indicates the molecular weight of the conjugate, from which the hapten number can be calculated.

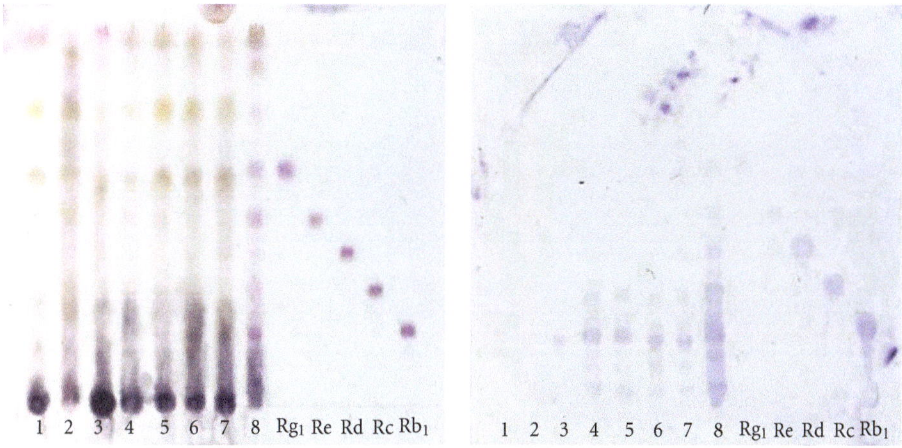

(a) TLC (b) Eastern blotting

FIGURE 2: Eastern blotting of ginsenosides in traditional Chinese medicine formulas using anti-ginsenoside Rb1 MAb. Samples (1) Jigengtang, (2) Dahuanggancaotang, (3) Nenshenyangyongtang, (4) Sijunzitang, (5) Nenshentang, (6) Buanxiaxiexintang, (7) Xiaocaihutang, and (8) Crude ginseng extract. Lanes 1 and 2 do not contain ginseng. Standard of ginsenosides indicated, ginsenosidr-Rg_1, -Re, -Rd, -Rc, and -Rb_1, respectively, from the upper spot.

Cross-reactivity is the most important factor in determining the value of an antibody. Since the ELISA for ginsenoside Rb_1 was established for phytochemical investigations involving crude plant extracts, the assay specificity was checked by determining the cross-reactivity of the MAb with various related compounds. The cross-reactivity data of MAb that was obtained were examined by competitive ELISA and calculated using picomole amounts of ginsenoside Rb_1. The cross-reactivity of ginsenoside Rc and Rd, which possess a diglucose moiety attached to the C-3 hydroxy group, was weak compared to ginsenoside Rb_1 (0.024 and 0.020%, resp.). Ginsenoside Re and Rg_1 showed no cross-reactivity (less than 0.005%). It is evident that the MAb reacted only with a small number of structurally related ginsenoside Rb_1 molecules, and very weakly, and did not react with other steroidal compounds like glycyrrhizin, digitoxin, tigogenin, tigonin, and solamargine.

In our ongoing studies on MAbs against ginseng saponins, anti-ginsenoside Rg_1 MAb [7] and Re [23] have been prepared and their ELISA was set up. Anti-ginsenoside Rg_1 MAb was also highly specific like anti-ginsenoside Rb_1. On the other hand, anti-ginsenoside Re MAb showed wide cross-reactivity. Therefore, the MAb can be used for the analysis for the total ginsenoside concentration.

2.3. Application of MAb in the Natural Products Field. Although western blotting is a common assay methodology for high-molecular-weight-substances, this method has not been employed for small molecules, as direct immunostaining of such compounds on a TLC plate is as yet unknown. Therefore, a new method for such small-molecular compounds is required. Moreover, if small molecules can be blotted to a membrane, fixing them also requires a new methodology. Previously, we succeeded in separating small-molecule compounds such as solasodine glycosides into a part of an epitope and fixing on the membrane [24], as follows.

2.3.1. New Staining Method for Ginsenosides, "Eastern Blotting". Figure 2 shows the H_2SO_4 staining and eastern blotting of ginsenoside standards and TCM using anti-ginsenoside Rb_1 MAb. It is impossible to determine the ginsenosides by TLC staining by H_2SO_4 as indicated in Figure 2(a). On the other hand, clear staining of ginsenoside Rb_1 occurred by eastern blotting, as indicated in Figure 2(b). Furthermore, it became evident that Jigengtang and Dahuanggancaotang prescriptions that did not contain ginseng, as indicated by the absence of a ginsenoside Rb_1band. The eastern blotting method was considerably more sensitive than that of H_2SO_4 staining. The H_2SO_4 staining detected all standard compounds. The eastern blotting indicated only limited staining of ginsenoside Rb_1, Rc, and Rd, whose cross-reactivities were under 0.02% as shown in Figure 2(b). We suggest that an aglycone, protopanaxadiol, and a part of the sugars may be of importance to the immunization and may function as an epitope for the structure of ginsenosides. In addition, it is suggested that the specific reactivity of sugar moieties in the ginsenoside molecule against anti-ginsenoside Rb_1 MAb may be modified by the $NaIO_4$ treatment of ginsenosides on the PVDF membrane, causing ginsenoside Rc and Rd to become detectable by eastern blotting.

Application of the eastern blotting method for the detection of glycyrrhizin in the serum samples was investigated. In general, it is difficult to detect glycyrrhizin in serum due to the large amount of impurities. The developed bands of impurities lapped over the band of glycyrrhizin, and we failed to find a proper developing solvent system that could separate glycyrrhizin and the impurities apart from each other clearly on the TLC plate. Figure 3 shows the detection of glycyrrhizin by the eastern blotting technique in the rat serum samples. Glycyrrhizin could not be identified on the TLC plate stained by H_2SO_4through many bands that were detected (Figure 3(a)). On the other hand, eastern blotting clearly shows the band of glycyrrhizin even after 1 h

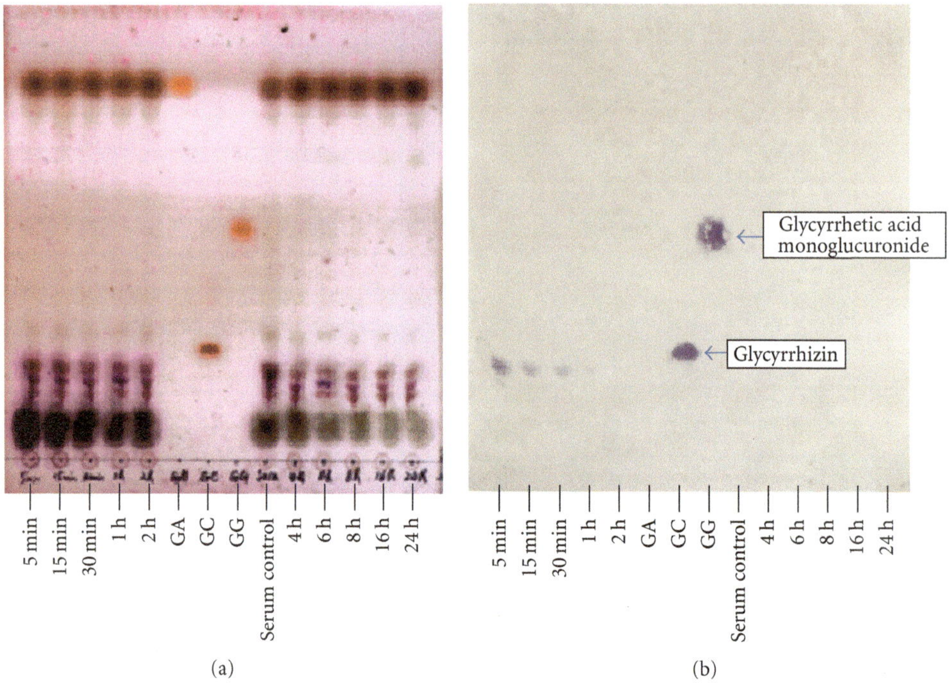

Figure 3: Eastern blotting profile of glycyrrhizin in rat serum after injection of glycyrrhizin (a) TLC of rat-serum-injected glycyrrhizin TLC was developed with *n*-BuOH-H_2O-AcOH (7 : 2 : 1). Spots were detected by 10% H_2SO_4. (b) Eastern blotting of rat-serum-injected glycyrrhizin; the band of glycyrrhizin was detected until 1 h after injection. GA: glycyhrrhetinic acid, GC: glycyrrhizin, and GG: glycyrrhetinic acid monoglucuronide.

(Figure 3(b)). Although the sensitivity of the eastern blotting method was greatly affected by the impurities, the detection limit was still at the nanogram level. The results proved that the eastern blotting technique could be a unique method for identifying glycyrrhizin against a background of a large amount of impurities.

When the mixture of anti-ginsenoside Rb_1 and Rg_1 MAbs and the pair of substrates were tested for staining for ginsenosides, all ginsenosides, ginsenoside Rb_1 , -Rc, -Rd, -Re, and -Rg_1 were stained blue although the purple color staining for ginsenoside Rg_1 was expected because 3-amino-9-ethylcarbazole and 4-chloro-1-naphotol might be different. Therefore, we performed successive staining of the membrane using anti-ginsenoside Rg_1 and then anti-ginsenoside Rb_1. Finally, we performed the double staining of ginsenosides indicating that ginsenoside Rg_1 and ginsenoside Re were stained purple and the other blue, as indicated in Figure 4. From this result, both antibodies can distinguish the individual aglycones, protopanaxatriol, and protopanaxadiol. For this application, the crude extract of various *Panax* species was analyzed by the newly developed double staining system. Major ginsenosides can be determined clearly by the double staining method, as indicated in Figure 4.

Therefore, it is suggested that the staining color shows the pharmacological activity. As shown in Figure 4(b), the purple bands indicate ginsenosides which have protopanaxatriol as an aglycone and stimulation activity for the central nervous system (CNS). On the other hand, the blue color indicates ginsenosides containing protopanaxadiol as an aglycone that possess a depression effect on the CNS. Moreover, the Rf value of ginsenosides roughly suggests the number of sugars attached to the aglycone. Both analyses make it possible to jointly identify which aglycone attaches and how many sugars it possesses, leading to the structure of the ginsenosides. In fact, three kinds of ginsenosides possessing protopanaxadiol-ginsenoside Rh_1, -Rf, and 20-*O*-glucoginsenoside Rf in *P. ginseng* root were determined by coloring and Rf value by comparing them with the structures reported in the previous paper [25].

Figure 5(a) indicates immunolocalization of glycyrrhizin in a licorice root slice using anti-glycyrrhizin MAb as another application of the eastern blotting method. The phloem (⟷) contained a higher concentration of glycyrrhizin than the xylem and cork part [26]. On the other hand, no staining occurred on ginseng (Figure 5(c)).

In the earlier experiments, we carried out the blotted staining on PVDF membrane using MAb on solasodine glycosides and called it western blotting [24]. Now we have applied this new methodology to licorice glycoside, glycyrrhizin and named it eastern blotting [12] for studying ginsenosides [23], saikosaponin [27], and so on.

2.3.2. Immunoaffinity Concentration and One-Step Purification of Ginsenoside Rb_1 by Immunoaffinity Column. A crude extract of *P. ginseng* roots was loaded onto the immunoaffinity column and washed with the washing solution of phosphate buffer. Figure 6 shows the fraction 1–8 containing

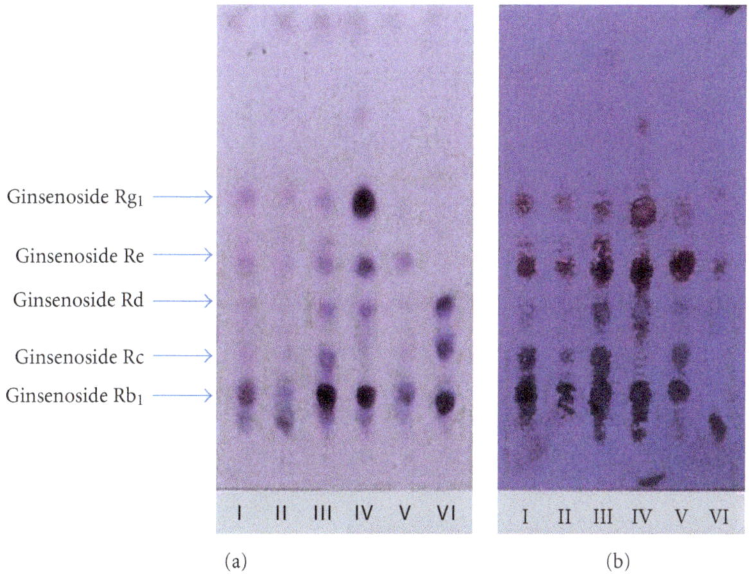

FIGURE 4: Double Eastern blotting staining of ginsenosides contained in various ginseng samples using anti-ginseenoside-Rb1 and anti-ginsenoside-Rg_1 MAbs. (a) TLC profile stained by sulfuric acid. (b) Eastern blotting by anti-ginsenoside-Rb_1 and anti-ginseside-Rg_1 monoclonal antibodies I, II, III, IV, V, and VI indicated white ginseng, red ginseng, fibrous ginseng (*P. ginseng*), *P. notoginseng*, *P. quinquefolius* and *P. japonicus*, respectively. Upper purple color spots and lower blue color spots were stained by anti-ginsenoside-Rg_1, and anti-ginsenoside-Rb_1 monoclonal antibodies, respectively.

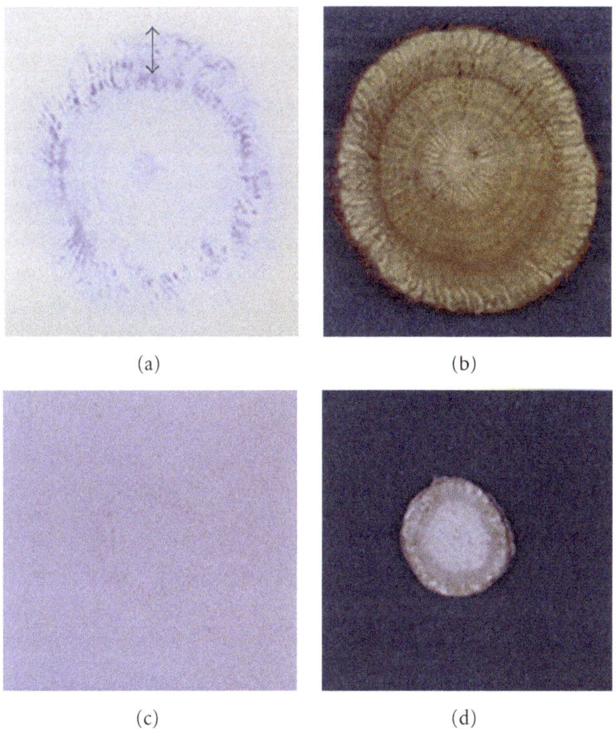

FIGURE 5: Immunocytolocalization of glycyrrhizin in fresh licorice root using anti-glycyrrhizin MAb. (a) Eastern blotting of fresh licorice slice, (b) fresh licorice slice, (c) Eastern blotting of fresh ginseng slice, and (d) fresh ginseng slice.

overloaded ginsenoside Rb_1. The other ginsenosides Rg_1, Rc, Re. and Rd were also detected in these fractions by eastern blotting (data not shown). A sharp peak appeared around fraction 20–24 eluted by acetate buffer containing KSCN and methanol to give pure ginsenoside Rb_1.

Overloaded ginsenoside Rb_1 was repeatedly immunoaffinity column choromatographed to separate ginsenoside Rb_1 completely [28]. The antibody was stable when exposed to the eluent, and the immunoaffinity column showed almost no decrease in capacity after repeated use more than 10 times under the same conditions, as was reported for a one-step separation of forskolin from a crude extract of *Coleus forskohlii* root [29]. This methodology is effective for the rapid and simple purification of ginsenoside Rb_1 and may open up a wide field of comparable studies with other families of saponins for which an acceptable method for one-step separation has not yet been developed. Furthermore, to separate the total ginseng saponins, a wide cross-reactive MAb against ginsenoside like anti-ginsenoside Re MAb could be designed, as was done for the total solasodine glycosides by an immunoaffinity column using an anti-solamargine MAb [30]. A combination of immunoaffinity column chromatography, Eastern blotting and ELISA could be used to survey low concentrations of ginsenoside Rb_1 of plant origin and/or in experimental animals and humans. In fact, we have succeeded in the isolation of ginsenoside Rb_1 from a different plant, *Kalopanax pictus* Nakai, which was not known previously to contain ginsenosides, using this combination of methods [31].

2.3.3. Newly Established Knockout Extract for Glycyrrhizin. Glycyrrhizin has been reported to possess numerous pharmacological effects like anti-inflammation [32], antiulcer [33], antitumor [34], antiallergy [35], and hepatoprotective

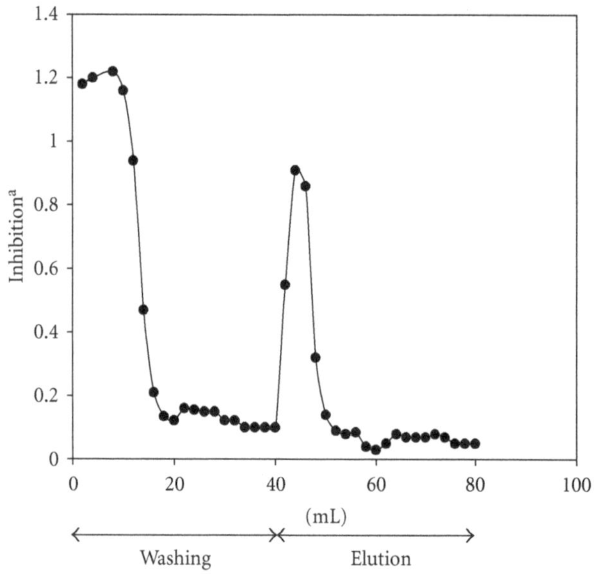

FIGURE 6: Elution profile of *Panax ginseng* crude extract on immunoaffinity column monitoring by ELISA using anti-ginsenoside-Rb_1 MAb. The column was washed by phosphate buffer and then eluted by HOAc buffer containing KSCN and MeOH. Individual fractions (2 mL) were assayed by ELISA using anti-ginsenoside-Rb_1 MAb.

activities [36]. Although glycyrrhizin is supposed to be a major active principle in licorice crude extract, a large number of studies have demonstrated that the licorice extract is rich in bioactive compounds other than glycyrrhizin such as triterpenes, flavonoids, and their aglycones [37]. In order to confirm the effect of glycyrrhizin in TCM, previously we purified glycyrrhizin from TCM using an immunoaffinity column conjugated with anti-glycyrrhizin MAb [38]. In this section, one-step purification of glycyrrhizin and its function in the licorice crude extract have been indicated as previously discussed about ginsenoside Rb_1.

To eliminate glycyrrhizin from licorice extract, 12 mg of licorice extract (glycyrrhizin content: 1275.0 μg) in loading buffer was applied on the anti-glycyrrhizin MAb immunoaffinity column, and then the loading buffer was continuously circulated through the column to enhance the binding efficiency. After overnight circulation at 4°C, the unbound column fraction was collected. The column was washed, and then bound column fraction was eluted by elution solvent. Glycyrrhizin concentration of bound column fraction was 1269.26 μg of glycyrrhizin (99.5% of the loading glycyrrhizin). On the other hand, glycyrrhizin content of unbound fraction was 3.50 μg (0.27% of the loading glycyrrhizin). These data indicated that anti-glycyrrhizin MAb immunoaffinity column could eliminate glycyrrhizin from licorice crude extract with high efficiency. From this evidence, we named this fraction knockout extract [16, 39].

To further characterize glycyrrhizin knockout extract, the TLC analysis and eastern blotting were performed. As shown in Figure 7(a), several spots including glycyrrhizin were detected in licorice extract. However, the spots of glycyrrhizin were undetected in glycyrrhizin-knockout extract, although all other spots were clearly detected (Figure 7(a), lane 2). eastern blotting by anti-glycyrrhizin MAb demonstrated that glycyrrhizin was detected in licorice extract, but the spot of glycyrrhizin was invisible in unbound fraction (Figure 7(b), lane 2). Therefore, these data suggest that glycyrrhizin was specifically eliminated from licorice extract by anti-glycyrrhizin MAb binding immunoaffinity column.

This technique is useful approach to clarify the multicomponent interaction with the specific protein such as enzymes, but it is unsuitable to investigate the interaction between one principal component and other components in herbal medicines. To investigate the role of one principal compound in the crude extract, we prepared knockout extract by removing target compound from the crude extract using high-specific MAb immunoaffinity column. By using this approach, it may become possible to determine the potential function of one natural compound on the crude extract or TCM by *in vitro* and *in vivo* assays.

Nitric oxide (NO) is free radical with multiple physiological functions, such as vasodilatation, neurotransmission, and inflammation [40]. During inflammatory process, large amount of NO is produced by inducible nitric oxide synthase (iNOS) by inflammatory cytokines and/or bacterial lipopolysaccharide (LPS) in various cell types including macrophages [41]. Overproduction of NO by iNOS triggers the pathogenesis of septic shock and organ destruction in certain inflammatory and autoimmune diseases [42–44]. Therefore, the inhibition of NO production by blocking iNOS expression may be useful strategy for the treatment of various inflammatory diseases.

To determine whether licorice crude extract inhibits NO production in LPS-stimulated mouse RAW264 macrophages, cells were treated with various concentrations of licorice extract for 30 min and then stimulated with LPS. LPS caused a dramatic increase in NO production, and this induction was inhibited in a dose-dependent manner by treatment with licorice extract.

We next investigated the inhibitory effect of glycyrrhizin-knockout extract and the combined treatment with glycyrrhizin-knockout extract and glycyrrhizin on NO production. Since 100 μg/mL of licorice extract contains 10.6 μg of glycyrrhizin, the cells were pretreated with licorice extract (100 μg/mL), glycyrrhizin-knockout extract (89.4 μg/mL), or combination of glycyrrhizin-knockout extract (89.4 μg/mL) and glycyrrhizin (10.6 μg/mL). Figure 8(a) indicated that treatment of licorice extract leads to a marked suppression of NO production as compared to LPS treatment [inhibition ratio (IR) = 57.7%]. Interestingly, although glycyrrhizin alone could not block NO production, the inhibitory effect of glycyrrhizin-knockout extract was decreased compared with licorice extract (IR = 17.8%). Moreover, the combined treatment with glycyrrhizin-knockout extract and glycyrrhizin significantly improved the inhibition of NO production (IR = 33.5%). To determine whether the combinational effect of glycyrrhizin-knockout extract and glycyrrhizin was related to iNOS expression, we performed western blotting (Figure 8(b)). Although the licorice extract strongly

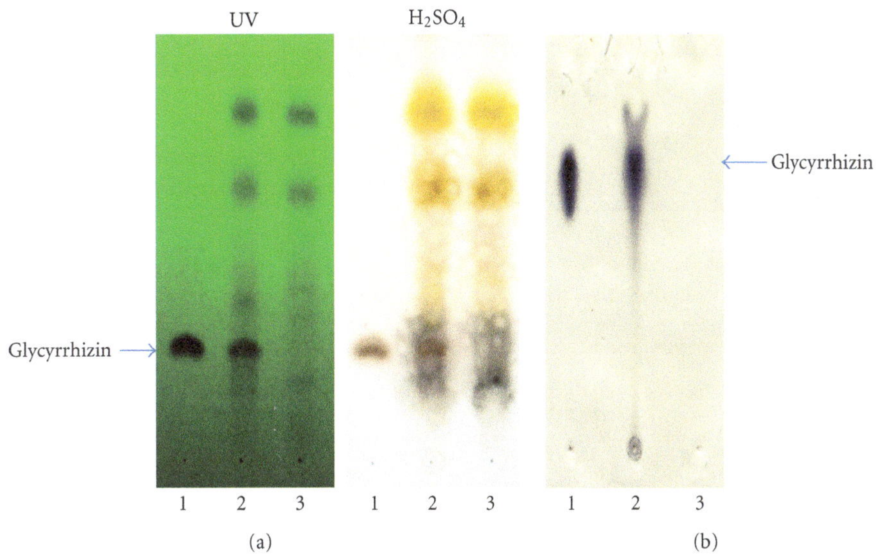

FIGURE 7: Preparation of knock-out extract eliminated glycyrrhizin from licorice crude extract using immunoaffinity column conjugated with anti-glycyrrhizin MAb. Lines 1, 2, and 3 indicate glycyrrhizin, crude extract, and knock-out extract, respectively.

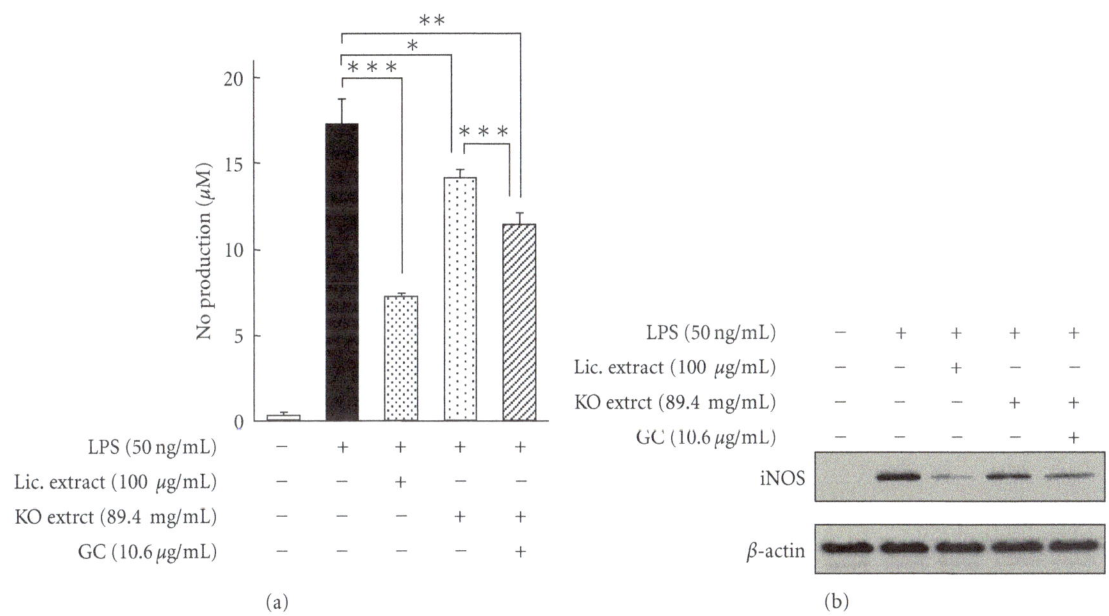

FIGURE 8: Effects of glycyrrhizin-knockout extract and the combination of glycyrrhizin-knockout extract and glycyrrhizin on LPS-induced NO production (a) and iNOS protein expression (b). $^{*}P < 0.05$, $^{**}P < 0.01$, $^{***}P < 0.001$.

suppressed LPS-induced iNOS expression, treatment of glycyrrhizin-knockout extract reduced this effect [45].

3. Conclusions

Two unique applications using MAb, eastern blotting and knockout extract have been introduced in this paper. The eastern blotting method has great potential applications for the wide range of natural products, especially glycosides like ginsenosides of ginseng and glycyrrhizin in licorice. When two kinds of MAbs, against ginsenoside Rb_1 and Rg_1 can be used, the double staining that enhanced the separate staining of ginsenosides having protopanaxatriol or protopanaxadiol in a molecule, occurred. The staining color can be used to monitor the pharmacological activity suggesting that the purple spots contain protopanaxatriol as an aglycone indicating central nervous system (CNS) stimulatory activity [46]. On the other hand, the blue color indicates ginsenosides having protopanaxadiol which possess a depression effect on the CNS [37]. Furthermore, the Rf value of ginsenosides suggests the number of sugars attached to the aglycone. Both evidences make it possible to confirm which aglycone is attached and how many sugars are combined with the aglycone.

We demonstrated that knockout extract prepared by anti-natural compound-specific MAb immunoaffinity column is a useful approach for determination of potential function of natural compound on *in vitro* and *in vivo* assays. The pharmacological analysis by knockout extract might be directly applicable to both other crude extracts and various TCMs to clarify the real pharmacologically active components.

A ginsenoside Rb_1 knockout extract can be prepared using an immunoaffinity column conjugated with anti-ginsenoside Rb_1 MAb. In this extract, all compounds except only ginsenoside Rb_1 are contained. Furthermore, glycyrrhizin-knockout extract prepared from licorice crude extract using an immunoaffinity column conjugated with anti-glycyrrhizin MAb. These knockout extracts may be able to support the pharmacological investigation for finding out a really active component in a crude drug and/or TCM. In fact, addition of glycyrrhizin to glycyrrhizin-knockout extract could improve the inhibition of iNOS expression resulting that glycyrrhizin may exert combinational inhibition of iNOS expression when coexisting with the other constituents contained in licorice extract.

References

[1] R. Sakata, Y. Shoyama, and H. Murakami, "Production of monoclonal antibodies and enzyme immunoassay for typical adenylate cyclase activator, forskolin," *Cytotechnology*, vol. 16, no. 2, pp. 101–108, 1994.

[2] M. Ishiyama, Y. Shoyama, H. Murakami, and H. Shinohara, "Production of monoclonal antibodies and development of an ELISA for solamargine," *Cytotechnology*, vol. 18, no. 3, pp. 153–158, 1996.

[3] L. Xuan, H. Tanaka, Y. Xu, and Y. Shoyama, "Preparation of monoclonal antibody against crocin and its characterization," *Cytotechnology*, vol. 29, no. 1, pp. 65–70, 1999.

[4] H. Tanaka, Y. Goto, and Y. Shoyama, "Monoclonal antibody based enzyme immunoassay for marihuana (cannabinoid) compounds," *Journal of Immunoassay*, vol. 17, no. 4, pp. 321–342, 1996.

[5] Y. Shoyama, T. Fukada, and H. Murakami, "Production of monoclonal antibodies and ELISA for thebaine and codeine," *Cytotechnology*, vol. 19, no. 1, pp. 55–61, 1995.

[6] H. Tanaka, N. Fukuda, and Y. Shoyama, "Formation of monoclonal antibody against a major ginseng component, ginsenoside Rb1 and its characterization," *Cytotechnology*, vol. 29, no. 2, pp. 115–120, 1999.

[7] N. Fukuda, H. Tanaka, and Y. Shoyama, "Formation of monoclonal antibody against a major ginseng component, ginsenoside Rg1 and its characterization: monoclonal antibody for a ginseng saponin," *Cytotechnology*, vol. 34, no. 3, pp. 197–204, 2000.

[8] J. S. Kim, H. Tanaka, and Y. Shoyama, "Immunoquantitative analysis for berberine and its related compounds using monoclonal antibodies in herbal medicines," *Analyst*, vol. 129, no. 1, pp. 87–91, 2004.

[9] O. Morinaga, S. Nakajima, H. Tanaka, and Y. Shoyama, "Production of monoclonal antibodies against a major purgative component, sennoside B, their characterization and use in ELISA," *Analyst*, vol. 126, no. 8, pp. 1372–1376, 2001.

[10] Z. Lu, O. Morinaga, H. Tanaka, and Y. Shoyama, "A quantitative ELISA using monoclonal antibody to survey paeoniflorin and albiflorin in crude drugs and traditional Chinese herbal medicines," *Biological and Pharmaceutical Bulletin*, vol. 26, no. 6, pp. 862–866, 2003.

[11] H. Tanaka and Y. Shoyama, "Formation of a monoclonal antibody against glycyrrhizin and development of an ELISA," *Biological and Pharmaceutical Bulletin*, vol. 21, no. 12, pp. 1391–1393, 1998.

[12] S. Shan, H. Tanaka, and Y. Shoyama, "Enzyme-linked immunosorbent assay for glycyrrhizin using anti-glycyrrhizin monoclonal antibody and an eastern blotting technique for glucuronides of glycyrrhetic acid," *Analytical Chemistry*, vol. 73, no. 24, pp. 5784–5790, 2001.

[13] P. Loungratana, H. Tanaka, and Y. Shoyama, "Production of monoclonal antibody against ginkgolic acids in Ginkgo biloba Linn," *American Journal of Chinese Medicine*, vol. 32, no. 1, pp. 33–48, 2004.

[14] K. Katsumi, K. Edakuni, O. Morinaga, H. Tanaka, and Y. Shoyama, "An enzyme-linked immunosorbent assay for aconitine-type alkaloids using an anti-aconitine monoclonal antibody," *Analytica Chimica Acta*, vol. 616, no. 1, pp. 109–114, 2008.

[15] K. Kido, O. Morinaga, Y. Shoyama, and H. Tanaka, "Quick analysis of baicalin in Scutellariae Radix by enzyme-linked immunosorbent assay using a monoclonal antibody," *Talanta*, vol. 77, no. 1, pp. 346–350, 2008.

[16] H. Tanaka, N. Fukuda, and Y. Shoyama, "Eastern blotting and immunoaffinity concentration using monoclonal antibody for ginseng saponins in the field of traditional Chinese medicines," *Journal of Agricultural and Food Chemistry*, vol. 55, no. 10, pp. 3783–3787, 2007.

[17] I. Wengatz, R.-D. Schmid, S. Kreissig et al., "Determination of the hapten density of immuno-conjugates by matrix-assisted UV laser desorption/ionization mass spectrometry," *Analytical Letters*, vol. 25, pp. 1983–1997, 1992.

[18] Y. Shoyama, T. Fukada, T. Tanaka, A. Kusai, and K. Nojima, "Direct determination of opium alkaloid-bovine serum albumin conjugate by matrix-assisted laser desorption/ionization mass spectrometry," *Biological and Pharmaceutical Bulletin*, vol. 16, no. 10, pp. 1051–1053, 1993.

[19] Y. Shoyama, R. Sakata, R. Isobe, H. Murakami, A. Kusai, and K. Nojima, "Direct determination of forskolin-bovine serum albumin conjugate by matrx-assisted laser desorption ionization mass spectrometry," *Organic Mass Spectrometry*, vol. 28, pp. 987–988, 1993.

[20] Y. Goto, Y. Shima, S. Morimoto et al., "Determination of tetrahydrocannabinolic acid-carrier protein conjugate by matrix-assisted laster desorption/Ionization mass spectrometry and antibody formation," *Organic Mass Spectrometry*, vol. 29, pp. 668–671, 1994.

[21] The Society for Korean Ginseng, In *Understanding of Korean Ginseng*, The Society for Korean Ginseng, Seoul, Korea, 1995.

[22] I. Kitagawa, T. Taniyama, M. Yoshikawa, Y. Ikenishi, and Y. Nakagawa, "Chemical studies on crude drug processing. VI. (1)) Chemical structures of malonyl-Ginsenosides Rb1, Rb2, Rc, and Rd isolated from the root of Panax ginseng C.A. Meyer," *Chemical and Pharmaceutical Bulletin*, vol. 37, no. 11, pp. 2961–2970, 1989.

[23] O. Morinaga, H. Tanaka, and Y. Shoyama, "Detection and quantification of ginsenoside Re in ginseng samples by a chromatographic immunostaining method using monoclonal antibody against ginsenoside Re," *Journal of Chromatography B*, vol. 830, no. 1, pp. 100–104, 2006.

[24] H. Tanaka, W. Putalun, C. Tsuzaki, and Y. Shoyama, "A simple determination of steroidal alkaloid glycosides by thin-layer chromatography immunostaining using monoclonal antibody against solamargine," *FEBS Letters*, vol. 404, no. 2-3, pp. 279–282, 1997.

[25] N. Fukuda, S. Shan, H. Tanaka, and Y. Shoyama, "New staining methodology: eastern blotting for glycosides in the field of Kampo medicines," *Journal of Natural Medicines*, vol. 60, no. 1, pp. 21–27, 2006.

[26] S. Yokota, Y. Onohara, and Y. Shoyama, "Immunofluorescence and immunoelectron microscopic localization of medicinal substance, Rb1, in several plant parts of Panax ginseng," *Current Drug Discovery Technologies*, vol. 8, no. 1, pp. 51–59, 2011.

[27] O. Morinaga, S. Zhu, H. Tanaka, and Y. Shoyama, "Visual detection of saikosaponins by on-membrane immunoassay and estimation of traditional Chinese medicines containing Bupleuri radix," *Biochemical and Biophysical Research Communications*, vol. 346, no. 3, pp. 687–692, 2006.

[28] N. Fukuda, H. Tanaka, and Y. Shoyama, "Isolation of the pharmacologically active saponin ginsenoside Rb1 from ginseng by immunoaffinity column chromatography," *Journal of Natural Products*, vol. 63, no. 2, pp. 283–285, 2000.

[29] H. Yanagihara, R. Sakata, H. Minami, H. Tanaka, Y. Shoyama, and H. Murakami, "Immunoaffinity column chromatography against forskolin using an anti-forskolin monoclonal antibody and its application," *Analytica Chimica Acta*, vol. 335, no. 1-2, pp. 63–70, 1996.

[30] W. Putalun, H. Tanaka, and Y. Shoyama, "Rapid separation of solasodine glycosides by an immunoaffinity column using anti-solamargine monoclonal antibody," *Cytotechnology*, vol. 31, no. 1-2, pp. 151–156, 1999.

[31] H. Tanaka, N. Fukuda, S. Yahara, S. Isoda, C. S. Yuan, and Y. Shoyama, "Isolation of ginsenoside Rb1 from Kalopanax pictus by eastern blotting using anti-ginsenoside Rb1 monoclonal antibody," *Phytotherapy Research*, vol. 19, no. 3, pp. 255–258, 2005.

[32] S. Amagaya, E. Sugishita, and Y. Ogihara, "Comparative studies of the stereoisomers of glycyrrhetinic acid on anti-inflammatory activities," *Journal of Pharmacobio-Dynamics*, vol. 7, no. 12, pp. 923–928, 1984.

[33] R. Doll and I.-D. Hill, "Triterpenoid liquorice compound in gastric and duodenal ulcer," *The Lancet*, vol. 280, no. 7266, pp. 1166–1167, 1962.

[34] Z. Y. Wang and D. W. Nixon, "Licorice and cancer," *Nutrition and Cancer*, vol. 39, no. 1, pp. 1–11, 2001.

[35] T. Kuroyanagi and M. Saito, "Effect of prednisolone and glycyrrhizin on passive transfer in experimental allergic encephalomyelitis," *Japanese Journal of Allergology*, vol. 15, no. 2, pp. 67–74, 1966.

[36] T. Nakamura, T. Fujii, and A. Ichihara, "Enzyme leakage due to change of membrane permeability of primary cultured rat hepatocytes treated with various hepatotoxins and its prevention by glycyrrhizin," *Cell Biology and Toxicology*, vol. 1, no. 4, pp. 285–295, 1985.

[37] M. N. Asl and H. Hosseinzadeh, "Review of pharmacological effects of glycyrrhiza sp. and its bioactive compounds," *Phytotherapy Research*, vol. 22, no. 6, pp. 709–724, 2008.

[38] J. Xu, H. Tanaka, and Y. Shoyama, "One-step immunochromatographic separation and ELISA quantification of glycyrrhizin from traditional Chinese medicines," *Journal of Chromatography B*, vol. 850, no. 1-2, pp. 53–58, 2007.

[39] C.-Z. Wang and Y. Shoyama, "Herbal medicine: identification, and evaluation strategies," in *Text Book of Complementary and Alternative Medicine*, C.-S. Yuan, E.-J. Bieber, and B.-A. Bauer, Eds., pp. 51–70, Informa Healthcare, Minnesota, Minn, USA, 2nd edition, 2006.

[40] S. Moncada, R. M. J. Palmer, and E. A. Higgs, "Nitric oxide: physiology, pathophysiology, and pharmacology," *Pharmacological Reviews*, vol. 43, no. 2, pp. 109–142, 1991.

[41] C. Nathan and Q. W. Xie, "Regulation of biosynthesis of nitric oxide," *Journal of Biological Chemistry*, vol. 269, no. 19, pp. 13725–13728, 1994.

[42] K. M. Naseem, "The role of nitric oxide in cardiovascular diseases," *Molecular Aspects of Medicine*, vol. 26, no. 1-2, pp. 33–65, 2005.

[43] T. J. Guzik, R. Korbut, and T. Adamek-Guzik, "Nitric oxide and superoxide in inflammation and immune regulation," *Journal of Physiology and Pharmacology*, vol. 54, no. 4, pp. 469–487, 2003.

[44] S. B. Abramson, A. R. Amin, R. M. Clancy, and M. Attur, "The role of nitric oxide in tissue destruction," *Best Practice and Research: Clinical Rheumatology*, vol. 15, no. 5, pp. 831–845, 2001.

[45] T. Uto, O. Morinaga, H. Tanaka, and Y. Shoyama, "Analysis of the synergistic effect of glycyrrhizin and other constituents in licorice extract on lipopolysaccharide-induced nitric oxide production using knock-out extract," *Biochemical and Biophysical Research Communications*, vol. 417, no. 1, pp. 473–478, 2012.

[46] K. Hashimoto, K. Satoh, P. Murata et al., "Components of *Panax* ginseng that improve accelerated small intestinal transit," *Journal of Ethnopharmacology*, vol. 84, no. 1, pp. 115–119, 2003.

7

Electrospray Ionization Mass Spectrometry: A Technique to Access the Information beyond the Molecular Weight of the Analyte

Shibdas Banerjee and Shyamalava Mazumdar

Department of Chemical Sciences, Tata Institute of Fundamental Research, Homi Bhabha Road, Colaba, Mumbai 400005, India

Correspondence should be addressed to Shyamalava Mazumdar, shyamal@tifr.res.in

Academic Editor: Troy D. Wood

The Electrospray Ionization (ESI) is a soft ionization technique extensively used for production of gas phase ions (without fragmentation) of thermally labile large supramolecules. In the present review we have described the development of Electrospray Ionization mass spectrometry (ESI-MS) during the last 25 years in the study of various properties of different types of biological molecules. There have been extensive studies on the mechanism of formation of charged gaseous species by the ESI. Several groups have investigated the origin and implications of the multiple charge states of proteins observed in the ESI-mass spectra of the proteins. The charged analytes produced by ESI can be fragmented by activating them in the gas-phase, and thus tandem mass spectrometry has been developed, which provides very important insights on the structural properties of the molecule. The review will highlight recent developments and emerging directions in this fascinating area of research.

1. Introduction

The basic concepts of chemistry originated from the quantitative estimation (e.g., weighing) of the constituents in a chemical reaction during the period of Lavoisier more than 200 years ago. Since then the analytical measurement of masses of the samples continuously evolved through the gravimetric analysis to weighing a single atom/molecule using the modern instrument called mass spectrometer. In mass spectrometry, a particular state of matter called gaseous ionic state is studied by transferring the analytes from condensed phase to the gas phase followed by their ionization. The success of the study of gas-phase ion chemistry and its application has been driven by the continuous advancement of the mass spectrometric technique since the studies were performed by Thomson [1]. As a result the mass spectrometry has become one of the most sensitive analytical methods for the structural characterization of molecules. Before the development of ESI-MS, there were several ionization methods (electron ionization, chemical ionization, etc.), but none of them could be able to overcome the propensity of the analyte fragmentation. In the mid 1980s, it became indispensable to precisely measure the molecular mass of the biologically important supramolecules like proteins [2]. But the proteins are polar, nonvolatile, and thermally labile molecules. So the ionization of the proteins by conventional ionization methods could lead to structural destruction. Although a technique called fast atom bombardment (FAB) [3] was available that time for the ionization of the biological samples, this technique produces predominantly singly charged ions of the analyte and the method works best for smaller species of mass below about 1000 Da. However, the available mass analyzers could not measure the high *m/z* value of the singly charged high molecular weight proteins during those days. So the only way to analyze the protein mass was to digest the protein and then the analysis of the digest mixture by FAB-mass spectrometry.

All those problems were overcome in 1989 when Fenn introduced electrospray ionization, a soft ionization technique, to ionize intact chemical species (proteins) by multiple charging [4]. The ionization is soft in the sense that a very little residual energy is retained by the analyte, and generally no fragmentation occurs upon ionization. Not only that but also very weak noncovalent interactions are preserved in the gas phase [5]. Because of the multiple charging, the *m/z* values of the resulting ions become lower and fall in the mass ranges of all common mass analyzers. Thus ESI became very useful in the production of gas-phase ions from large biologically important macromolecules like proteins and nucleic acids, and their subsequent mass spectrometric analysis for structural characterization as well as their rapid identification on the basis of molecular mass, a very specific property of the analyte. Gradually a systematic analysis of proteins with the mass spectrometry as the central tool led to a discrete subject called "Proteomics," one of the fastest growing research areas in the chemical sciences [6].

In 2002 Fenn, the inventor of ESI-MS shared the 4th Nobel Prize in mass spectrometry along with Koichi Tanaka (for the development of MALDI mass spectrometry, another soft ionization technique) and Kurt Wuthrich (for the work in NMR spectroscopy). "A few years ago the idea of making proteins or polymers "fly" by electrospray ionization (ESI) seemed as improbable as a flying elephant, but today it is a standard part of modern mass spectrometers" as stated by the Professor Fenn in his Nobel lecture [7]. Nowadays ESI-MS is not only being used as a balance to weigh protein molecules but also to gain a deeper understanding of the protein three-dimensional structures, noncovalent interaction, posttranslation modification, and amino acid sequence. Soft landing of the mass-selected multiply charged gaseous protein ions into liquids (after the mass spectrometric separation) was recently shown to retain the native structures and even the biological activities of some proteins [8, 9]. Although the development of ESI-MS has had a major impact in biology and proteomics, its application has extended to a broad range of analytes including polar organic [10], inorganic [11], and metal-organic complexes [12]. Recently ESI efficiency scale of the different organic molecules with different polarities has been developed [13, 14]. The best ESI response has been observed for the analytes with ionizable basic/acidic polar functional groups. High-performance liquid chromatography has been coupled with the ESI-MS for the molecular fractionation prior to mass-spectrometric analysis. Thus, HPLC/ESI-MS has become a very powerful technique capable of analyzing both small and large molecules of various polarities in a complex biological sample mixture.

Here we would briefly review the development of the ESI-MS technique in last two and half decades not only for the mass access but also for the detailed understanding of the structural properties of the analyte in the different aspects of chemistry and biology including the fundamentals of the ionization mechanisms.

2. The Historical Perspective

There is an interesting history behind the development of ESI-MS. Although the process electrospray was known more than hundred years ago [15], the actual thought process on ESI-MS was initiated by Professor Dole, a physical chemist at Northwestern University. Much of the Dole's research focused on the polymerization chemistry. In the 1960s he was trying to characterize the size as well as mass distribution of some synthetic polymers (originally polystyrene) by mass spectrometric technique. But that time the troubles he encountered were the lack of a suitable ionization system which can produce molecular ions (without decomposing their structures) in the gas phase from the highly nonvolatile synthetic polymers and also the unavailability of the suitable detector system which can probe the appearance of the large molecular ions with high *m/z* value. Accidently he discovered the existence of electrospray while visiting a car manufacturer, and he observed the car painting by a process called electrospray. Then he applied the electrospray process in the production of gas-phase polystyrene ions and their subsequent collection using a Faraday cage detector [16, 17]. Although their experiments showed that the electrospray is a very promising soft ionization (no fragmentation of the analyte) technique, no mass spectrometer was available that time to separate and detect the ions of polystyrene molecules.

However, Dole's report [16, 17] on electrospray got the attention of Professor Seymour Lipsky and Professor Csaba Horvath at Yale Medical School [2]. That time (1970s) Professor Lipsky was thinking about the alternate ways of ionizing biopolymers like proteins, and Professor Horvath was to work on the development of HPLC known as high-pressure liquid chromatography. They noticed that two of the Dole's reports referred the work of Fenn who was a professor in the Yale Engineering Department that time. Fenn was a specialist in the field of molecular beams and their production by nozzle-skimmer systems. Through those references Lipsky got in touch with Fenn. Fenn accepted the challenge of the production of biomolecular ions in the gas phase using his molecular beam apparatus even though he was approaching 65, typical retirement age [2]. Finally Fenn group's ground breaking discovery on the ionization and characterization of large biomolecules in the gas phase by electrospray ionization mass spectrometry [4] created a new dimension in the field of proteomics. After the Fenn's discovery of ESI-MS technique, the uses of the electrospray ionization continue to grow at an unprecedented rate (see Figure 1), and every day new applications are developed as the instrument continues to advance as fast as the need.

3. Basic Architecture of the ESI-Mass Spectrometer

Like other mass spectrometers, ESI-mass spectrometer is also composed of three basic components, for example, ion source, mass analyzer, and detector (see Figure 2). The intact molecular ions (not truly ions, see later) are produced in the ionization chamber where the ion source is kept, and then they are transferred in the mass analyzer region via several ion optics (electromagnetic elements like skimmer, focusing lens, multipole, etc.), which are basically kept to focus the ion stream to maintain a stable trajectory of the ions. The mass analyzer sorts and separates the ions according to their mass to charge ratio (*m/z* value). The separated ions are then passed to the detector systems to measure their concentration, and the results are displayed on

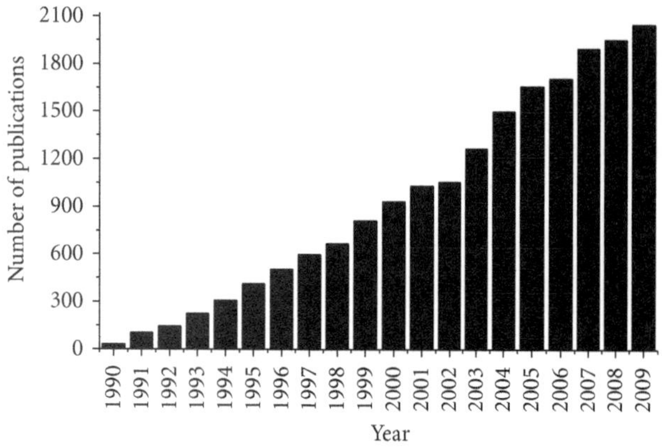

Figure 1: Yearly histogram of the papers dealing with the use of electrospray ionization after the Fenn's introduction of electrospray ionization mass spectrometry to ionize the biomolecules in 1989. The information was obtained by searching ISI Web of Knowledge on 26.06.2011 for the term "Electrospray ionization."

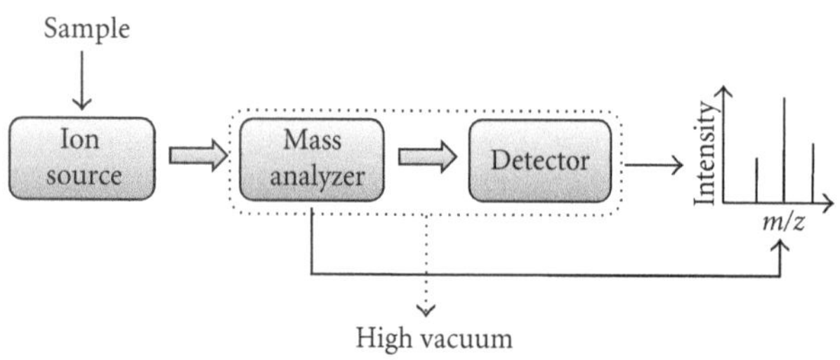

Figure 2: The basic components of the ESI-mass spectrometer.

a chart called a mass spectrum (see Figure 2). Since the ions in the gas phase are very reactive and often short lived, their formation and manipulation should be conducted in high vacuum. For this reason the ion optics, analyzer, and also the detectors are kept at very high vacuum (typically from 10^{-3} torr to 10^{-6} torr pressure). Mass spectrometers typically use either oil diffusion pumps or turbomolecular pumps to achieve the high vacuum required to operate the instrument. Generally the ion source is kept at atmospheric pressure, and a continuous pressure gradient and voltage gradient are used from source to the detector to help pump out the ions from source to the detector through the analyzer.

3.1. Ion Source. A suitable ESI source for the mass-spectrometric analysis was designed by the Fenn group in the mid 1980s [4, 18–20]. Later on it was modified by different research groups to improve the system's robustness [21–25]. Generally a dilute (less than mM in polar volatile solvent) analyte solution is injected by a mechanical syringe pump through a hypodermic needle or stainless steel capillary (~0.2 mm o.d and ~0.1 mm i.d) at low flow rate (typically 1–20 μL/min). A very high voltage (2–6 kV) is applied to the tip of the metal capillary relative to the surrounding source-sampling cone or heated capillary (typically located at 1–3 cm from the spray needle tip). This strong electric field causes the dispersion of the sample solution into an aerosol of highly charged electrospray (ES) droplets (see Figure 3). A coaxial sheath gas (dry N_2) flow around the capillary results in better nebulization. This gas flow also helps to direct the spray emerging from the capillary tip towards the mass spectrometer. The charged droplets diminish in size by solvent evaporation, assisted by the flow of nitrogen (drying gas).

Finally the charged analytes are released from the droplets, some of which pass through a sampling cone or the orifice of a heated capillary (kept in the interface of atmospheric pressure and the high vacuum) into the analyser of the mass spectrometer, which is held under high vacuum. The heated capillary (typically 0.2 mm inner diameter, 60 mm in length and heated to 100–300°C) causes the complete desolvation of the ions passing through it. The use of drying gas and the heated capillary can influence the system's robustness and reduce the degree of cluster ion formation [24]. The transfer of analyte ions from solution to gas phase is not an energetic process, but rather the desolvation process effectively cools the gaseous ions. So the analyte ions with low internal energies are allowed to enter into the mass spectrometer from the electrospray probe, and the structure of the analytes generally remain intact (no fragmentation) when appropriate instrumental conditions (e.g., no activation of the ions in gas phase) are used. Nowadays a number of sprayer modifications like pneumatically assisted electrospray [26–28], ultrasonic nebulizer electrospray [29, 30], electrosonic spray [31], and nanoelectrospray [23, 32] have been developed to expand the range of ESI applications. Among them the most popular one is nanoelectrospray.

Nanospray ionization is a low flow rate (20–50 nL/min) version of electrospray ionization [32]. A very low sample concentration (nanomole/mL) and low volume are required for nanospraying. Such downscaling has been achieved by replacing the spray needle with borosilicate glass capillary of some microliters volume to which a fine tip (1–4 μm inner diameter) is pulled with a micropipette puller. The spray voltage of 0.7–1.1 kV is normally applied via an electrically conducting coating (usually a sputtered gold film) on the outer surface of the spray capillary. When the high voltage is switched on, the analyte solution flow is solely driven by capillary forces refilling the aperture as droplets are leaving the tip. While conventional ESI generates initial charged droplets of 1-2 μm in diameter, the nanospray produces the charged droplets of the less than 200 nm diameter; that is, their volume is about 100–1000 times smaller than the droplets produced by a conventional microemitter. The nano-ESI has an increased tolerance to high aqueous solvents and salt contamination [23, 32]. In this technique not only less analyte sample is consumed than with the standard electrospray ionization, but also a small volume of sample lasts for several minutes, thus enabling multiple experiments to be performed.

3.2. Mass Analyzer. The mass analyzer is the heart of the mass spectrometer. The mass analyzer can be compared with the prism. The component wavelengths of a light are separated by a prism, and then they are detected by an optical receptor. Similarly in the mass analyzer, the different types of ions (m/z) of an ion beam are separated, and then they are passed to the detector. There are many types of mass analyzers

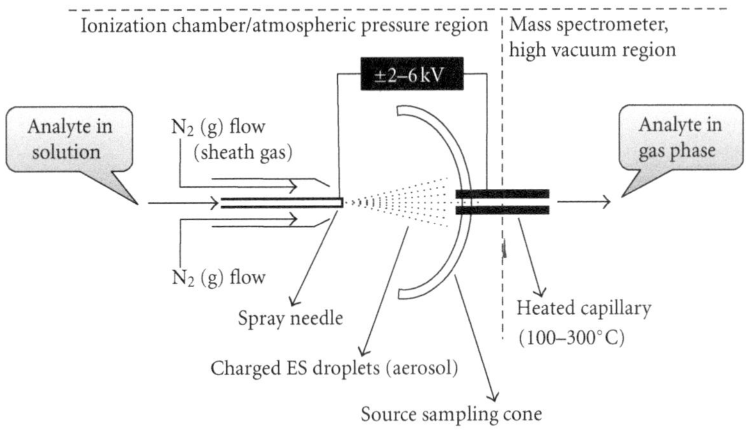

FIGURE 3: A schematic representation of the ESI-ion source.

[33, 34], for example, magnetic (B)/electric (E) sector mass analyzer, linear quadrupole ion trap (LIT), three-dimensional quadrupole ion trap (QIT) [35], orbitrap, time-of-flight mass analyzer (TOF), and ion cyclotron resonance mass analyzer (ICR), all of these which use the static or dynamic magnetic/electric field, and all operate according to two fundamental laws of physics, for example, Lorentz force law and Newton's second law of motion. Proper selection of the mass analyzer depends on the resolution, mass range, scan rate, and detection limit required for an application. Although the detailed discussion of different types of mass analyzers is beyond the scope of this paper, interested readers can go through any standard mass spectrometry text book for this issue.

3.3. Detector. The simplest way to detect the ion is the use of Faraday cup, where the ion is allowed to be neutralized and the resulting current is measured. This Faraday cup is used when the ion flux is relatively large. But a more modern way to measure the low ion flux is the use of electron multiplier (just like photomultiplier tube to measure the photon flux) [36, 37]. The energetic ions are allowed to strike a metal or semiconductor plate (e.g., copper/beryllium alloy oxide) called a conversion dynode that emits secondary electrons (SE). This conversion dynode is held at very high voltage (order of kV). The secondary electrons emitted are accelerated and focused onto the second and subsequent dynodes (kept in positive potentials), which are set at potentials progressively closer to earth. At each dynode there is an increase in the number of electrons emitted (electron avalanche), such that at the end of the multiplier a gain of approximately 10^6 is achieved. The output current is converted to a voltage signal, which finally can be translated to an intensity value (the ordinate axis of the mass spectrum) by means of an analog-to-digital (ADC) converter. There are several types of multipliers like discrete dynode electron multipliers [36, 37], channel electron multipliers (CEMs) [34], microchannel or multichannel plates (MCPs) [34], and so forth. Unlike other mass spectrometers, ions are not detected by hitting a detector such as an electron multiplier in FT-ICR-MS [38], but ions are detected by the measurement of the image current produced by ions cyclotroning in the presence of a magnetic field. A detector is selected according to its speed, dynamic range, gain, and geometry. Some detectors are sensitive enough to detect a single ion. Although there has been a revolution in the mass spectrometer development in the last twenty years by several researchers and companies, the question regarding the response of the detector haunts the researchers till now. How does the detector respond to the large multiply charged ions produced by ESI? No precise information is available regarding the fact whether the observed peak height corresponding to a multiply charged macroion reflects the number of incident ions, the number of charges they carry, the conformation of the ion, the energy of incidence, its velocity, or an unknown combination of these factors [39]. Though it is assumed that a peak height in a mass spectrum is directly proportional to the number of corresponding incident ions to the detector, this issue still remains suspicious as the detector response has not been characterized appropriately in those aspects mentioned above [39].

Figure 4 shows some hybrid mass spectrometers (commercially available) which are constructed by combining different types of *m/z* separation devices or mass analyzers. Different types of detectors and spray (ion source) geometries are also noticeable in those instruments (Figures 4(a)–4(c)). These instruments are specially designed for the tandem mass analysis (see later). The trajectory of the ions from ion source to the analyzer is linear (on-axis/line-of-sight) in conventional electrospray sources (Figure 4(a)). But, recently the spray geometries have been modified to orthogonal spray (Figure 4(b)) or z-spray (Figure 4(c)) where the ion trajectory from source to the analyzer is, respectively, orthogonal and z shaped. These off-axis spray geometries circumvent the problem of the clogging of heated capillaries and skimmers by neutral molecules, nonvolatile materials, and so forth.

4. The ESI-Mass Spectrum

Generally the ions derived by ES process are multiply charged, and the analyte remains intact (no fragmentation) when

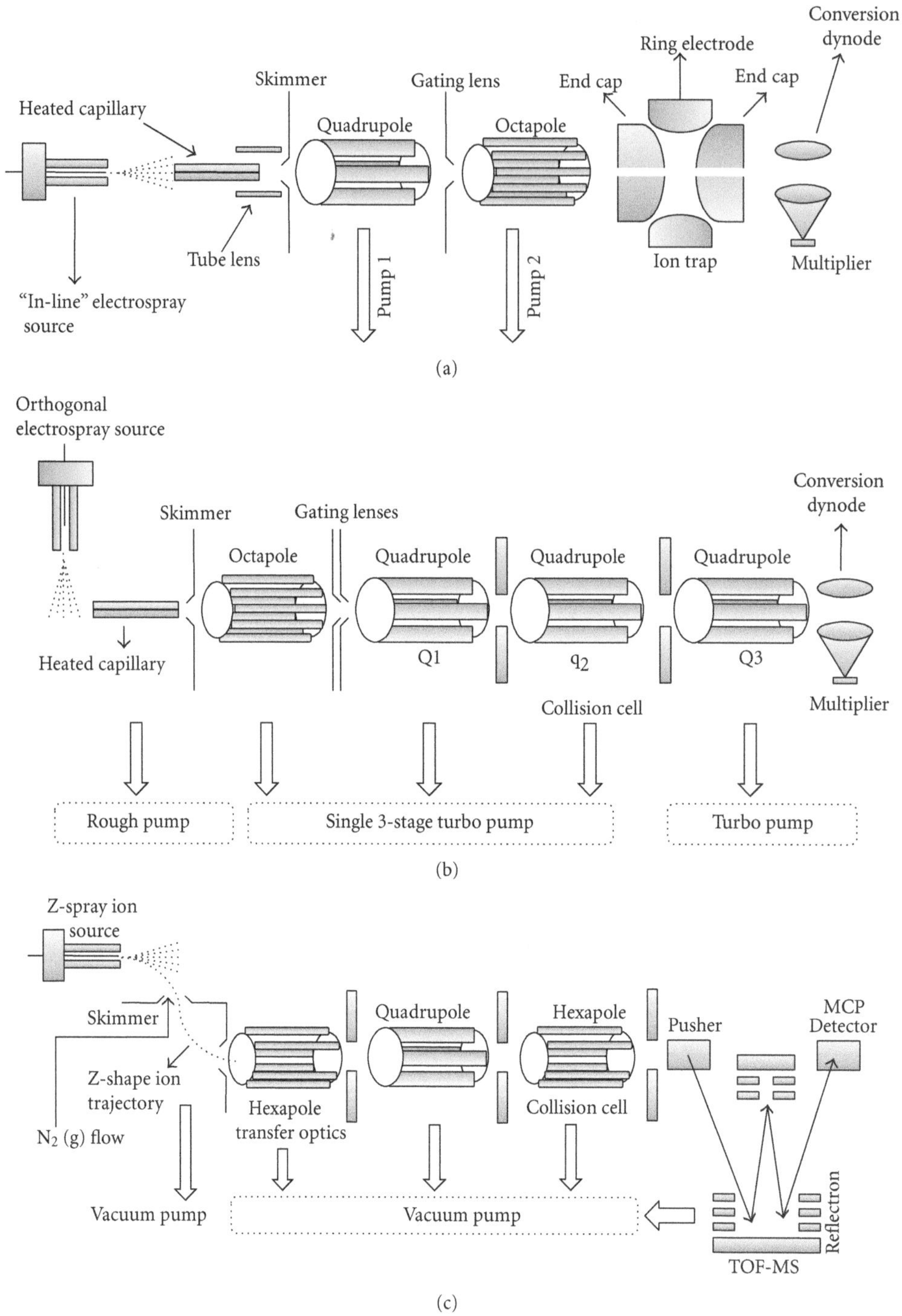

FIGURE 4: Schematic of the (a) thermo-Finnigan LCQ Deca mass spectrometer (on-axis spray), (b) Agilent 6410 Triple Quad LC/MS system (off-axis/orthogonal spray), and (c) Waters Micromass Q-TOF Ultima ESI-MS (z-spray).

appropriate instrumental conditions are used. In positive ion mode (when the spraying nozzle is kept at positive potential) the charging generally occurs via protonation (sometimes metalation also), but in negative ion mode (when the spraying nozzle is kept at negative potential) charging occurs via deprotonation of the analyte. The mechanism of charging has been discussed later. Since the charging of the analyte occurs by transfer of protons, the ionic species detected are not the true molecular ions (which are formed by the loss or gain of the electron), but they are more preferably protonated or deprotonated molecules [40]. When a pure analyte solution is electrosprayed all the peaks appearing in the corresponding ESI-mass spectrum represent the intact molecular species with variable charging as shown in Figure 5. The ordinate or vertical axis represents the relative abundances of multiply charged species of the same analyte, and the abscissa

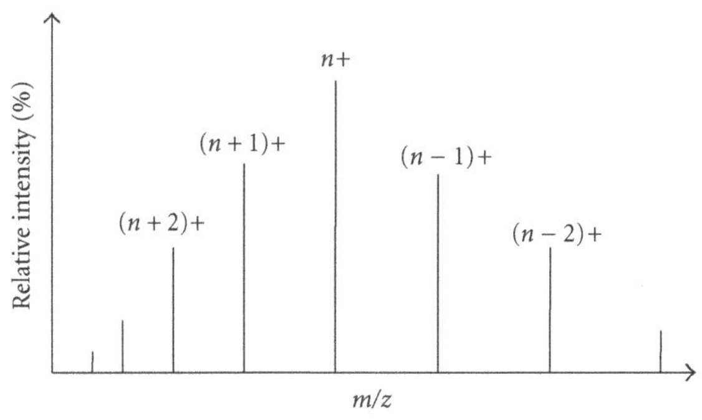

Figure 5: A typical cartoon representing the nature of ESI-mass spectrum in positive ion mode.

or horizontal axis represents the mass to charge ratio (*m/z*) of the multiply charged analyte. As usual the most intense or the tallest peak (peak of *n*+ charge in Figure 5) with 100% relative abundance is called the "base peak." The quantity *m/z* is dimensionless since it is the ration of two dimensionless quantities mass number (*m*) and the elementary charges (*z*). Sometimes the unit *Thomson* [Th] is applied to the *m/z* value to honor Thomson. Very often the symbol "u" (unified atomic mass) or Da (Dalton, in biological mass spectrometry field to honor J. Dalton) is also used as the molecular mass unit.

Now the question is how one can determine the molecular mass of the analyte with the help of the ion signals (different *m/z* values) related to the same molecule? The procedure of finding the charge and mass of the analyte through the ESI-mass spectrum is called "Mathematical Charge Deconvolution" [41–43]. The process is straightforward and based on the assumption that two adjacent peaks in an ESI-mass spectrum of a single analyte have charge state difference one as has been shown in Figure 5. If it is assumed that the molecular mass of the analyte is *M* and the observed *m/z* values for two neighboring charge states *n*+ and $(n-1)+$ are m_1 and m_2, respectively, then two mathematical equations can easily be drawn as $(M+n)/n = m_1$ and $[M+(n-1)]/(n-1) = m_2$. Since there are only two unknown parameters *M* and *n*, a minimum of two equations are required to find out the values of *M* and *n*. So the above two equations can easily be employed to determine the value of *n* (charge state) and *M* (deconvoluted mass). One can think the observed multiple peaks as the multiple mass assessment of the same molecule. A more accurate mass of the analyte can be derived by averaging the calculated mass for each peak. The above algorithm works well for the pure analyte and when the analyte produces the successive charge states of the difference unity. But problems arise when the analyte produces metaliated as well as protonated species in gas phase or skips some charge state in between two observed charge states. Therefore, several refined procedures have been developed to cope with these requirements [44]. Nowadays all the modern ESI-MS instruments are equipped with elaborate software for charge deconvolution.

Proteins, the most popular analytes for ESI-MS study exhibit different charge state distribution profiles in their ESI-mass spectra. To quantify the charge state distribution of the analyte, the term "average charge state" (Z_{av}) was introduced [45] and defined as $\sum z_i \cdot I_i/\sum I_i$, where I_i is the mass spectrometrically detected signal intensity of a given charge state (z_i) carried by the *i*th ion. Latter on several correlations between this average charge state and the protein structures were found (discussed later).

5. The Mechanism of Electrospray Ionization

After the development of ESI-MS, many different assumptions and hypotheses were made in the early 1990s to interpret the multiple charging of the analyte by the ES process [46–49]. That time it was thought that the distribution of the charge states in the ESI-MS spectra actually reflects the degree of charging of the analyte (say proteins) in the neutral solution (as determined by solution phase equilibrium) [41, 50, 51]. But later it became evident that there is no correlation between solution charging and electrospray charging after the report of Kelly and coworkers [47]. Their report showed that in positive ion mode ESI produced protonated (positively charged) analyte (myoglobin) though the analyte is overall negatively charged (deprotonated) in basic aspirating solution (pH 10), and in negative ion mode it produced deprotonated (negatively charged) analyte (myoglobin) though the analyte is overall positively charged (protonated) in acidic aspirating solution (pH 3). These observations implied that the charging process in ESI might occur in an entirely different manner than that it was thought.

When an analyte is transferred from solution to the gas phase via ESI, the analyte solution undergoes three major processes. These are (a) production of the charged droplets from the high-voltage capillary tip where the analyte solution is injected; (b) repeated solvent evaporation (from the charged droplet) and droplet disintegration, resulting a very small charged droplet, which is able to produce the charged analyte; (c) finally a mechanism by which the gas-phase ion is formed. Since the electrospray process existed long before its application in the mass spectrometry and earlier it was used for the electrostatic dispersion of liquids and the creation of aerosol, the first two processes were mostly studied by the researchers in the aerosol science, and the processes are well understood nowadays [15, 52]. But the last process, that is, the mechanisms of the ion formation from vary small highly charged droplet is still under controversy, and the exact process happening at this stage is not unambiguously known. Over the last two decades, the issue has hotly been discussed and debated, and several hypotheses based on the theory and experimental evidences were invoked [22, 53, 54], and here we would discuss the most popular models of the gas-phase ion formation.

5.1. Production of Charged Droplets. When the analyte solution is pumped through the high-voltage capillary (emitter), an electrochemical reaction of the solvent occurs which causes an electron flow to or from the metal capillary depending on its polarity [55]. In absence of any redox active analyte [56, 57], the oxidation of the solvent occurs in the positive ion mode and reduction of the solvent occurs in the

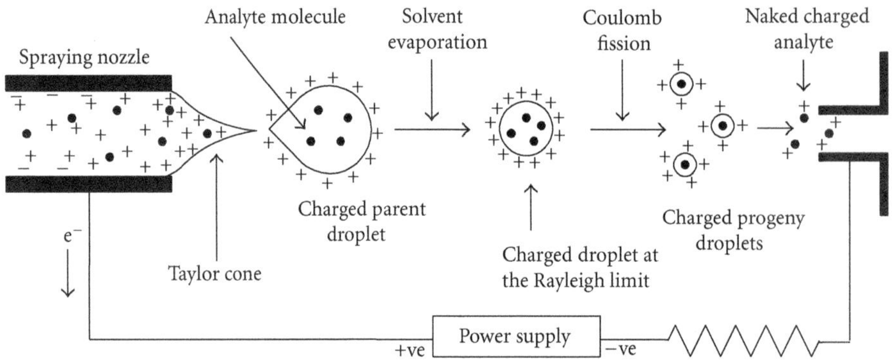

Figure 6: Schematic representation of the electrospray ionization process.

negative ion mode as has been shown in Table 1 [58]. So the ES ion source could be "viewed as an electrolytic cell of a special kind" [56]. These redox reactions (Table 1) supply positive or negative ions in the solutions depending on the polarity of the emitter electrode. Generally polar solvents (e.g., water, methanol, acetonitrile, etc.), which easily undergo electrochemical reactions in the spraying nozzle, are used in the ESI-MS experiments [58]. The accumulated charges (positive or negative) would be repelled by the high-voltage capillary (of the same polarity), and then they would drift towards the liquid surface at the capillary outlet (see Figure 6). However, the accumulated like charges at the surface are destabilized, and finally the meniscus would be drawn out and deformed into a cone under the influenced of a very high electric field (typically ~10^6 V/m) [22, 58]. This cone is called a Taylor cone [59] after the name of the scientist Taylor who first theoretically described the conditions under which a stable liquid cone can exist with the competing forces of an electric field and the surface tension of the liquid [59]. This Taylor cone is the zone of high turbulence. At high field strength, it immediately starts ejecting a fine jet of liquid from its apex towards the counter electrode (heated capillary) [60]. This charged jet easily breaks up into small droplets. The droplets produced at low flow rate (typically ~5 μL/min) have a narrow distribution of sizes with a most abundant radius ~1.5 μm [22]. Such an electrospray droplet was shown to have a charge of ~10^{-14} C, which corresponds to ~60,000 singly charged ions. All those charged droplets are driven away from each other by Coulombic repulsion and move along the direction of the electric fields (towards the heated capillary) (see Figure 6). Overall, under positive ion mode positively charged aerosol is formed and in negative ion mode negatively charged aerosol is formed. The positive charges are mostly contributed by protons, and negative charges are contributed by some negative ions (e.g., OH^-) (see Table 1) [58].

5.2. Coulomb Explosion and Disintegration of the Charged Droplets. The charges (say protons in positive ion mode) in the droplets are distributed on its surface with equidistant spacing to minimize the potential energy [39]. There are two forces acting in opposite directions in the charged droplets. One is surface tension of the charged droplet, which tries to retain the spherical shape of the droplet, and the other is Coulomb force of repulsion between the like charges on the surface, which tries to break down the spherical shape of the charged droplet [22, 39, 54, 58]. The solvent evaporation occurs when the droplets traverse the space between spraying nozzle and the heated capillary (see Figure 6). As a result the size of the droplet decreases until it reaches the point (Rayleigh limit) [62] where the surface tension can no longer sustain the Coulomb force of repulsion, and at this point "Coulomb explosion" or "Coulomb fission" occurs; that is, the parent droplet disintegrates into much smaller offspring droplets. The emitted stream of the offspring/progeny droplets holds about 2% mass and 15% charge of the parent droplet [22]. So the offspring droplets are not only much smaller than their parent but also have much higher charge-to-mass ratio. The process of solvent evaporation and Coulomb fission occurs repeatedly to generate smaller and smaller progeny droplets and finally the charged nanodroplets from which the gas-phase charged analyte molecule is formed [22, 63]. Figure 7 describes the detail evolution of the initial droplets, formed from the Taylor cone, to the droplets that are the precursor of the gas-phase ions. Typically the flying microdroplets were observed to vibrate alternately from oblate to prolate shapes, and this elastic vibration causes the parent droplet to emit a tail of much smaller offspring droplets (droplet jet fission) [22, 53, 64] as has been shown in the inset of Figure 7. This disruption pattern is quite similar to the disruption at the Taylor cone. Therefore, the charge density on the droplet surface is not homogeneous, but significantly increased in the region of sharper curvature, and it has been shown the charged microdroplets fission somewhat before, at ~80% of the Rayleigh limit [22]. The concept of this charged droplet jet fission is based on theoretical as well as experimental evidence (flash microphotographs) [64–66]. The life time of an ES droplet largely dependent on several parameters like ion spray voltage, nature of the solvents, sheath gas flow rate, distance between the spraying nozzle and heated capillary and temperature of the heated capillary, and so forth, [22, 28] and that would in turn affect the charging of the analyte. However, the average life time of a charged droplet produced by the ES process is around one to a few milliseconds for typical interfaces that one would use in ESI-MS [22, 67].

TABLE 1: Typical redox reactions of the solvents expected to occur in the electrospray emitter and their corresponding standard potentials.

Solvent systems	Positive-ion mode Oxidation reactions	E^0 (V)	Negative-ion mode Reduction reactions	E^0 (V)
Water	$2H_2O = O_2 + 4H^+ + 2e^-$	1.23	$2H_2O + O_2 + 4e^- = 4OH^-$	0.40
	$2H_2O = H_2O_2 + 2H^+ + 2e^-$	1.77	$H_2O + O_2^- + e^- = HO_2^- + OH^-$	0.20
	$H_2O = HO^* + H^+ + e^-$	2.72	$H_2O + HO_2^- + e^- = HO^* + 2OH^-$	0.18
			$2H_2O + 2e^- = H_2 + 2OH^-$	0.07
			$2H_2O + O_2 + 2e^- = H_2O_2 + 2OH^-$	−0.13
			$H_2O + O_2 + 2e^- = HO_2^- + OH^-$	−0.83
Methanol	$CH_3OH = HCHO + 2H^+ + 2e^-$	0.23	$CH_3OH + H_2O + 2e^- = CH_4 + 2OH^-$	−0.25
	$CH_3OH + H_2O = HCOOH + 4H^+ + 4e^-$	0.10	$CH_3OH + 2H^+ + 2e^- = CH_4 + H_2O$	0.58
	$HCOOH = CO_2 + 2H^+ + 2e^-$	−0.20		

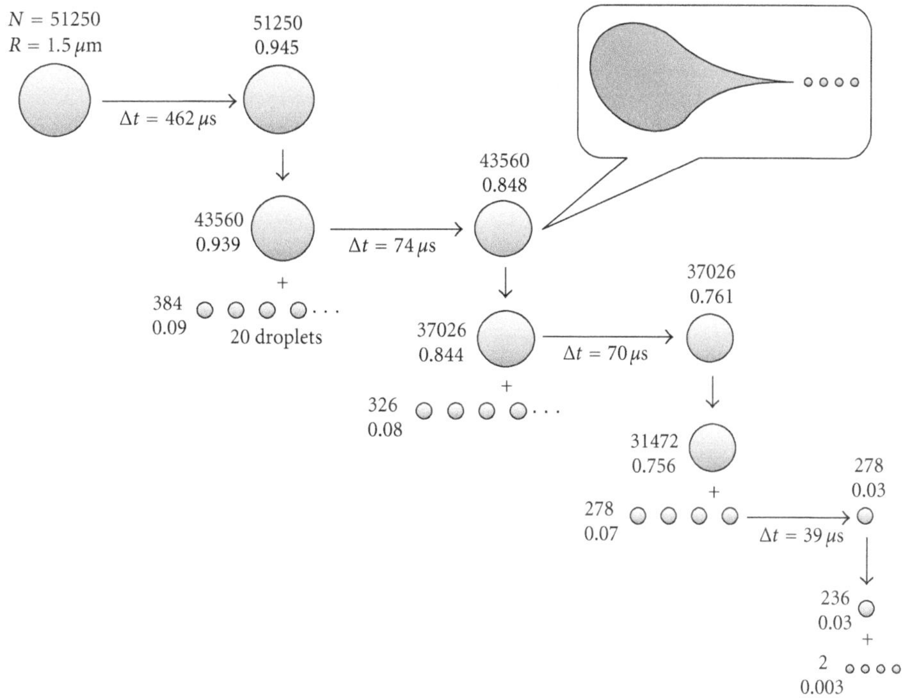

FIGURE 7: Time history of the charged methanol droplet produced by microelectrospray process. The droplet at the top left is a typical parent droplet created at the ES capillary tip. The successive solvent evaporation and Coulomb fission leads to the charged nanodroplets that are the precursors of the gas-phase analyte ions. The numbers beside the droplets give radius R (μm) and number of elementary charges N on the ES droplet; Δt corresponds to the time required for evaporative droplet shrinkage to size where fission occurs. Only the first three successive fissions of a parent droplet are shown. At the bottom right, the fission of the offspring droplet to produce the charged nanodroplets is shown. The inset shows a drawing of droplet jet fission based on actual flash microphotograph. (Adapted with permission from [22], Copyright 1993, American Chemical Society).

5.3. Formation of Gas-Phase Analyte Ions from the Charged Droplets. Two principle mechanisms have been proposed to account for the formation of gas-phase analyte ions from very small highly charged ES droplets, and those are *charge residue model* (CRM) [16, 68, 69] and *ion evaporation model* (IEM) [63].

5.3.1. Charge Residue Model. As a result of a series of solvent evaporation and Coulomb fission, an extremely small charged droplet ($R \approx 1$ nm) is formed which contains only one analyte molecule [16, 22]. Desolvation of this droplet causes its charges (on the surface) to land on the analyte molecules. Since the "residual" droplet charge at the last stage of the solvent evaporation in the ES process is retained by the analyte molecule in the gas phase, this mechanism is called charge residue mechanism/model. Originally this mechanism was hypothesized by Dole et al. [16] and a more detail consideration as well as support of this mechanism was given by Schmelzeisen-Redeker et al. [68].

A Study of Winger and coworkers showed that generally large macromolecules like proteins follow CRM [70]. They determined one interesting empirical correlation (see (1)) between the analyte molecular mass (M) and the observed average charge state (Z_{av}) in the ESI-mss spectra of the

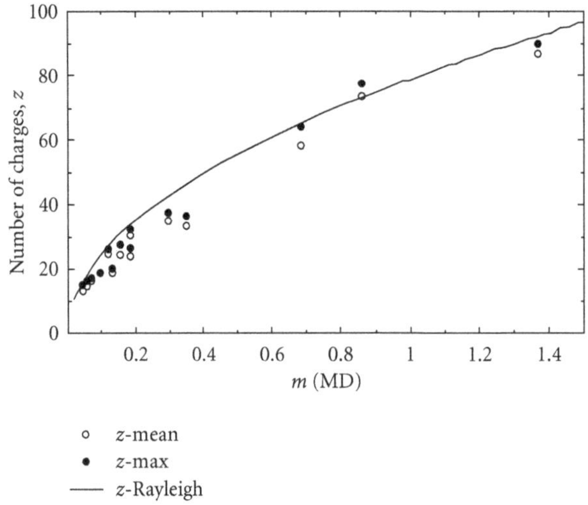

FIGURE 8: Plot of the average charge state (z) against molecular mass (m) for various proteins that follow CRM. [(•) Observed highest charge state; (◦) observed lowest charge state; solid curve corresponds to the average charge state predicted by (4)]. The molecular weights of the proteins given on the x-axis are in units of 10^6 Da. (Reprinted with permission from *Analytica Chimica Acta* [61], Copyright 2000, Elsevier).

corresponding analyte (starburst dendrimers) [71]. The "a" and "b" are constants and $b = 0.53$ gave the best result

$$Z_{\text{av}} = aM^b. \tag{1}$$

Later on Fernandez De La Mora [61] showed that the above equation (1) could be theoretically derived for the analytes like proteins that follow CRM. He assumed that after the final Coulomb fission event that produces the ultimate charged water droplet containing one neutral protein molecule would be a little bit larger in size than that of the protein. Then the final solvent evaporation would completely transfer the droplet charges to the protein. He also assumed that the density of the globular protein is the same as that of water ($\rho = 1$ g/c.c) [61, 72], and then he calculated the radius of the protein as

$$\left(\frac{4}{3}\pi R^3\right)\rho N_A = M. \tag{2}$$

The Rayleigh limit charging (which deals with the Rayleigh stability limit for coulomb fission of the charged droplet) of the water droplet having the same radius (R) as of the analyte protein can be given by the Rayleigh equation (3) [62], and then combining (2) and (3) he found the (4) [61] which is similar to (1) empirically obtained by Winger and coworkers. Notably that the exponent for M equals to 0.5 is very much close to the exponent 0.53 of the empirical equation (1)

$$Z_R \cdot e = 8\pi \left(\varepsilon_0 \gamma R^3\right)^{1/2}, \tag{3}$$

$$Z_R = 4\left(\frac{\pi \gamma \varepsilon_0}{\rho e^2 N_A}\right)^{1/2} \times M^{1/2} = 0.078 M^{1/2}, \tag{4}$$

where e is the elementary charge, Z is the charge number, ε_0 is the permittivity of the surrounding medium, N_A is the Avogadro's number, ρ is the density of the water, M is the molecular weight of the protein, and γ is the surface tension of the solvent (water). A plot (solid line) shown in Figure 8 is based on the above theoretically derived equation (4), and that fits well with the empirical data (shown by circles) of several proteins investigated by Fernandez De La Mora. So the observed agreement between the charges on the proteins and the charges on water droplets of the same size, at the Rayleigh limit, is not a coincidence but a consequence of the multiply charged proteins being formed by the CRM.

5.3.2. Ion Evaporation Model. After repeated solvent evaporation and Coulomb fission, the radii of the charged droplets decrease to a given size when the electric field due to the charges at the surface of the droplet is strong enough to cause direct emission of the solvated ions [22, 63, 73]. Typically when the droplet reaches the size $R \leq 10$ nm, the ion emission dominates over Rayleigh fission. Thus unlike CRM, this mechanism does not require the production of very small droplets ($R \approx 1$ nm) that contains only one analyte molecule. This mechanism is called ion evaporation mechanism/model as proposed by Iribarne and Thomson [22, 63, 73]. They proposed that the rate constant k_1 for ion emission from the highly charged droplet surface can be given by the following equation (5) where k_b is the Boltzmann constant, T is the temperature of the droplet, h is the Planck constant, and $\Delta G^{\dagger}$ is the free energy of activation:

$$k_1 = \frac{k_b T}{h} \exp\left(-\frac{\Delta G^{\dagger}}{RT}\right). \tag{5}$$

The activation energy is influenced by several parameters like (a) attraction between the escaping ion and the solvent by which the droplet is composed, (b) Coulomb repulsion of the escaping ion by the remaining charges on the droplet surface, and (c) ion desolvation energy. Although the above Iribarne and Thomson equation (5) can successfully interpret the

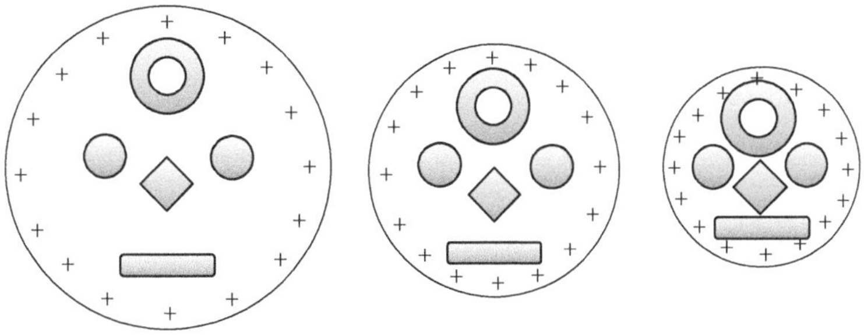

Figure 9: Hypothetical picture of a charge droplet containing five analytes of different shapes and sizes at three different stages of the solvent evaporation before Coulomb fission.

ion evaporation process, it cannot predict the observed relative intensities of ions in the ESI mass spectra, which is mostly influenced by the structural and physicochemical nature of the analyte and the solvent used. However, IEM is experimentally well supported for small inorganic and organic ions [53, 61, 74, 75].

5.3.3. Fenn's Model of Ion Formation: An Extension of IEM. Unlike IEM, the Fenn's model [39, 69] states that the analytes remain neutral inside the neutral core of the charge droplet (charges lie on the surface). If the analyte is intrinsically charged (e.g., proteins), this charged neutrality could be maintained either by a nearby counter anion as its "shadow" (i.e., ion pair formation) or by a proton transfer process between analyte and the solvent. The overall neutral analyte would have a rapid Brownian diffusion motion inside the charged droplet. The successive solvent evaporation and Coulomb fission would continuously increase the surface charge density of the droplet. In positive ion mode these charges are basically protons. Protons are pretty small, and they can have a strong attractive interaction with the molecules of polar solvents. When a given part of the analyte encounters the proton-rich surface by Brownian thrashing (energy on the order of k_bT), the basic residues located at that portion of the analyte would be attached to the surface protons. Thus the protons would serve to anchor the analyte molecule nearer to the surface. Thermal activation (Brownian oscillation, internal vibration, and rotation, with amplitudes constrained by k_bT) may provide the energy for the analyte to move some distance outside the droplet. This separation between the protonated analyte residues and the charges on the droplet surface leads to repulsion, which facilitate the escape of a remaining portion of the molecule that carries other charge sites. With increase in the charged residues (outside the droplet surface) of the escaping analyte, the Coulomb repulsion increases and that facilitate the escape of the whole analyte with a varying degree of charging. The degree of charging varies not only for the size, shape, orientation, and fugacity or escaping tendency of the analyte but also for the continuously variable charge spacing and electric field at the droplet surface during the process of repeated solvent evaporation and Coulomb fission. Thus, the number of charges that are attached to the analyte molecule when it desorbs from the surface is the number of charges that it can span on the surface during its emission. The situation is illustrated in Figure 9, which shows three hypothetical snapshots of a charged droplet at different stages during its evaporation before undergoing Coulomb fission. The droplet contains five analytes of different shapes and sizes. The droplet size decreases with solvent evaporation and the distance between equally spaced positive charges decreases since the number of charges remains constant. Clearly, at the evaporation stage III, the desorption of the analytes from the droplet surface would lead more charging than that at the stage I. Again at each stage the charging will vary according to the conformation and size of the analyte. But in practice the ejection of the charged analyte would continuously lower the number of charges and number of analyte species in the host droplet from stage I to stage III. As a result, above and below some critical charge states, the relative ion abundance decreases and that is reflected by a bell-shaped distribution of the ion abundances (see Figure 5). Of course the scenario is much more complicated because of the involvement of the Coulomb fission since it also contributes the charge and analyte concentration in a droplet.

6. Factors That Influence the Charge State Distribution (CSD) of Proteins

Multiple charging is an intrinsic feature of electrospray ionization of macromolecules like proteins [41–43]. Several factors affect the charge state distributions (CSDs) of proteins in vacuum generated by ESI [76]. Mostly the physical dimension of the protein molecules in solution are the major determinant of this CSD [76, 77]. As discussed earlier, in positive ion mode the basic sites in the proteins can accommodate protons and in the negative ion mode the acidic sites in the proteins can give up protons (see Figure 10) in the vanishing charged droplet and thus responsible for the gas-phase ion formation mostly via CRM [78]. Here we would discuss some critical factors known to significantly influence the protein CSD.

6.1. Solution pH. The change of pH of the protein solution can change the protein conformation (tertiary structure), and the degree of the conformational change depends on the

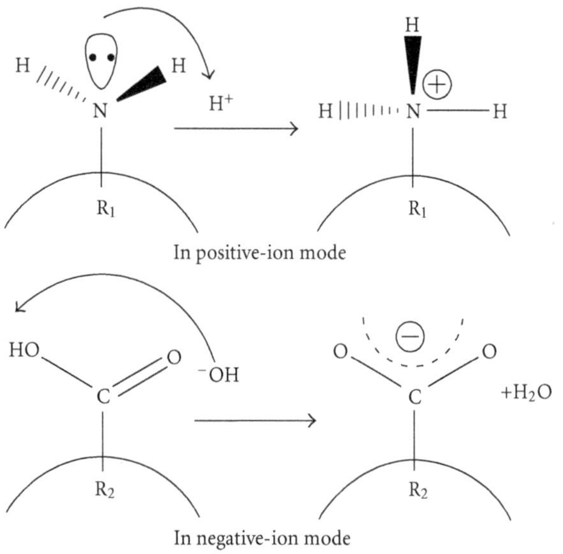

FIGURE 10: The proton transfer reaction responsible for protein charging in gas phase via ESI.

nature of the protein and the acidity/basicity of the solution. According to the CRM, it is the geometry or size of the protein that actually determines the average charge state (CS) of the electrosprayed protein in vacuum [76]. Relatively small number of charge states (low average CS) is produced when a natively folded protein is transferred to the gas phase upon ESI than that when an unfolded/denatured protein undergoes ESI. This is because the native proteins maintain a compact structure by its tightly folded polypeptide chain, and this compactness is lost upon denaturation. After denaturation, the solvent accessibility as well as the size of the protein increases, which in turn leads to increase the charge state distribution over a broad range (m/z) with the high average CS [45, 48, 76, 79].

Figure 11 shows a typical example of ESI-CSD of cytochrome c at different pH in positive ion mode [80]. At pH 6.4, where cyt c is known to adopt native, tightly folded conformation, the positive ion ESI-MS (Figure 11(a)) shows relatively narrow CSD with +8 as the most intense CS (base peak). But at pH 2.3, where a substantial unfolding of cyt c occurs, the positive ion ESI-MS (Figure 11(d)) shows a broad CSD with +17 as the most intense CS (base peak). Native and nonnative protein states often coexist at equilibrium under mildly denaturing conditions. In such conditions, the CSD becomes bimodal, reflecting the presence of both native and denatured states. At pH 2.6 cyt c exists in both native and unfolded states as is evidenced by the bimodal CSD in the ESI-mass spectrum (Figure 11(c)).

In certain cases proteins, for example, ubiquitin unfolds in both acidic (pH 2) and basic solution (pH 11.7) and when the protein was electrosprayed from both acidic and basic solution, the corresponding positive ion mass spectra showed characteristic broad CSD with maxima at +11 and +8, respectively, [81]. But the ESI-mass spectrum of ubiquitin in neutral solution (pH 7.2) [81] showed narrow CSD with most intense charge at +6.

The native environment of pepsin (in gastric juice) is extremely acidic, and at native pH (pH $\leq$ 2) it remains in folded active form. So when it is electrosprayed from the acidic solution (pH 1.6), it shows remarkably narrow CSD [82], and the CSD does not change until the solution pH is raised above 2.5, a threshold pH of pepsin deactivation [77]. The large scale unfolding of pepsin occurs in neutral and basic solutions. Therefore, a dramatic change of pepsin CSD in ESI-mass spectra acquired under these conditions [76, 77] is observed, and the corresponding most intense CS increased to above +30.

So the analysis of the CSD of proteins in ESI-MS provides reliable information on the protein structure and stability at different solution pH.

6.2. Number of Acidic and Basic Functional Groups in the Proteins. Conformation-dependent neutralization theory (CDNT) was proposed to interpret the appearance of main CS in ESI of proteins [83, 84]. The difference between acidic (aspartic acid, glutamic acid, etc.) and basic (arginine, lysine, histidine, etc.) residues in the protein sequence was used to determine the predominant charge state in the ESI-mass spectrum. However, the CDNT could not explain the appearance of the positive ion mass spectra of several proteins although they contain overall negative charge and vice versa. Later it was proposed that only the surface-exposed basic or acidic groups of a protein could hold charge in the gas phase [85]. However, some surface-exposed residues may form salt bridges, ion pair, or hydrogen bonding within the protein, reducing the effective charge density on these residues [86] and thereby decreasing the charge on the protein. Hence, we reported a more refined model to interpret the CS based on the three-dimensional structure (crystal structure) of the protein [78]. In view of this, we have identified the free basic and acidic groups (which do not form any ion pair or salt bridge) on the protein surface from the crystal structure analysis of a series of the proteins. Then we investigated the ESI-mass spectra of those native folded proteins in both positive and negative ion mode [78]. Interestingly it was observed that the maximum charge state of the gaseous protein ion actually corresponds to the number of surface-exposed free basic (in positive ion mode) or free acidic residues (in negative ion mode). This model was further supported by the ESI-mass spectra of the denatured proteins [78]. Upon denaturation all the acidic and basic residues become surface exposed, and the number of the basic and acidic residues in the protein sequence corresponded to the maximum CS in positive and negative ion mode, respectively.

Some exceptional examples of the above model were reported later. One particularly interesting example is the ESI-mass spectra of native folded pepsin (33 kDa) which contains total 41 acidic and only four basic residues [82]. Despite the small number of basic sites, the native pepsin (pH $\leq$ 2) can accommodate maximum 11 positive charges (most intense CS is +10) in gas phase upon positive ion ESI [76, 77]. This maximum charge state (+11) is higher than that predicted by our model and lower than that predicted by CRM ((4), which predicts +14 as the average CS). Since most proteins possess a high number of basic sites, the requisite number of

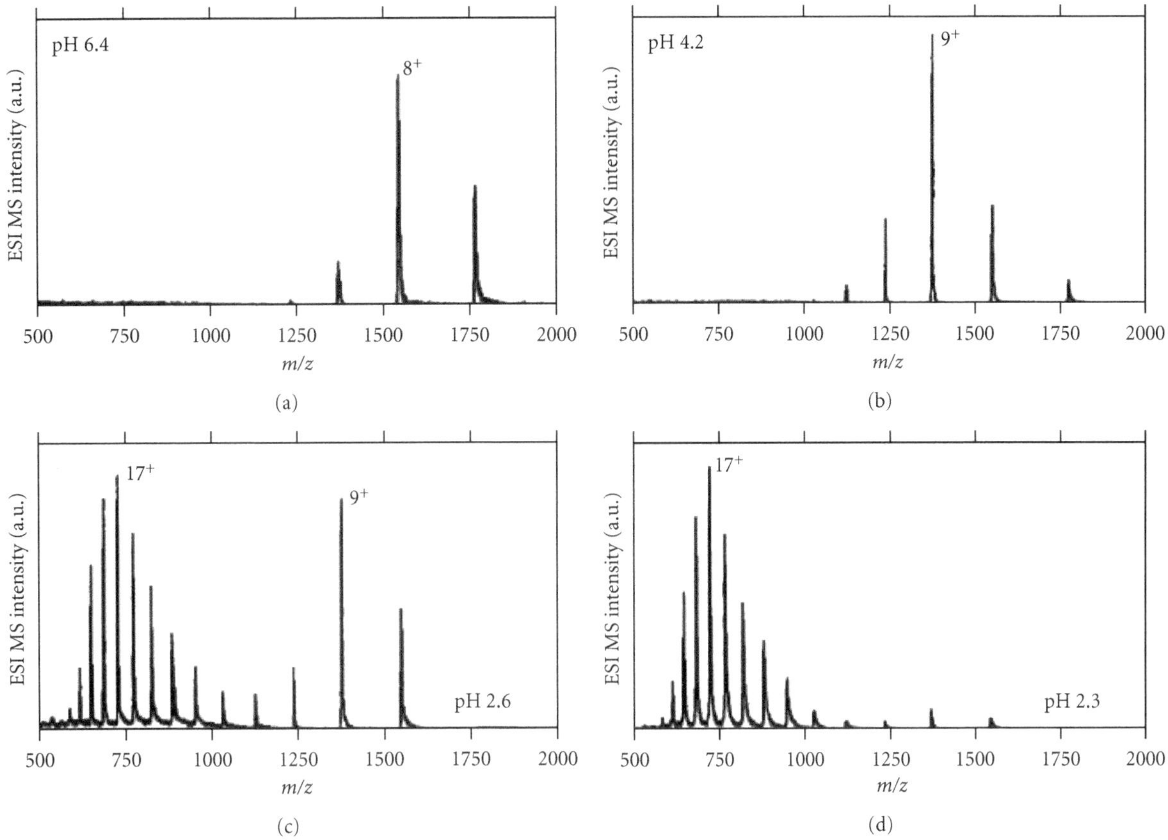

FIGURE 11: ESI-mass spectra of cyt c in water containing 3% methanol and 0.5 mM ammonium acetate at (a) pH 6.4, (b) pH 4.2, (c) pH 2.6, and (d) pH 2.3. The pH was adjusted by the addition of hydrochloric acid. (Reprinted with permission from *Biochemistry* [80], Copyright 1997, American Chemical Society).

positive charges as predicted by CRM can almost always be distributed among the available free basic sites on the surface of the native proteins. But, the ability of gaseous pepsin to accommodate a number of protons far exceeding the number of available basic sites might be due to the participation of the other mild basic functional groups (e.g., amide moiety, etc.) or collectively a number of polar functional groups over the large flexible surface of the protein to hold (solvate) protons in gas phase [87]. The lack of the required number of strong basic functional groups in pepsin possibly inhibits the extensive protonation as predicted by CRM. Moreover, it has been proposed that the proton affinity of the solvent molecules provides a "cut-off" level for the amino acid residues that are protonated in gas phase by ESI [87].

6.3. Protein Surface Area in Solution. Despite the fact that the extent of multiple charging in ESI is influenced by the number of surface-exposed free acidic/basic functional groups in the proteins (i.e., the gas-phase ion chemistry), the physical dimension or the structural integrity of the protein also largely influence the CSD. Kaltashov and Mohimen [77] demonstrated that the CSD of protein ions in ESI can be correlated to the solvent-exposed surface area in solution. In their study on 22 proteins ranging from the small protein insulin (5 kDa) to the large protein ferritin (500 kDa), they showed that the observed average charge states (Z_{av}) in the ESI-mass spectra could be correlated more precisely to the solvent exposed surface area rather than the radius or molecular weight of the proteins as suggested by CRM ((3) and (4)). The spherical approximation of the proteins which was used by Fernandez De La Mora [61] could provide erroneous results in CRM ((3) and (4)) since in practice the protein shape can deviate from spherical. Therefore, they extended (3) to (6) as given below to correlate the Z_{av} with the protein surface area S (Since, $S = 4\pi R^2$) [77].

$$Z_{av} \cdot e = 4\left(4\pi\varepsilon_0^2\gamma^2 S^3\right)^{1/4}. \tag{6}$$

Their empirical data is shown in Figure 12, which clearly indicates that the surface area (S) evaluated from the known crystal structure and the charge state Z_{av}, obtained from the ESI-mass spectra of the proteins, are correlated and fitted well by (7) or (8) with the value $a = 0.69$. The fit of the data points (Figure 12) is much better than the fit in the Fernandez De La Mora plot (Figure 8)

$$Z_{av} = AS^a,$$
$$\ln Z_{av} = \ln A + a \ln S. \tag{7}$$

Thus, they showed that the average CS-molecular mass correlation holds only for the tightly packed globular protein.

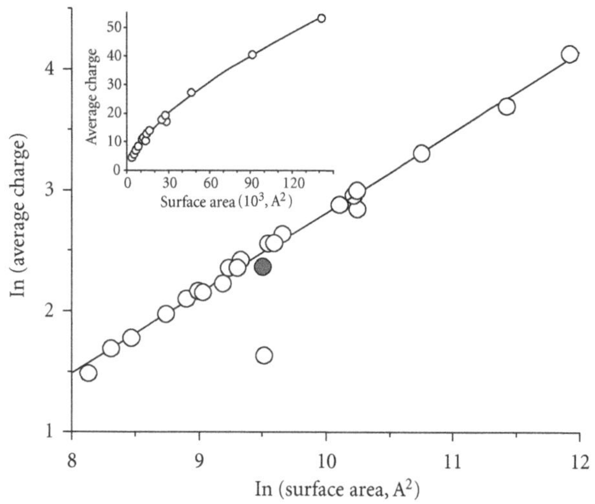

FIGURE 12: Correlation between the average charge state of protein ions generated by ESI under near-native conditions (10 mM ammonium acetate, pH adjusted to 7) and their surface areas in solution, whose calculation was based upon their crystal structures. The data are plotted in the logarithmic scale (a graph plotted in the normal scale is shown in the inset). A gray-shaded dot represents a pepsin data point. An open circle underneath represents the highest charge of pepsin if the extent of multiple charging was limited by the number of basic residues within the protein molecule. (Reprinted with permission from *Analytical Chemistry* [77], Copyright 2005, American Chemical Society).

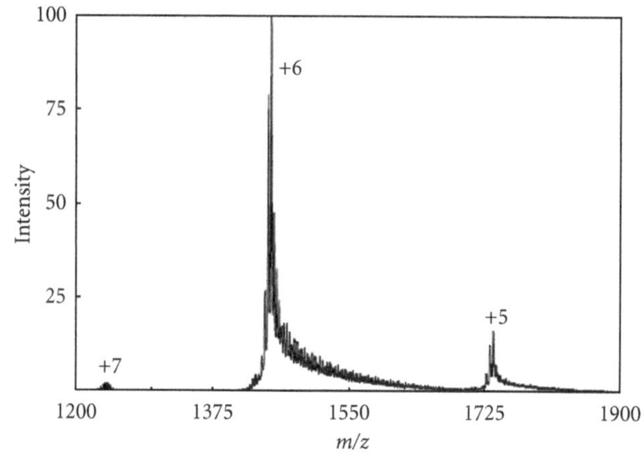

FIGURE 13: Nano-ESI-mass spectrum of 25 μM ubiquitin in 1 mM NaCl solution. The spectrum was obtained at low nozzle-skimmer potential, so that there was little collisional activation of the protein. It is obvious that each charge state is not a single peak but consists of multiple peaks due to the adduct formation of the salt ions (e.g., Na^+ & Cl^-) with ubiquitin. (Reprinted with permission from *Journal of the American Society for Mass Spectrometry* [91], Copyright 2005, Elsevier).

But, there are some proteins (e.g., ferritin) that are not tightly packed in their native conformations and it is actually the solvent exposed surface area of the proteins that is the major determinant of the average CS of the gaseous proteins produced by ESI process [77].

6.4. Presence of Nonvolatile Salts in Protein Solution. It is often the case that the proteins are purified and stored in some buffer solution containing nonvolatile salts (e.g., sodium chloride, potassium chloride, sodium phosphate, potassium phosphate, etc.). So if the protein solution is not properly desalted before injecting it in the ESI probe, the ions such as Na^+, K^+, and so forth, make strong ion pairing with the ionized acidic residues of the protein in the gas phase [88]. The nonvolatile metal ion impurities not only form strong bonds with carboxylate functional groups but also with other polar functional groups including the amide groups in the peptide backbone [89, 90]. This ion pairing would be driven by large loss of the solvent from the droplet and thus concentrating the nonvolatile salt component along with the proteins in the vanishing charged droplet. The binding of those metal ions (Na^+ and K^+) with the protein in the gas phase is so strong that they even survive in the clean-up stages (heated capillary, nozzle-skimmer dissociation potential) of the ESI mass spectrometer [91]. But the adduct formation of the metal ions (M^+) and/or their salts (MA) with the proteins is highly undesirable because of the mass shifting in variable extent that generates multiple peaks corresponding to each charge state (see Figure 13) [91]. The observed multiple peaks corresponding to each charge state can be rationalized by (a) replacement of H^+ with M^+ and (b) ion pairing of M^+ with the ionized acidic residues (aspartic acid/glutamic acid/C-terminal carboxylic acid, etc.) and/or ion pairing of A^- (counter anion of the non volatile salt) with the ionized basic residues (Arginine, Lysine, Histidine, N-terminal amine, etc.) [91]. If the mass spectrometer is not of sufficiently high resolution, then a complicated and "messy" spectrum is produced for the adduct formation of the component ions of the salt impurities with the protein in the gas phase.

In our recent study we have showen that the relative volatility of the charged particles (from salts) and the surface accessibility of the polar residues of the protein towards the charged particles are important factors in the CSD of the protein in gas phase [92]. The surface accessibility of surface-exposed free basic residues (SEFBRs) and surface-exposed free acidic residues (SEFARs) were determined from the crystal structure and compared with the most probable charge-state (most-intense peak obtained in the positive ion and the negative ion mode ESI-MS spectra of the proteins). The results indicated that the number of the SEFBRs that have surface accessibility above the mean surface accessibility of all the SEFBRs in the protein corresponds to the most-intense charge state of the protein in the positive ion mode. Analogously, the number of the SEFARs that have surface accessibility above the mean surface accessibility of all SEFARs in the protein would give the most-intense negative ion charge state of the protein. Hence, the most-intense charge states both in the positive ion as well as in the negative ion modes for a large number of proteins could be predicted based on this simple model from surface accessibility of the protein.

6.5. Use of Ammonium Acetate Salt Additive. Ammonium acetate buffer solution is very often used in the mass spectrometric analysis of proteins (see Figure 11) since unlike other buffer of nonvolatile salts it produces very clean mass spectra solely due to the protonated proteins in the gas phase [88, 91]. Even the use of ammonium acetate prevents the formation of sodium/potassium (if present as impurities) adduct with the gaseous protein. When ammonium acetate is added to the protein solution, following ion pairing reaction (8) occurs on the solvent accessed surface of the protein [91]. Ammonium cations pair with acidic functional groups, and acetate anions pair with basic functional groups in the protein in the gas phase upon ESI. The larger concentration of ammonium acetate relative to sodium/potassium ion impurities reduces the possibility of ion pairing involving sodium/potassium because of the predominance of ammonium acetate interaction with proteins (8). When the resulting protein ions containing the ion pairs (8) fly through the heated capillary and nozzle skimmer in the intermediate pressure region of the ESI interface, they encounter collision with the background gas and solvent vapor, which causes loss of volatile acetic acid and ammonia by the proton transfer reaction (9). The process is entropically favorable, and the dissociation is also facile because of the low bond (proton transfer) energy for the dissociation (15 kcal/mol for acetic acid and 11 kcal/mol for ammonia) [91]. The net effect of the complete process is nothing but the proton transfer from the protonated basic residues to deprotonated acidic residues. Though the positive and negative groups are neutralized, the molecular weight of the protein does not change. Thus, a clean mass spectrum of the protein is obtained without the mass change

$$\begin{aligned} &{}^{+}H_3N\text{-(protein)-}COO^{-} + CH_3COONH_4 \\ &\quad = CH_3COO^{-}\,{}^{+}H_3N\text{-(protein)-}COO^{-}\,NH_4^{+}, \end{aligned} \tag{8}$$

$$\begin{aligned} &CH_3COO^{-}\,{}^{+}H_3N\text{-(protein)-}COO^{-}\,NH_4^{+} \\ &\quad = CH_3COOH + H_2N\text{-(protein)-}COOH + NH_3. \end{aligned} \tag{9}$$

Felitsyn et al. [93] and Peschke et al. [94] also suggested that, in presence of the volatile salt like ammonium acetate, the ammonium cations can also bind to the basic residues of the protein, and there is proton transfer reaction from the ammonium ion to the basic residues of the protein with formation of volatile ammonia (base) and the protonated protein in the gas phase. Thus, ammonium acetate can also enhance the propensity of protonation of the protein in the gas phase.

The above equations ((8) and (9)) suggest that the use of ammonium acetate not only suppresses the metal ion adduct formation but also prevents the adduct formation involving the anion of strong acids (e.g., phosphate, trifluoroacetate anions, etc.) and ionized basic residues. Though (trifluoroacetic acid) TFA is frequently used in the mobile phase of the liquid chromatography mass spectrometry (LC-MS), the presence of the TFA anion in the effluent causes the loss of sensitivity, and the use of ammonium acetate prevents that sensitivity loss [95].

6.6. The Nature of Solvent. Several physical properties of the solvent like polarity (dielectric constant) [96], gas-phase basicity [97–99], and the surface tension [61, 100, 101] are also known to be responsible for the CSD of the protein. It is expected that the solvent with higher dielectric constants will tolerate high charge densities in the evaporating droplets produced by the electrospray process. The excess charges can also be confined to a thin layer at the surface because of the increased conductivity of the polar solvent with higher dielectric constant. At the last stage of solvent evaporation or gaseous ion emission from the charged droplet, high polarity solvents are more effective in stabilizing the multiply charged species, whereas the low polarity solvent may disfavor that. So the solvent with higher dielectric constant would boost higher charge states in the gaseous analyte ions generated from those ES solvent droplets. Loo et al. also showed that the denaturing capacities of different solvents also affect the protein CSD, which in turn reflect the conformation (solvent accessibility) of the protein in that solution [102].

Apart from the apparent gas-phase basicities (calculated based on the intrinsic gas-phase basicities and Coulomb energy of the interacting charges) of the basic sites in the protein, the gas-phase basicity of the solvent is also an important factor to significantly affect the CSD of the protein [87, 103]. The effect of the solvent composition on both maximum CS and the CSD of the proteins were investigated by Iavarone and coworkers [96]. They showed that CSD of cytochrome c and myoglobin (see Figure 14), formed from 47%/50%/3% water/solvent/acetic acid solutions, shifts to lower charge (higher *m/z*) when the 50% solvent fraction is changed from water to methanol, to acetonitrile, to isopropanol. This is also the order of increasing gas-phase basicities of these solvents. However, the effect is relatively small for these solvents, possibly due to their limited concentration (in gas phase) inside the electrospray interface. But the addition of small amounts of diethylamine (0.4%) in the aspirating solution results in dramatic shifts to lower charge, presumably due to preferential proton transfer from the higher charge state ions to diethylamine. These results clearly show that the maximum charge states and charge state distributions of ions formed by electrospray ionization are influenced by solvents that are more volatile than water and the gas-phase basicity of the solvent too [96].

It is also not surprising that along with different physical properties of the solvent the surface tension plays an important role to the CSD of the protein, since the surface tension parameter (γ) is involved in the Rayleigh equation (3) [62], based on which the CRM is proposed [61]. The increase in surface tension of the solvent should increase the average charge state of the analyte. Iavarone and Williams showed that, in absence of other factors, the surface tension of the ultimate electrospray droplet (formed after repeated solvent evaporation and Coulomb fission), from which the charged gaseous analyte is formed, is a significant factor in determining the overall analyte charge [100].

6.7. The Supercharging Effect. Iavarone and coworkers reported interesting additives, for example, m-nitrobenzyl alcohol (m-NBA) and glycerol to the aqueous solution of the protein and peptide, which can dramatically increase both the maximum charge state and the most intense charge state

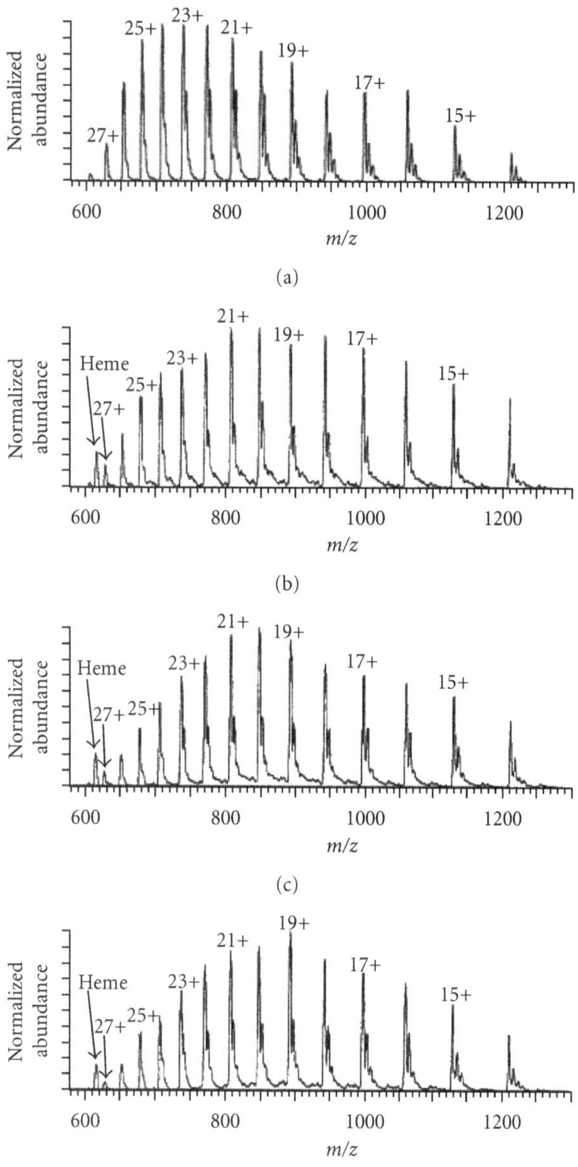

Figure 14: Myoglobin (10^{-5} M) electrosprayed from 47%/50%/3% water/solvent/acetic acid solutions, where the "solvent" was (a) water, (b) methanol, (c) acetonitrile, or (d) isopropanol. (Reprinted with permission from *Journal of the American Society for Mass Spectrometry* [96], Copyright 2000, Elsevier).

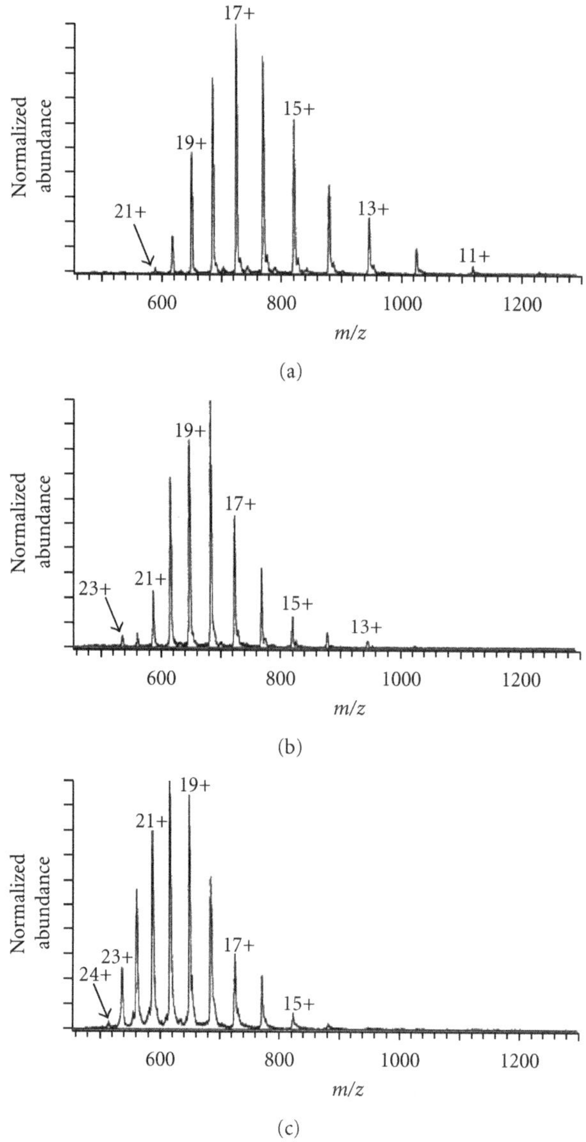

Figure 15: Electrospray ionization mass spectra of cytochrome c (10^{-5} M) from solutions containing (a) 0, (b) 0.3, and (c) 0.7% m-NBA. The base solution is 47/50/3% water/methanol/acetic acid. (Reprinted with permission from *Analytical Chemistry* [101], Copyright 2001, American Chemical Society).

[101]. They termed this phenomenon "supercharging effect." Figure 15 shows that the addition of just <1% m-NBA to the electrospray solution, the maximum charge state as well as average charge state increases significantly [101]. Similarly the addition of small amount of m-NBA in the peptide (KKKK) solution substantially increased the most intense and average CS [101]. The exact mechanism of this enhanced charging is still not clear. Recently they have suggested that these supercharging reagents have low vapor pressures (less volatile), and aqueous droplets are preferentially enriched in these reagents as evaporation occurs [100]. Less evaporative cooling will occur after the droplets are substantially enriched in the low volatility supercharging reagent, and the droplet temperature should be higher compared with when these reagents are not present. Protein unfolding induced by chemical and/or thermal denaturation in the final electrospray droplet, from which the charged gaseous protein is produced, appears to be the primary origin of the enhanced charging [104, 105]. The fact is further supported by the arrival time distributions obtained from traveling wave ion mobility spectrometry [105], which showed that the higher charge state ions that are formed with the supercharging reagents are significantly more unfolded than lower charge state ions. Though this model successfully explained the supercharging of proteins and their noncovalent complexes, it fails

to explain the supercharging of small molecules like small peptides [101], which do not have any tertiary structure. A detail investigation of the physical parameters of the evaporating and disintegrating droplet containing this supercharging reagent is required in the future to understand the exact mechanism of supercharging.

Recently Lomeli group has reported other supercharging agents, for example, benzyl alcohol, m-nitroacetophenone, m-nitrobenzonitrile, o-NBA, m-NBA, p-NBA, m-nitrophenyl ethanol, sulfolane (tetramethylene sulfone), and m-(trifluoromethyl)-benzyl alcohol [106]. Among these sulfolane displayed a greater charge increase (61%) than m-NBA (21%) for myoglobin in aqueous solutions based on average charge state. All these reagents that promote higher ESI charging appear to have low solution-phase basicities and relatively low gas-phase basicities and are less volatile than water [106].

Thus the increased multiple charging by the supercharging reagents promises its implication in the accurate mass measurement of the large macromolecules and the structural characterization of the proteins by tandem mass spectrometry and MS^n experiments, where the multiple charging is important for efficient fragmentation (discussed later) [107–110].

6.8. Modification of the Protein Charge State Distribution. Several approaches like acid/alcohol-induced proteins denaturation [102], protein disulphide bond reduction using chemical reducing agent dithiothreitol, and so forth, [50] were undertaken to modify the protein CSD in last two decades. When the disulfide bonds are reduced, much higher charge states can be obtained because the reduction of the disulfide bonds allows the protein to unfold further and expose additional acidic/basic sites for protonation/deprotonation [50].

Though the protein CSD can be altered by manipulating the solution conditions (pH, reduction of disulphide bonds, use of different solvents, denaturing agent and supercharging agent, etc.), it can also lead to deleterious effects such as poor ESI response, dissociation of noncovalent complexes, and so forth. But, sometimes it is desirable to alter CSDs independent of initial solution conditions. The CS alteration by gas-phase proton transfer reaction (post-ESI process) has been demonstrated via ion/ion [111, 112] or ion/molecule [113] reaction. But that resulted in significant reduction of the absolute charge of the gaseous analytes. Recently Kharlamova and coworkers have established a method for the manipulation of the CSD of the electrosprayed proteins by exposing the nano ES droplets to the volatile acid/base vapors in the atmospheric pressure region of the mass spectrometer [114, 115]. Exposure of the positive ES droplets to the acid vapors [115] and the negative ES droplets to the base vapors [114] increases the propensity of high CS formation of the analyte (protein). On the basis of changes in protein CSDs, protein folding and unfolding phenomena are implicated in their studies. Since no change in the CS distributions of buffered proteins exposed to reagent acid/base vapors was observed, the charge state changes are attributed largely to a pH effect. They suggested that, when the acid/base vapor interacts with the nanoelectrospray generated protein droplets, it changes the pH of the droplets and thereby affects the structure of the protein present in the ES droplets during the very short nanodroplet lifetime (few tens of microseconds). This change of the protein structure is attributed by the change of the CSD. They also showed that the species bound by relatively weak interactions can be preserved, at least to some extent, allowing for the observation of high charge states of protein-ligand complexes [114, 115].

Recently Krusemark et al. used chemical derivatization method to control the CSD, which eventually gave the answer about the role of protein functional groups in the CSD [116]. When the free carboxylic acid function groups of the denatured proteins were modified by amidation (neutral functional group), minimal change in the CSD in positive ion mode was observed compared to the unmodified proteins, indicating that the carboxylic acid functional groups do not play significant role in charging in the positive ion mode. The modification of proteins with additional basic sites or fixed positive charges generated substantially higher charge states in positive ion mode, providing evidence that the number of ionizable (basic) sites determines ESI charging for denatured proteins irrespective of their shape and sizes. Combining this chemical modification and the ion/molecule or ion/ion reactions in gas phase, the authors also illustrated some unique approaches to alter/control the CSD [117].

6.9. Analyte Concentration and the Presence of Electrolyte Impurities. Apart from the analyte structure and conformation, the analyte concentration in the initial spray solution is also an important factor that significantly affects the ESI-MS signal intensity [22, 39, 118]. For multiply charged analyte, ions of discrete charge state exhibit a linear response of the ion signal with the analyte concentration in solution until a certain concentration limit is exceeded. This concentration limit is dependent on the nature of the analyte. Beyond that concentration the signal intensity loses that linear response and may even decrease at extremely high concentration. This departure from the linearity as the aspirating solution concentration increases can be attributed to the analyte concentration having reached its saturation value during the droplet evaporation so that further loss of solvent results in precipitation, rather than any further increase in its concentration in solution [39]. At elevated concentration the relative abundances of the gaseous analyte ions of higher charge state also decrease due to the increase in competition between the analyte molecules inside the ES droplet for the limited number of available charges on the droplet surfaces [119].

The presence of any ionic species (i.e., electrolytes) other than analyte in the solution (except for small amounts of electrospray friendly acids, bases, or buffers as discussed before) is usually avoided when possible, because their presence in the ES solution has been found to suppress the formation of gas-phase ions of the analyte of interest [118, 120–122].

6.10. Instrumental Parameters. Very often it is observed that the CSD in the ESI-mass spectra of a particular analyte acquired from different instruments is significantly different though a specific protocol for the sample preparation has been followed. Moreover, ES-mass spectra acquired from the same instrument at different points in time can show some degree of variability. This is because there are several instrumental parameters which are responsible for the ion production, filtration, and detection. The instrumental parameters like spray tip orifice diameter (e.g., microspray and nanospray [32, 123]), ion spray voltage (different onset voltages for different solvents [22]), source geometry [98, 124] (e.g., on-axis, off-axis, etc., [125]), source gas pressure [39, 126], source temperature [58], heated capillary temperature (thermal dissociation may occur here) [58, 126–128], sheath/auxiliary gas flow rate [39] that (affects the evaporation and cooling rate of the ES droplets), nozzle-skimmer voltage (declustering potential) [129–132], different ion-optics voltages [34, 58] (that guide the ion filtration and trapping) [34, 58] and nature of the detectors (detector response) [34, 58] are all known to influence the ESI-MS response of the analyte.

7. Is Substantial Conformational Change of the Protein Possible inside the ES Droplet?

There are still several aspects of ESI not fully understood though it is almost two and half decades elapsed after the discovery of ESI-MS. One such major issue is the conformational and other subtle chemical properties of the analyte during the transfer from solution to gas phase. The issue becomes further complicated due to the complexity of the processes affecting the ES droplets including the charge separation [22], solvent electrolysis at the emitter tip [133, 134], droplet evaporation then subdivision [22] the increase of the droplet acidity/basicity, and so forth. As discussed earlier the droplets produced at low flow rate (typically $\sim 5\,\mu L/min$) have a narrow distribution of sizes with a most abundant radius $\sim 1.5\,\mu m$ [22]. Such an electrospray droplet was earlier shown to have a charge of $\sim 10^{-14}$ C, which corresponds to ~60,000 singly charged ions [16]. If we assume that all these singly charged ions are protons (in positive ion mode) and they are homogeneously distributed in the droplet, then the estimated pH of the initially produced droplet would be ~5.2. The repeated solvent evaporation and Coulomb fission of the droplets would increase the proton density and thereby continuously lower the pH of the droplets. According to the IEM, the gas-phase ions are supposed to be formed when the droplet reaches the radius ≤ 10 nm [63]. The Rayleigh charging [62] at this limit ($r = 10$ nm) corresponds to 125 elementary charges (protons) in the droplet. So the corresponding pH of the droplet at this limit is ~1.3. The acidity of the ES droplets was also experimentally measured [135]. Overall the chemical composition in the final small droplets from which the naked analyte ions are formed may be significantly different from that of the original sample solution [22]. For example, the acid/base equilibrium, the conformational states, and/or noncovalent bonding in the analyte may all be affected by the subtle changes in the composition and properties of the droplets during the electrospray. There have been no reports addressing the possible effects of the microenvironment of the solvent droplets on the analyte during the ES process. Generally it has been assumed that the solute molecules residing at the neutral core of the charged droplet remain "oblivious" to the harsh conditions on the droplet surface and the environment beyond it until the last of the solvent molecules is gone. As a consequence it is presumed that the protein charge state distribution in the gas phase actually represents the protein conformation in the initial solution. In some reports it has been also assumed that the violent nature of charged jet breakup and subsequent Coulombic explosion of the ES droplets encapsulating the proteins can cause repacking of the supramolecular assemblies, forcing them to hold a minimal volume [77, 136]. But no direct experimental evidence is still available to support this assumption on the molecular rearrangement inside the charged ES droplets. So the molecular behavior inside the charged ES droplets still remains like a "black box" and the exact processes that happen to a molecule residing inside the charged droplets are far from being understood.

If the physical and chemical properties of the ES droplet continuously change during the lifetime of the droplet then what happens to the conformation of the proteins residing inside the charged droplet? Does a protein change its tertiary structure or conformation inside the violently behaved (charged jet breakup) charged ES droplet? The preservation of the overall protein integrity via ES process does not necessarily mean the retention of the initial solution structure intact inside the charged droplets. Some recent works [137, 138] suggested that after the ES process the new gaseous environment can ultimately cause dramatic structural alteration of the protein (post-ESI process) and the existence of gaseous macromolecules (proteins) with a number of different conformations [138]. Though in the solvent-free environment the protein conformational change is not surprising, but the issue regarding the conformational change inside the ES droplet relative to the initial solvent is still far from being understood. If the protein molecule is susceptible to change their three-dimensional conformation inside the ES droplet, then the CSD in the corresponding ESI-mass spectrum should obviously reflect the three-dimensional structure of the protein inside the droplet not in the initial solution.

In view of this we have recently conducted some experiments [139] on some model analytes (small molecules) that are labile to be transformed inside the charge droplet. It is often argued that the charges on an ES droplet are distributed over its surface with equidistant spacing and that they are locked into this pattern by the forces of Coulomb repulsion [39], and the bulk or core of the ES droplet essentially remains free of charge to minimize the potential energy of the droplet [140]. But when we put a highly stable intrinsically charged molecule inside the charged ES droplets, we observed a drastic transformation of their structure due to huge Coulomb force of repulsion on the positively charged analyte imparted by the surface protons of the ES droplet (see Scheme 1). This result unambiguously suggests that the violent nature of the charged ES droplets can affect the analyte residing inside it depending on the nature of the analyte [139].

Recently we have also induced the protein conformational change inside the charge droplet by perturbing the droplet solvent composition during the lifetime of the

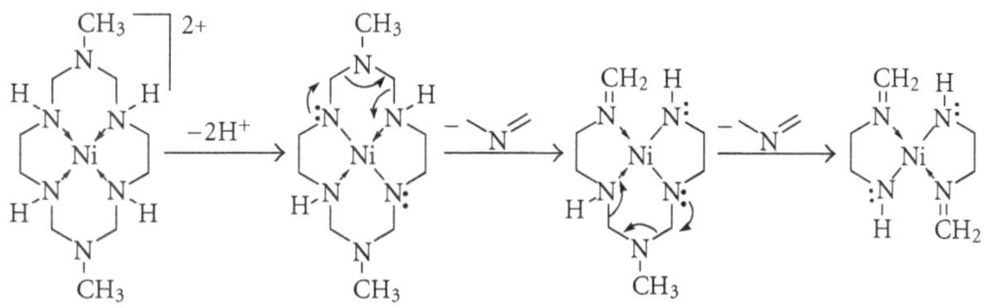

1 A B C

Scheme 1: The fragmentation pathway of the cationic complex **1** inside the charged droplet.

droplet (unpublished work). But it is not clear to us whether without perturbing the solvent composition of the charge droplet the protein conformational change is possible or not in the natural way of ES process.

8. Tandem Mass Spectrometry: Collision-Induced Dissociation

The word "tandem" means arrangement of two or more objects/persons one behind another, [141]. Similarly tandem mass spectrometry (MS/MS) is a method where the gaseous ions are subjected to two or more sequential stages of mass analysis (which may be separated spatially or temporally) according to quotient mass/charge [40, 142]. For example, in a tandem mass spectrometric experiment, a precursor ion is mass selected by a mass analyzer Q_1 (see Figure 4(b)) and then focused into a reaction cell q_2 (collision cell) where it undergoes a gas-phase chemical reaction. The reaction gives different product ions with different masses, which are then passed to another mass analyzer Q_3 (see Figure 4(b)). This last mass analyzer scans the masses of the product ions and generates the product ion spectrum. The mass analyzers are set up in series either in space (sector, triple quadrupole, and hybrid instruments) or in time (trapping instruments).

ESI transforms the analyte in gas phase without rupturing any covalent bonds and provides the information about the molecular weight (MW). This MW alone can not determine the structure of an unknown analyte. So, the fragmentation of the gaseous analyte is deliberately required for the structural elucidation. The fragment ion spectrum is basically the fingerprint of the presence of a particular analyte (precursor ion). By investigating the fragmentation reaction (fragment-ion spectrum), one can actually determine the correct chemical structure of the unknown analyte sample. The task is like a jigsaw puzzle, where the players have been given a broken plate and asked to join them together in a sensible way to find the actual shape of the plate. Now the question is how one can fragment a charged molecule in gas phase? The mass-selected ion (precursor analyte) is activated in the collision cell (e.g., q_2 in Figure 4(b)) by increasing its internal energy. This activation causes the homolytic or heterolytic cleavage of the chemical bonds. As a result different fragment ions (product ions) are produced. There are different methods of precursor ion activation, which include collisional activation (collision-induced dissociation or CID [143], surface-induced dissociation or SID [144, 145], etc.), photon-induced activation (infrared multiphoton dissociation or IRMPD [146], blackbody infrared radiative dissociation or BIRD [147, 148], ultraviolet photodissociation or UVPD [149, 150], etc.) and electron-mediated activation (electron capture dissociation or ECD [151], electron transfer dissociation or ETD [152], electron detachment dissociation or EDD [153], electron-induced dissociation or EID [154]), and so forth. Here we would discuss about the most popular and most common ion activation method in the ESI-MS system, which is collision-induced dissociation (CID).

In collision-induced dissociation, the gaseous precursor ion is allowed to collide with a gaseous target in the collision cell. The gaseous target is inert and neutral, generally inert gases such as nitrogen, helium, or argon [155]. As a result of collision, the energy is gained and redistributed among different vibrational degrees of freedom (internal energy) within the precursor ion. Thus, an unstable excited state of the precursor ion is populated, which causes the precursor ion to decompose into the product ions in a process termed "collision-induced dissociation" (CID) [143]. So the CID is basically a unimolecular fragmentation reaction of the mass-selected precursor ion.

8.1. Multistage CID/Tandem Mass Spectrometry. But it is often the case that the single-stage collision-induced dissociation provides a lot of confusing fragmentation information or does not provide enough information for the structural elucidation. In both of these cases multistage tandem mass spectrometry (MS^n)/multistage CID is performed. Generally MS^n experiments are performed in ion-trap or in FT-ICR instrument, where each MS step is separated in time [58]. Those trapping instruments allow the refragmentation of the product ions and thus produce next generation of product ions and so on. For example, in the first stage, the normal ESI-mass spectrum (MS^1) is produced without fragmentation. Then one precursor ion is isolated in the trap and fragmented to produce the first generation product/fragment ions, and the corresponding spectrum is called MS/MS or MS^2 spectrum. Again one precursor is isolated from the first generation fragment ion (from MS^2 spectrum) and then fragmented to produce the second generation product ions, which are scanned to record the MS^3 spectrum (MS/MS/MS). Thus process of isolation and fragmentation can be repeated a number of times, resulting in a series of MS^n spectra where "n" represents the number of times, that is,

$(n - 1)$ times, the isolation-fragmentation cycle has been carried out.

8.2. High- and Low-Energy CID. The appearance of the CID spectrum is dependent on several parameters like nature of the projectile ion and target gas, target gas number density, ion activation time, nature of the collision cell, and above all on the collision energy between gaseous projectile ion and the target atom/molecule [58]. This collision is controlled by the kinetic/translational energy of the projectile. The fraction of the kinetic energy that can be converted to the internal energy of the precursor ion in a collision is determined by the centre of mass collision energy (E_{cm}) [143]

$$E_{cm} = \frac{m}{m+M}E_K, \tag{10}$$

where "m" is the atomic/molecular weight of the target gas "M" is the molecular weight of the precursor ion, and "E_K" is the precursor ion kinetic energy. E_{cm} is the maximum amount of the precursor ion's kinetic energy, which can be converted into the internal energy for fragmentation. The E_{cm} above 25 eV is defined as high collision energy and that below 20 eV as low collision energy [58, 158, 159]. Since a very high kinetic energy (KeV) is gained by the projectile ion in the traditional sector instruments, the high energy CID spectrum is produced in those instruments [58]. But the projectile translational energy is in the range tens to hundreds of electronvolts in the triple-quadrupole instruments and thus those instruments provide low energy CID spectrum [58]. In the ion-trapping instruments (e.g., QIT), the ions are trapped within a three-dimensional electric field inside the ion trap, and a stable ion trajectory is guided by Mathieu equation. So in QIT the precursor ion kinetic energy can only be raised to a level where stable ion motion still occurs. Consequently the achievable kinetic energy of the precursor ion falls in the range of low energy CID regime [58].

In low energy CID, the internal energy increase is sufficient to vibrationally (not electronically) excite the precursor ion, which results in the cleavage of the most labile bonds when a certain threshold energy is reached. But in high energy CID, electronic excitation of the precursor ion becomes possible if the ion-neutral interaction time is very short ($<10^{-14}$ s). But this generally does not occur for medium-to-large size molecules undergoing high-energy CID [58]. In those cases a curve-crossing mechanism more likely occurs [58]. For example, a crossover from the first electronically excited state to the higher vibrational level of the ground state occurs. But both direct electronic excitation and curve-crossing are likely to result, a different fragmentation pattern compared to that in the low-energy CID. Although the high-energy CID spectrum is typically much more complex than low-energy spectrum, it contains maximum structural information. For example, generally backbone cleavage of the peptide is observed upon low-energy CID but side-chain cleavage including, backbone cleavage of the peptide is observed upon high-energy CID [159–161]. High-energy CID is known to promote charge-remote fragmentation [162, 163], while low-energy CID is often guided by charge-directed fragmentation mechanism (see later) [161].

Here we would discuss the low- and/or high-energy fragmentation of some biologically important molecules like peptides, oligosaccharides, oligonucleotides, and lipids, which are very often analyzed by ESI-CID-MS/MS experiments.

8.3. CID of Peptides. When a peptide is electrosprayed in positive ion mode, gaseous protonated peptide is produced. The protons are mostly localized on the most basic sites (e.g., Arg, Lys, and his side-chain or N-terminal α-NH_2 group) in the peptides prior to activation [164, 165]. Upon collisional activation, the ionizing protons are transferred from the unreactive sites of higher gas-phase basicity (e.g., Arg, Lys, and his side-chain or N-terminal α-NH_2 group) to several backbone amides to form energetically less favored but reactive backbone-amide-protonated species [161]. The amount of activation energy that has to supply to the precursor-protonated peptide to make the proton sufficiently mobile is dependent on the gas-phase proton affinities of the different amino acids (Arg > Lys > His > Try > Glu > Pro > Gln > Met > Tyr > Asn > Phe > Thr > Ile > Leu > Val > Asp > Ser > Ala > Cys > Gly) [166, 167] present in the sequence. For example, when a doubly protonated peptide, which contains a strongly basic amino acid residue (e.g., Arg/Lys/His), is collisionally activated one proton would mostly be anchored on the basic residue, and the second one would be mobile in the backbone leading to the heterogeneous population of the ions varying the location of the second proton. This fast proton "dancing"/migration process to various backbone amide bonds is called "mobile proton model" as proposed by Wysocki (see Figure 16) [156, 168, 169]. This mobile proton model has been verified by hydrogen/deuterium scrambling studies [170] and theoretically also [171, 172]. Since amide nitrogen protonation removes the resonance stabilization of the amide bond, the protonated amide bonds become weak and labile to fragment in low-energy CID. As a result of heterogeneous population of the activated precursor ions, a series of peptide bonds are cleaved [161, 173]. Apart from the amide bond fragmentation, other backbone bond fragmentation is also possible via mobile proton model. A notation has been developed to indicate the peptide fragments that arise from the CID of the protonated peptides. The N-terminal fragment ions (charge is retained on the N-terminus) are indicated by a_n, b_n, and c_n, and C-terminal fragment ions (charge is retained on the C-terminus) are indicated by x_n, y_n, and z_n [174, 175] (see Figure 17). The subscript "n" indicates the number of amino acid residues in the fragment. As shown in Figure 17, the N-terminal a_n, b_n, and c_n ions are complimentary to the C-terminal $x_{(n-m)}$, $y_{(n-m)}$, and $z_{(n-m)}$ ions, respectively, where "m" is the total number of constituent amino acids in the precursor peptide. Under low-energy collision condition, the peptide dissociation mainly leads to the formation of N-terminal a, b-fragments, and C-terminal y-fragment [58, 161]. As a qualitative basis, the fragment with higher gas-phase basicity (GB) will remain protonated with higher probability than the fragment with lower GB [161]. New fragment can further be created, under

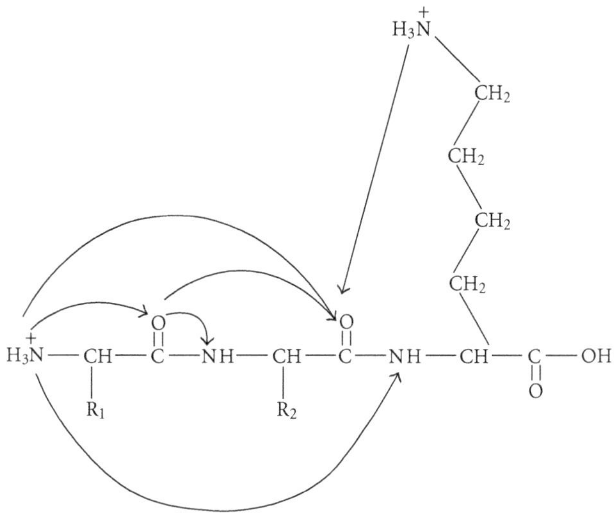

FIGURE 16: Mobile proton model. (Reprinted with permission from *Journal of the American Society for Mass Spectrometry* [156], Copyright 2010, Elsevier).

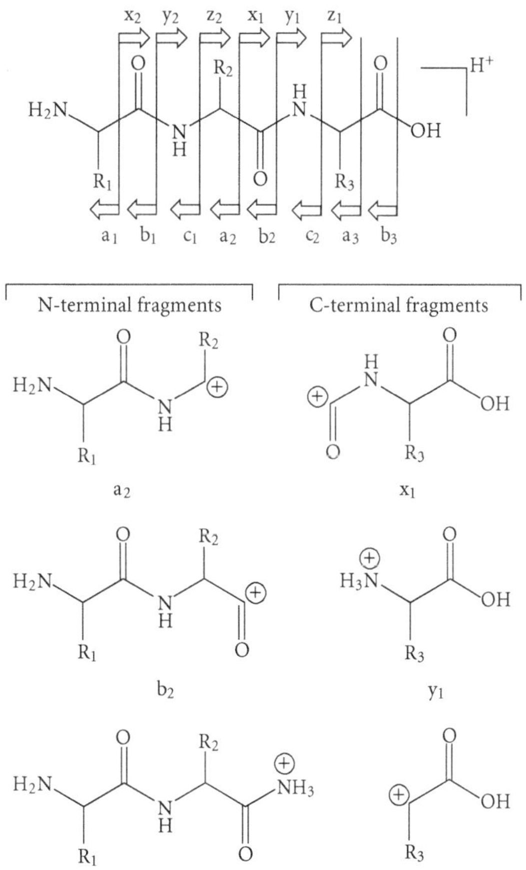

FIGURE 17: Nomenclature of the peptide fragments showing some typical qualitative structure of the fragments.

multiple collision condition, by the primary fragment dissociation leading to the secondary ion products (satellites) such as internal fragments [176], amino acid-specific immonium ions [177], neutral-loss a/b/y-ions (e.g., b_n-NH_3, b_n-H_2O, etc.) [178, 179], and smaller member of b, a, and y-ion series. All these ions provide peptide sequence information directly and that is why they are called "direct sequence ion" [180]. Figure 18 shows a typical example of the CID-MS/MS

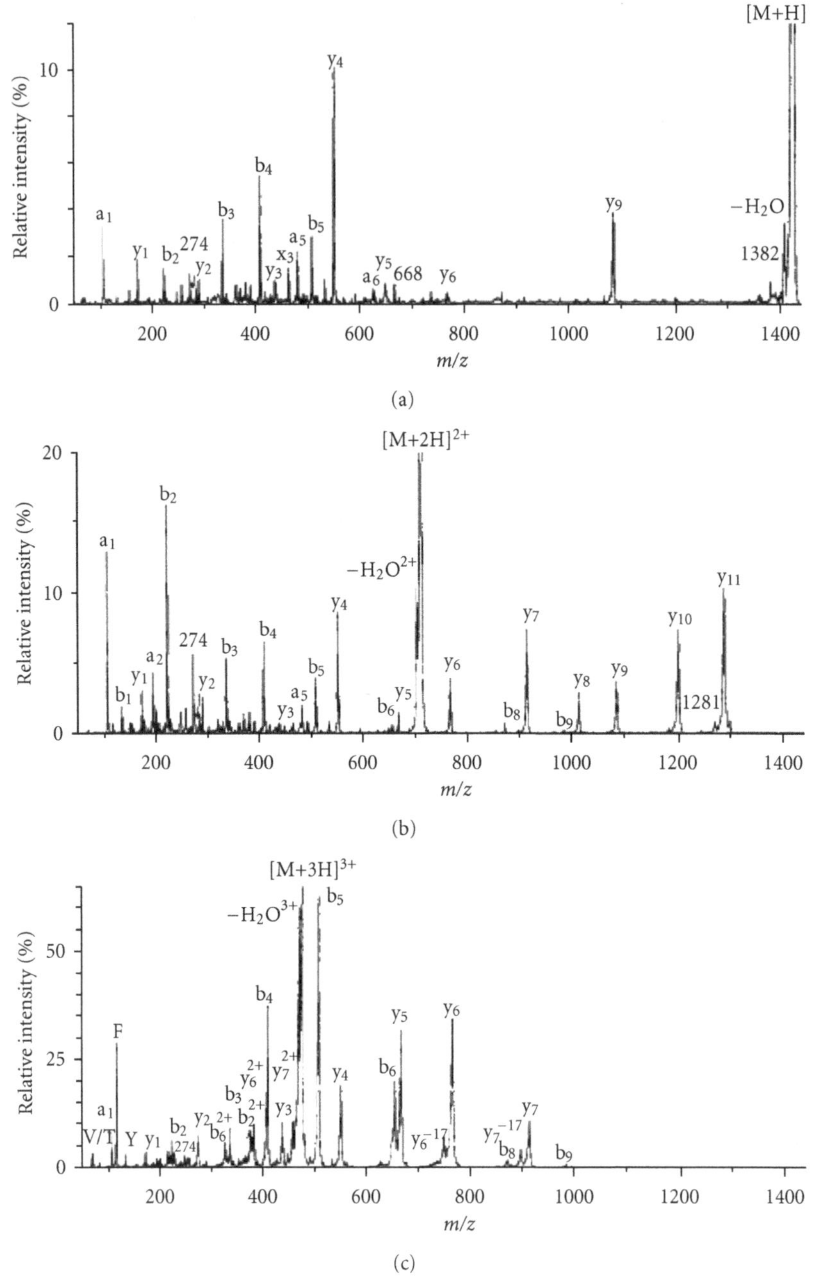

FIGURE 18: CID-MS/MS spectra of different protonated forms of the peptide HSDAVFTDNYTR: (a) CID of singly protonated peptide $[M + H]^+$, (b) CID of doubly protonated peptide $[M + 2H]^{2+}$, and (c) CID of triply protonated peptide $[M + 3H]^{3+}$. (Reprinted with permission from *Analytical Chemistry* [157], Copyright 1993, American Chemical Society).

spectra of the different protonated forms of the peptide HSDAVFTDNYTR representing different "direct sequence ions" [157].

The actual mechanism by which the peptide backbone is cleaved follows complicated pathways. Although the peptide fragment ions (product ions) were identified long ago [107, 174], it is the recent trend to understand the mechanism of the fragmentation of gaseous protonated peptides. For example, recently it has been understood how a singly protonated peptide can fragment to produce the C-terminal/N-terminal fragment ions [181]. The b_n-ion does not exist as acylium cation as shown in Figure 17, rather they form a cyclic oxazolone structure in the gas phase. Scheme 2 shows the proposed mechanism of the amide bond cleavage of a singly protonated peptide [181]. Before the cleavage of a protonated amide bond, the corresponding amide carbonyl is attacked by the nucleophilic oxygen of the previous amide carbonyl group and thus makes an oxazolone ring followed by a proton-bound complex between the oxazolone (N-terminal) and a truncated peptide or amino acid (C-terminal). Then the fragmentation of the proton-bound complex can occur by the proton transfer process leading to either b_n-ion

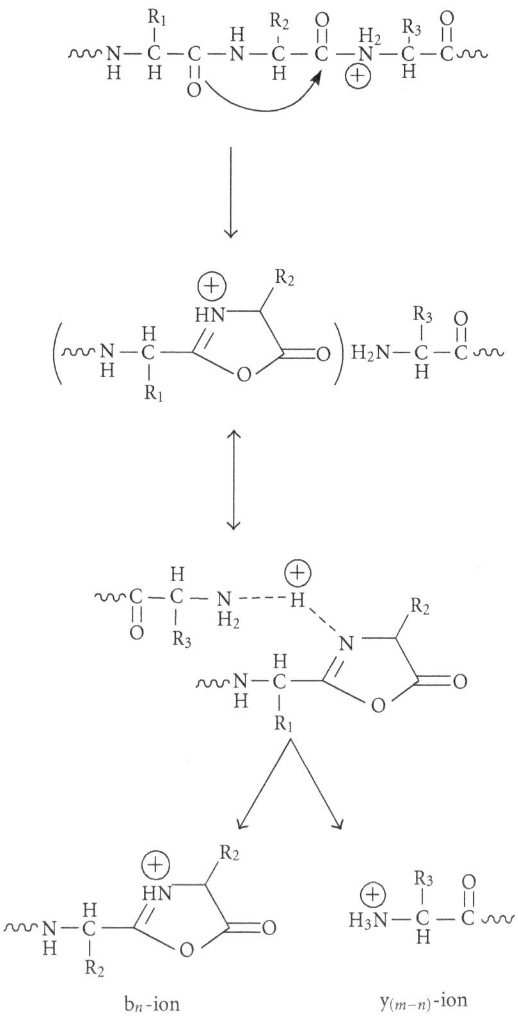

SCHEME 2: A mechanism for the production of b_n and y_n ions.

or y_n-ion. Thus, the propensity of the formation of b_n or y_n ion is dependent on the relative proton affinities of the N-terminal oxazolone and the C-terminal truncated peptide/amino acid. The N-terminal oxazolone (b_n-ion) can undergo a neutral loss of carbon monoxide (CO) to form a_n-ion. It has also been proposed that the N-terminal a, b-fragments are terminated with five membered [182] (oxazolone) ring, via rearrangement type reaction. Generally the amide bonds N-terminal to proline and C-terminal to Asp/Glu are more prone to fragmentation [183]. This is because the tertiary amide nitrogen of the proline residue is more basic and thus more likely to be protonated than other backbone amide bonds due to the mobile proton mechanism and thereby leading to enhanced peptide bond cleavage. The acidic residues Asp/Glu provide "local" mobile proton, which catalyzes the cleavage at the peptide bonds C-terminal to these residues [183]. It should be noted that the dissociation of the doubly or multiply charged peptide is more advantageous compared to that of a singly protonated peptide [58]. The population of more than one charge on the peptide backbone results in more dissociation pathways by mobile proton mechanism and thus produces more fragment ions to provide much more sequence information compared to that derived from the singly protonated peptide. Furthermore, the presence of more than one charge of similar polarity can impart Coulomb repulsion in gas phase rendering the precursor ion less stable and thereby further contributes to more facile dissociation. In general the peptide fragmentation is mostly influenced by the charge (e.g., protons), and thereby the fragmentation is called "charge directed fragmentation" [159, 161].

However, this conventional knowledge of fragmentation pathways and rules cannot reasonably explain many anomalous fragmentations because it is often the case [157, 184–187] that abundant fragment ions appear in the MS/MS spectra of peptides that do not belong to the above-discussed direct sequence ion series. These fragments are formed in complex rearrangements and are termed as "nondirect sequence ions" [180]. The nondirect sequence ions discovered so far include scrambled fragments of the b/a type ions, which is equivalent to the loss of amino acid residue from the interior

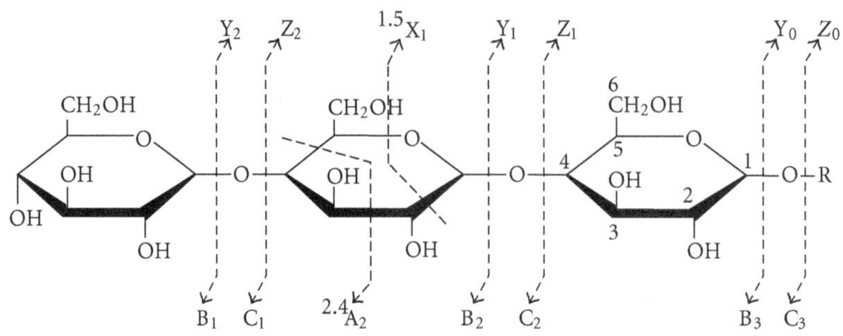

Figure 19: Nomenclature of the product ions formed from the precursor oligosaccharide.

of the peptide chain [180]. As reported in the recent literatures [180, 187–192], the first step of the mechanism concerned to the internal amino acid loss (sequence-scrambling fragmentation pathway) is a nucleophilic attack by N-terminal α-NH_2 group on the acylium carbocation of the b-ion (a primary fragment) forming a protonated cyclic-peptide intermediate (CPI). This CPI reopens at other sites, producing linear fragment ions where residues are relocated at the C/N-termini. These linear fragment ions further dissociate via conventional fragmentation pathways, forming daughter ions with neutral losses of internal residues from the parent ions. These types of ions are called the nondirect sequence ions of the protonated parent peptides. At the first stage of CID, this unique fragmentation of singly charged b-ions is difficult to be observed. However, under multi-stage CID, almost all of the investigated b-ions (containing three to seven residues) notably displayed the above unique fragmentation patterns. The propensity of the sequence scrambling increases with the increase in the chain length and the charge states of the peptides [193]. A recent report [194], on the basis of the analysis of 15,897 low-energy and 10,878 high-energy CID mass spectra of doubly protonated tryptic peptides, is suggesting that the rate of sequence scrambling due to b-ion cyclization is negligible (<1%) and can be safely ignored as a possible source of erroneous sequence assignment in shotgun proteomics. Recently we have also found the loss of a specific amino acid residue from the interior of the protonated peptide in gas phase, which follows a degradation mechanism other than the above sequence scrambling pathway (unpublished work).

Although historically the peptide fragmentation is performed on the gaseous protonated peptides to derive the sequence information [173], fewer efforts have been invested on the dissociation of the gaseous deprotonated peptides. Hence the dissociation pathways of the deprotonated peptides are not well established [173]. It should be noted that in negative ion mode the –OH functional groups in serine, threonine, and tyrosine side-chains and –COOH functional groups in glutamic and aspartic acid side-chains are the potential deprotonation sites [195, 196]. Fragmentation of deprotonated peptides has been shown to provide complementary structural information to their positive counterparts [196–198]. Therefore, sequencing unknown peptides can benefit from the interpretation of both positive and negative mass spectra.

Several aspects of CID of the protonated peptides are still under controversy. It is continuously being debated where exactly the proton is located before the amide bond cleavage, whether the location is carbonyl oxygen or amide nitrogen [107, 164, 168, 171, 199, 200]. How exactly the fragmentation pattern is dependent on the amino acid sequence of the precursor peptides and how different fragments get stabilized? Apart from sequence informative ions, there are lots of anomalous fragments, which are observed in the MS/MS spectra of the peptides. As a future outlook the detail understanding of the formation of those anomalous fragment ions is required and that continuously drawing the attention of the mass spectroscopists.

8.4. CID of Oligosaccharides. The analytical and structural characterization of carbohydrates or saccharides (highly abundant biological compounds) is quite a challenging problem because of the diversity of the linkage types, linkage position, and anomeric configuration [201]. Analogous to the peptide cleavage in the gas phase, the gaseous protonated/metallated oligosaccharides also undergo facile cleavages in certain specific positions upon CID. A systematic nomenclature (see Figure 19) of the different product ions produced from a precursor oligosaccharide was introduced [202] by Domon and Costello in 1988. Under low-energy collision conditions, the abundant product ions (B, C, Y, and Z) are generated by the glycosidic bond cleavages. But the high-energy collision conditions tend to favor cross ring cleavages, which produces "A" or "X" type fragments (see Figure 19) [203]. These cross-ring cleavage preferably produced by charge remote fragmentation from the sodium adduct of the precursor analyte [203]. The dominant pathway for the ion decomposition is highly dependent on the nature of the precursor ion, that is, protonated or metallated, or the cationic or anionic adduct species [203–207]. Generally for protonated oligosaccharides, the protons are mainly located on the glycosidic oxygen and thereby induce the cleavage of either C_1–O bond or C_4–O bond (see Figure 19) [202, 204]. The C_1–O and C_4–O bond cleavages give rise to the formation of B_n/Y_m and C_i/Z_j complementary ions, respectively, via "ion-dipole complex" mechanism involving an oxonium ion [58] (see Schemes 3 and 4).

8.5. CID of Oligonucleotides. Low energy CID is also useful for the structural elucidation of nucleic acids like DNA, RNA,

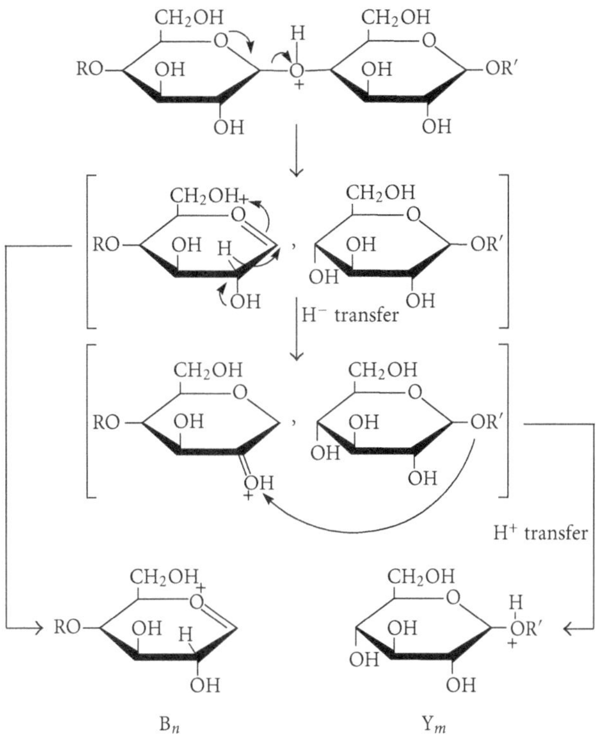

SCHEME 3: The proposed mechanism of the formation of B_n and Y_m product ions via ion-dipole complex.

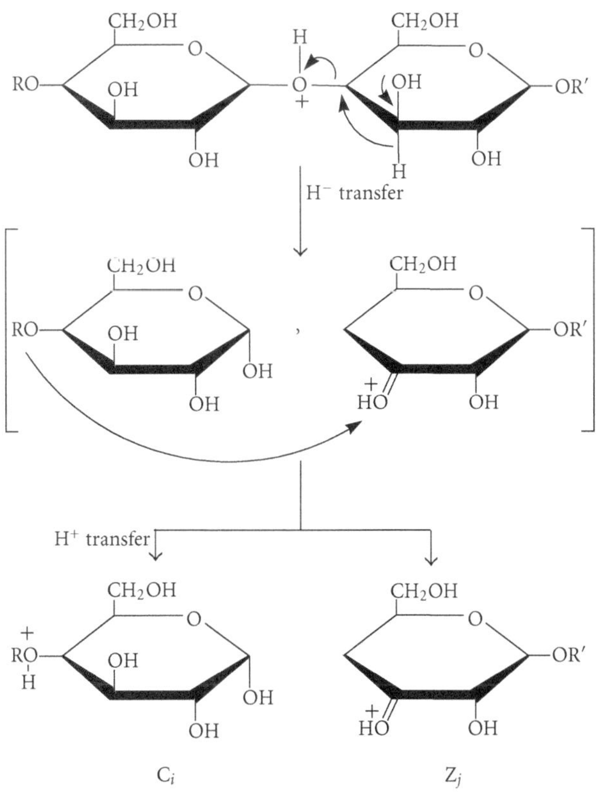

SCHEME 4: Proposed mechanism of the formation of C_i & Z_j product ions via ion-dipole complex.

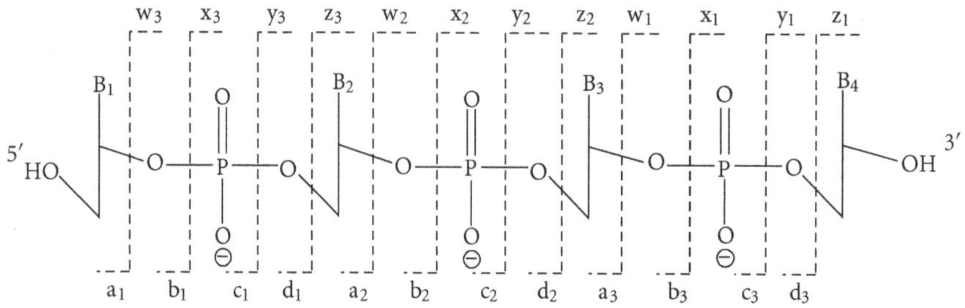

Figure 20: Nomenclature of the fragment ions produced by the CID of oligonucleotides.

and so forth, [208]. Polyanionic oligonucleotides derived from DNA/RNA can be efficiently transferred to the gas-phase by ESI. Those oligonucleotides undergo selective fragmentation in CID producing ladder-like product ion spectra similar to that produced by peptides. A systematic nomenclature was proposed by McLuckey et al. [208] to describe the product/fragment ions generated by CID of the oligonucleotide (see Figure 20). Cleavage of the four types of phosphodiester bond yields eight types of fragments for example, a_n, b_n, c_n, and d_n, which contain the 5′-OH group, and w_n, x_n, y_n, and z_n, which contain 3′-OH group. The subscript "n" represents the number of residues contained in the fragments and thus designates the position of cleavage. Sometimes additional loss of a purine/pyrimidine base (B_n) is indicated by parenthesis with the identity of the base (if possible). For example, the fragment a_4-B_4 (A) indicates the cleavage of the bond between ribose carbon and the oxygen of the phosphodiester group at position 3, with the additional loss of adenosine base at the same position. ESI-MS/MS spectra of the oligonucleotides are generally dominated by a_n, a_n-B_n, d_n, and w_n-ions. Other types of fragments are also possible by some complex rearrangement type reaction. Several fragmentation pathways have been proposed to explain the formation of the complimentary ions in gas phase by the precursor oligonucleotides [209–213]. Scheme 5 shows the typical mechanisms [213] of the formation of a_n, a_n-B_n, d_n, and w_n-ions from the oligonucleotide anion upon CID.

8.6. CID of Lipids. Lipids are a broad group of naturally occurring hydrophobic organic molecules, which includes fatty acyls, glycerolipids, glycerophospholipids, sphingolipids, sterol lipids, prenol lipids, saccharolipids, and polyketides [215]. Mass spectrometry plays a key role in the structural elucidation and quantification of these molecules. It is beyond the scope here to discuss details about the CID of all classes of lipids. Here we would only discuss about the CID of two typical lipid molecules, for example, fatty acids and bile acids. Fatty acids are long hydrocarbon chain of varying length and varying degree of unsaturation, terminated by a carboxylic acid functional group. Bile acids are derived from cholesterol, and they contain a 5β steroid ring made up of four fused cycles bearing a side-chain attached to the C-17 carbon atom of the cycle, terminated by a carboxylic group (see Figure 22). They differ from each other by the number and position of the hydroxyl/keto groups and by the presence of unsaturation in the cycle. The bile acids can exist as free carboxylic acids or amide conjugates of the carboxylic groups with glycine ($NH_2CH_2CO_2H$) or taurine ($NH_2CH_2CH_2SO_3H$).

The fragmentation of fatty acids and bile acids is believed to occur via mechanisms that do not involve the charged group directly and thus termed as charge-remote fragmentation (CRF) mechanisms [162, 214, 216, 217]. A typical example of the CRF of a pseudomolecular anion has been shown in Scheme 6, which either involves a simple homolytic cleavage [218] or a 1,4-H_2 elimination [216] to produce distonic radical anion or terminally unsaturated anion.

When fatty acids are electrosprayed in negative ion mode, they produce $[M-H]^-$ ions in gas phase. Upon collisional activation, the CRF of the chain in different positions occurs. According to the nature of the bond, which is broken, different symbols of the product ions are used [214]. When a regular C–C bond is broken, this is represented by C, a vinyl bond by V, an allyl bond by A, a homoallyl bond by H, and a double bond by D. A subscript to the right of the capital letter (e.g., C_n) indicates the number of the carbon atoms remaining in the corresponding charged fatty acid fragment. A prime symbol to the left of the capital letter (product ion symbol) indicates that the product ion is deficient in one hydrogen atom relative to a fragment ion formed by homolytic cleavage at the same site of a hypothetical precursor molecule, and $M^{-\bullet}$ ($[M-H]^-$ is deficient one hydrogen relative to $M^{-\bullet}$). Multiple hydrogen deficiencies are denoted by multiple prime symbols to the left of the letter [214]. Figure 21 shows the fragmentation characteristic of a saturated fatty acid (stearic acid) and an unsaturated fatty acid (oleic acid) under high energy collision conditions [214]. Both the fatty acids show their unique backbone fragmentation pattern.

Similarly the charge remote ring cleavage of the bile acids also occurs, resulting in ring opening, loss of neutral ethylene, and double-bond formation on the product ions. Griffiths et al. have developed a systematic nomenclature (see Figure 22) to interpret the fragment ions produced by the CID of the bile acids [217, 219]. The rings are labeled as (a), (b), (c), (d) from left to right (see Figure 22), and the corresponding capital letter is used to indicate which ring is opened and broken in the cross-ring cleavage. The subscript (1, 2, and 3) to the right side of the capital letter indicates the cross-ring cleavage site on that ring. Again the "prime" symbol to the left of the letter is used to keep track of hydrogen deficiencies as discussed above.

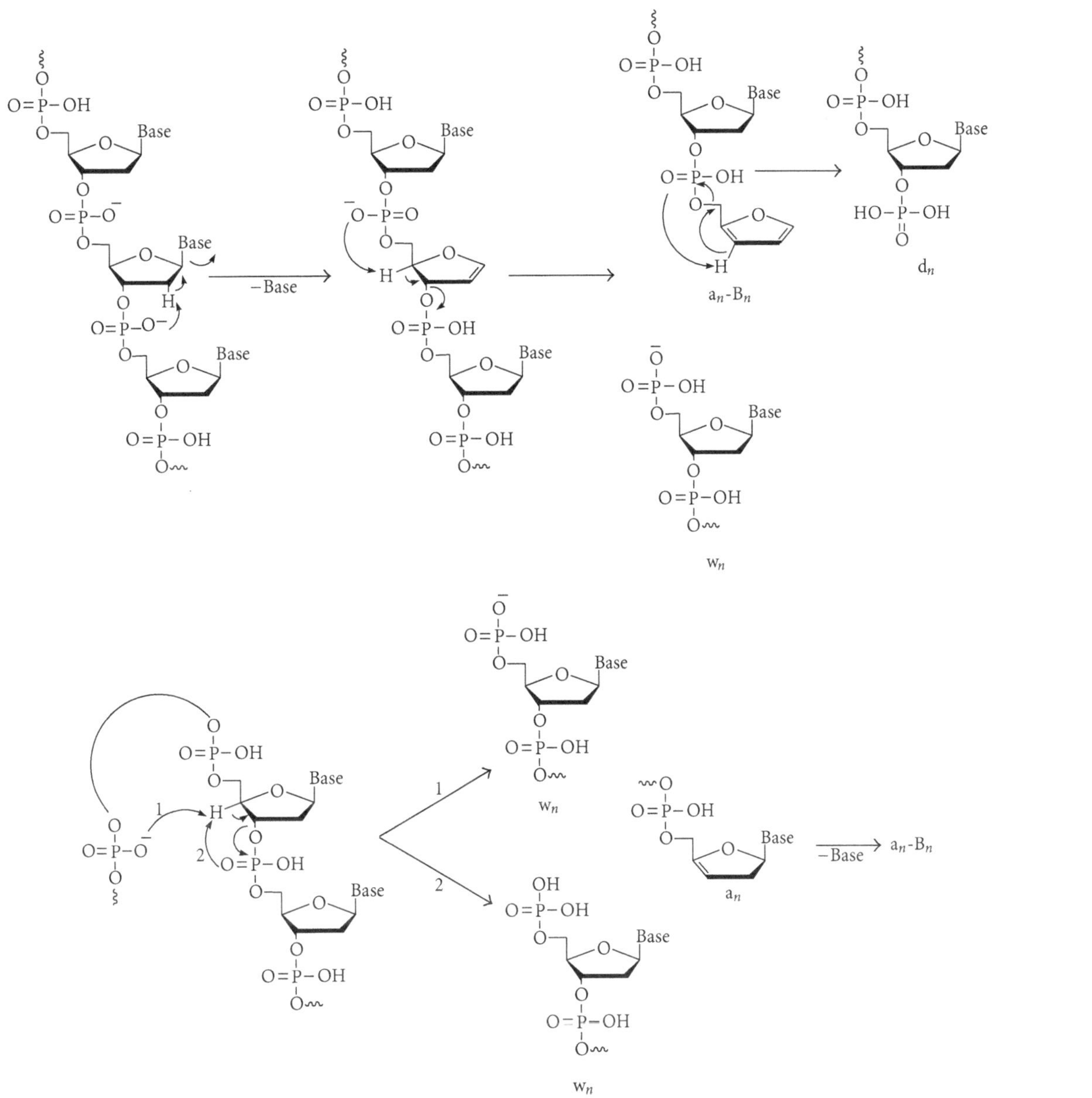

SCHEME 5: The fragmentation mechanism of the oligonucleotide in two different ways. The mechanism in the upper panel shows the necessity of a charge proximate to the cleavage site. The mechanism in the lower panel does not require the charge proximate.

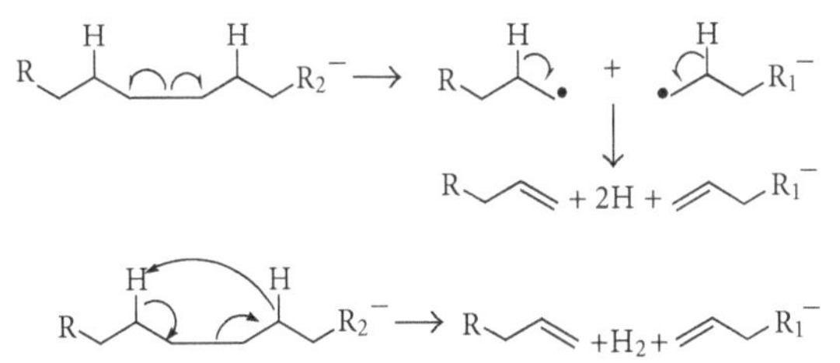

SCHEME 6: The charge remote fragmentation (CRF) mechanism for the formation of distonic ions (anions) and terminally unsaturated ions (anions).

9. Application of ESI-MS

After the discovery of the ESI-MS, its application areas rapidly expanded from large macromolecules to small organic and inorganic molecules. Nowadays the ion signals in the ESI-mass spectra are providing a clear and deep perception about the analytes properties far beyond the conventional mass and structural properties of the analytes. Here we will discuss in brief about certain applications of the ESI-MS.

9.1. Protein Identification and Characterization. Since mass is a very specific property of a molecule, determination of the molecular weight with high precision allows to solve many problems in proteomics. Over the past twenty years, ESI-mass spectrometry has emerged as a powerful tool in the life science to determine the identity [220–222], quantity [223, 224], and structural properties [6, 225–227] of the protein molecules. Mass spectrometry is a very fast process to verify the structure and purity of the proteins and peptides

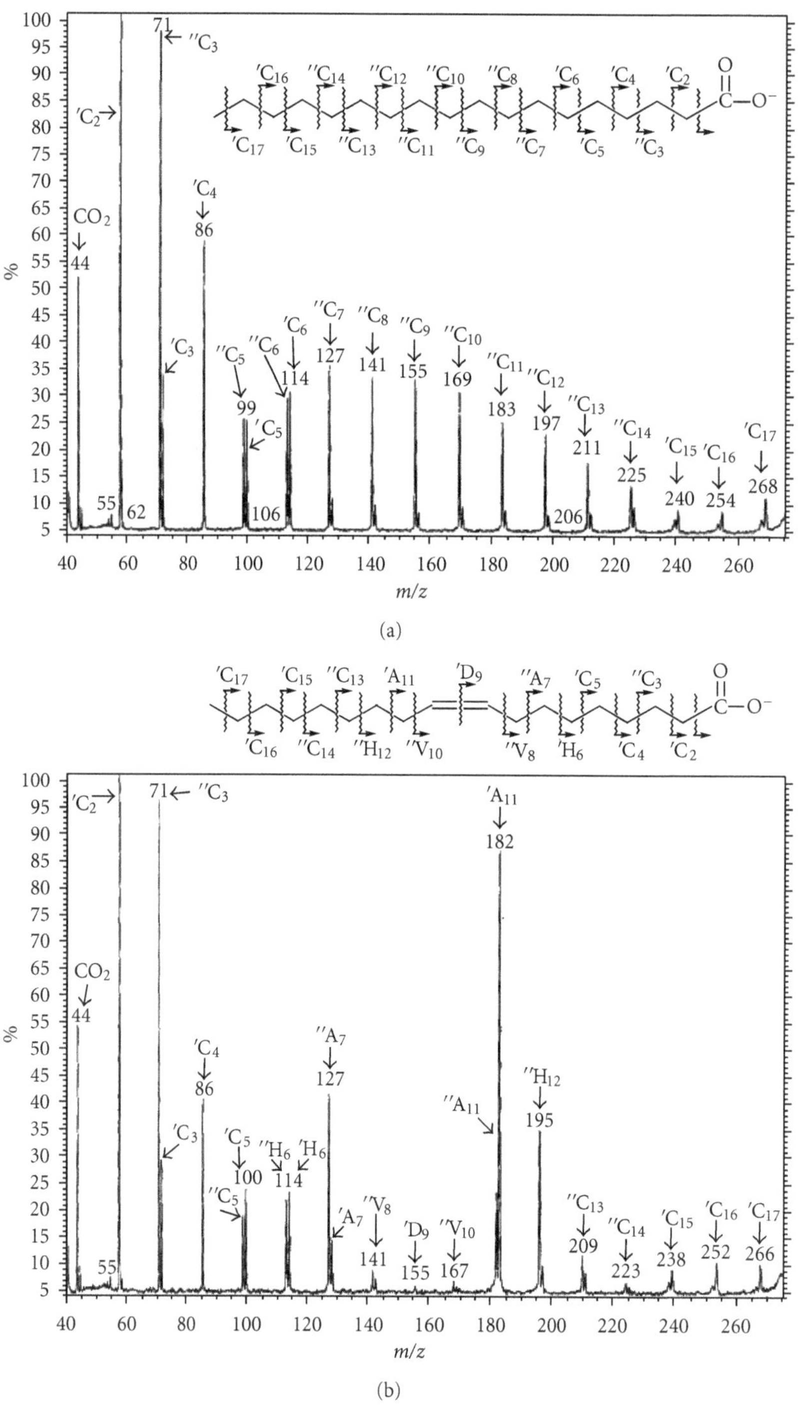

Figure 21: 400 eV (E_{lab}) CID spectra of stearic acid (a) and oleic acid (b) pseudomolecular anion. (Reprinted with permission from *Rapid Communication in Mass Spectrometry* [214], Copyright 1996, John Wiley and Sons).

[222, 228]. Classically the determination of the primary structure of the protein (i.e., the amino acid sequence) requires a chemical approach called Edman degradation method [229]. But this technique is time consuming, and a large quantity of the protein sample is also required. But after the discovery of LC-ESI-MS, the mass spectrometry is the most efficient way of sequencing a protein [107, 160, 230, 231]. In this MS method, the protein is digested by an endopeptidase (e.g., trypsin, chymotrypsin, pepsin, etc.), which cut the protein into small peptide/polypeptide fragments very specifically. Then the digest mixture is passed through the high-performance liquid chromatography where the collection of peptide fragments is separated. At the end of the column the individual peptide fragment is electrosprayed in vacuum, where they are trapped and allowed to undergo collision activation to produce MS/MS spectra as discussed

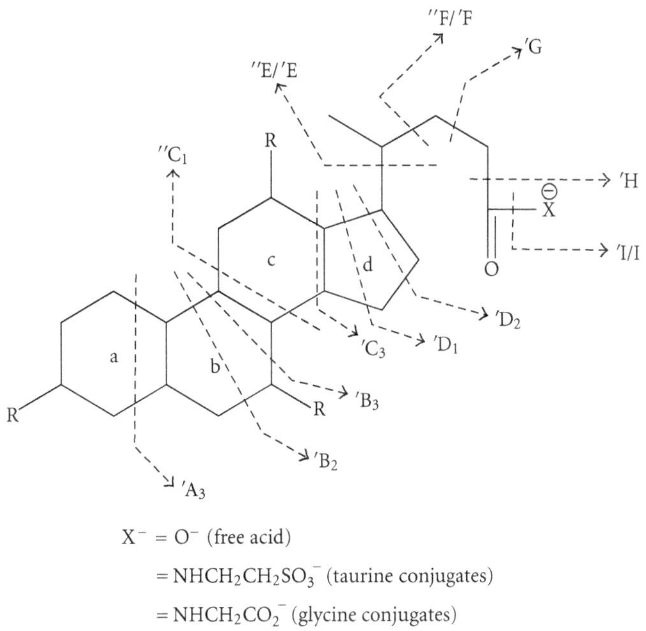

Figure 22: The charge remote fragmentation of bile acids (metabolites of cholesterol) with the nomenclature of different fragments.

above [160, 231]. The resulting MS/MS spectra are nothing but the mass fingerprints of the ladder-like product ions of the peptide of a particular sequence. The MS/MS spectra of the peptides are usually assigned with the help of various bioinformatic tools that implement sequencing algorithms based on peptide fragmentation chemistry as discussed before [160, 231, 232]. Thus the peptide mass fingerprinting (PMF) helps to identify the amino acid sequence of the whole proteins. The process is repeated using different digestion enzymes, and the overlaps of the resulting sequences are used to construct a sequence of the protein. When the identification of the proteins is solely based on the sequence data obtained from the tandem mass analysis, it is called "de novo sequencing" [232], and the corresponding procedure (as discussed above) of the protein analysis is called "bottom-up" approach [233, 234].

Sometimes intact charged protein generated by ESI is introduced into the mass analyzer and are subjected to gas-phase fragmentation, and this type of approach is referred to as the "top-down" strategy of protein analysis [235].

Although the ESI-MS technique is not sensitive to probe the local conformation of the polypeptide chain, that is, the secondary structure of the protein, the technique can successfully interpret the three-dimensional conformation of the proteins [76]. As discussed already that the charge state distribution actually represents the three dimensional folding or the tertiary structure of the proteins [76]. This property makes ESI-MS an excellent method, complimentary to CD (circular dichroism), to characterize the protein conformational change. Hence an increasing interest based on ESI-MS is continuously developing to study the protein folding process.

ESI-MS can also successfully detect the posttranslational modifications (PTMs) [236, 237] and the mutation [238] of the proteins as those processes lead to the change of protein masses. Not only does the total mass change, but also the position or point of the PTM or mutation can be determined by tandem mass spectrometric experiments [222, 239]. Recently, using tandem mass spectrometric technique, we have identified the reactive lysine residues of cytochrome c [240]. We found that the lysine residues, which are in the turn or loop region of the protein, are more reactive to succinylation compared to those which are in the helical region. It has been proposed that the lower reactivity of the lysine residues present in the helical regions might be due to the higher rigidity of the helical region than that of the turn or loop region [240]. In another study [241], using tandem mass spectrometry we have determined the molecular basis of the PTM involving covalent attachment of the heme with a glutamic acid of the protein matrix in the Cytochrome P450 enzyme.

Disulphide bonds (a kind of PTM) play important roles in the structure and biological activity of the cystinyl proteins. So the determination of the disulphide bond linkage between two adjacent or closely spaced cysteine residues is an integral part of the structural characterization of the proteins. Several mass spectrometry-based strategies have been developed to map the disulphide bond linkages in the proteins [242–247]. Most of the time, the protein of interest is cleaved enzymatically in its nonreduced states, and then the resulting disulphide-linked peptides are separated, identified, and characterized by LC-MS technique. Then the data is compared with the similar experiments performed on the disulphide-reduced protein of the interest [243]. This comparison study can locate the disulphide linkages in the folded proteins. Sometimes chemical cleavage at disulphide residues followed by chemical derivatization is also performed for the mass spectrometric study to identify the disulphide linkages in the proteins [247].

9.2. Studying Noncovalent Interaction. Since ESI is a sufficiently soft ionization technique, the noncovalent complexes of the analytes formed in solution can representatively, transferred to the gas phase when appropriate instrumental conditions are used [225, 248–250]. Mainly a collection of weak interactions such as Van der Waals forces, hydrophobic forces, and hydrogen bonding or salt bridges (electrostatic interactions) are responsible for the analyte association in solution. When the analyte is transformed in the gas-phase via ESI, probably most of those interactions are retained, and some of those interactions become more prominent in the gas phase compared to solution and thus provide the structural integrity in the gas phase [251–253]. Generally the propensities of the ionic interactions become more in the gas phase compared to solution [251–253]. For example, a recent ESI-MS study of the protonated and deprotonated gaseous ions of a single chain antibody-trisaccharide complex has showen several specific intermolecular H-bonds in the gas phase [254]. Likewise a recent report, shown that, in the presence of K^+, Rb^+, and Cs^+, uracil, thymine, and their homologues form self-assembled quintet structures that are stabilized by hydrogen bonding and ion-dipole interactions in the gas phase [255]. ESI-MS has allowed the observation of a large number of biomolecular noncovalent complexes such as protein-protein [256, 257], protein-metal ion [258], protein-drug [259], and protein-nucleic acid [260] complexes. Since proteins provide a large number of functional groups for the noncovalent interaction with the partner molecule(s), the resultant noncovalent forces are large enough to retain the association during their transfer from condensed to the gas phase. But, in small molecules for example, in small peptides, this number of noncovalently interacting atoms with proper orientation for the intermolecular interaction is less and it is sometimes very difficult to probe their association in solution by ESI technique. Recently we have shown the formation of noncovalent dimers of the lysine containing basic peptides by ESI-MS [253], which demonstrated that the intermolecular electrostatic/H-bonding network is mainly responsible for the dimer formation of those small peptides in the gas phase. ESI-mass spectrometry also evolved as a valuable tool for the determination of the association/dissociation constants for several protein-ligand complexes [261]. Although the gas-phase studies on the proterin-ligand complexes have been primarily focused on the complexes, which are stabilized by the ionic interaction or H-bonding, the nonpolar intermolecular interaction between protein (bovine β-lactoglobulin) and ligand (fatty acids) in vacuum has also been reported lately [262]. The corresponding interaction strength of the nonpolar protein-ligand complexes has also been quantified.

It is very difficult to ionize the hydrophobic proteins by ESI because of their inherent insolubility in the buffers compatible with electrospray. For this reason the ESI-mass spectrometry has not been applied to intact membrane protein complexes. But recently Barrera et al. have been successful to transfer some hydrophobic membrane protein complexes in vacuum via ESI by encapsulation in a solution phase detergent micelle [263].

The ESI-MS has also been used to study the complexes between polyether (e.g., crown ethers) and protonated peptides [264, 265]. The protonated amine functional groups make hydrogen bonds with the oxygen atoms of the crown ethers in those complexes. The energy-variable collision-induced dissociation was carried out to analyze the strengths of noncovalent interactions of protonated peptide/polyether complexes [264, 265].

9.3. In Clinical Laboratory. Since ESI-MS is a sensitive, robust, and reliable tool for studying the femtomole samples in microlitre volumes, it has become an increasingly important technique in the clinical laboratory for structural study or quantitative measurement of metabolites in a complex biological sample [266]. For example, HPLC/ESI-MS is useful to a great extent than other conventional techniques in screening for inborn errors of amino acid [267, 268], fatty acid [269], purine [270], pyrimidine [270] metabolism, and diagnosis of galactosaemia [271] and peroxisomal [272, 273] disorders. Because of the preservation of the noncovalent interaction in gas phase, ESI-MS has nurtured a new and improved approach (versus electrophoresis) for identification and quantification of haemoglobin variants [274]. With the understanding of glycohaemoglobin structure, an IFCC reference method for glycohaemoglobin assay has been established using ESI-MS [275]. It also represents a promising strategy for the standardisation of HbA1c in diabetic monitoring [276]. With its other applications such as in therapeutic drug monitoring and identification of biomarkers [277, 278], ESI-MS will continue to exert a more important influence in the clinical laboratory in near future.

9.4. Probing Molecular Dynamics: H/D-Exchange Experiments. The application of the ESI-MS is not only restricted to the structural characterization, but it has also been extended recently to the study of the molecular dynamics. For this type of application, the analyte ions produced by the ESI-ion source are trapped and subjected to the ion-molecule reaction (H/D exchange) with some gaseous deuterated molecules possessing exchangeable deuterium (e.g., CH_3OD, ND_3, etc.) for well-defined reaction intervals in the collision cell [279]. As a result the H/D-exchange between analyte and the deuterated molecules occurs in the gas phase and thus provides some critical information regarding the molecular motion in the noncovalent complexes. For example, the H/D exchange reactions have been applied to uncover the tumbling motion of ammonium guests bound inside the cavity of resorcinarene hosts [280]. Very recently a highly dynamic motion of the crown ethers along the oligolysine peptide chains has been probed by H/D-exchange experiments in the gas phase [281]. The authors suggested this phenomenon as the wire dance on the molecular level. It has been observed that the crown ethers (guest molecules) can move directly between different binding sites of the oligolysine (a multitopic host) without intermediate dissociation. Furthermore, the exchange experiments unambiguously revealed the zwitterionic structure of the crown ether/oligolysine complexes, highlighting the success of the gas-phase experiments for investigating noncovalent interactions [281].

The gas-phase folding and unfolding of the protein (protein dynamics) can also be monitored by the H/D-exchange experiments [282, 283]. Valuable information regarding the protein conformation in vacuum has also come out by the gas-phase isotope exchange experiments [138, 284, 285]. Similarly several ion-molecule reactions in gas-phase have been used for the covalent modification of the gaseous analyte [286].

9.5. Monitoring Chemical Reactions and Studying Reactive Intermediates. The applications of ESI-MS have also been explored in synthetic organic and organometallic chemistry to study the reactive intermediates and the mechanisms [287–292]. For example, catalytic intermediate of the Suzuki coupling reaction [289] and the Heck reaction mechanism [291] has been studied by ESI-MS. Raney nickel-catalyzed coupling reaction of 2-bromo-6-methylpyridine was studied, and a reactive dimer of the intermediate Ni[II] complex of 6,6′-dimethyl-2,2′-bipyridine was detected [293]. Desorption electrospray ionization mass spectrometry (DESI-MS) has been used for monitoring solid-state organic reaction in ambient air, specifically the Bayer-Villiger type reactions involving the oxidation of ketones by m-chloroperbenzoic acid in solid state [294]. The 1-adamantyl radicals have been identified as a reactive intermediate in several organic syntheses. Recently a tert-adamantyl peroxyl radical has been trapped in gas phase using ESI-MS, and its unusual structure and reactivity has been investigated [295]. Many other types of the reaction mechanisms investigated by the ESI-MS technique have recently been reviewed by Eberlin [296].

9.6. Chemical Imaging. Desorption electrospray ionization mass spectrometry (DESI-MS) is relatively a new ambient surface analysis method. The development of imaging mass spectrometry by DESI has been described recently [297–299], and its application to high-throughput biological tissue imaging was also demonstrated [298]. In this technique the spatial distribution (on the tissue surface) as well as structural identification of the molecule of interest can be accomplished successfully [298]. Chemical imaging by DESI has been applied to label-free detection of drugs and metabolites in tissue [298]. DESI-MS imaging is advantage in a number of ways over the conventional whole-body autoradiography approach: (1) no radioactive label is required and (2) it allows simultaneous detection of the parent drug compound and metabolites in tissue [300]. DESI also holds the advantages of speed and specificity inherent in the mass spectrometric experiment. Chemical imaging by DESI-MS is still, however, in its early childhood. The ongoing research activities in this area address the questions concerning sensitivity, compound-specific ionization yields, tissue-specific ion suppression, and effects of solvent composition on ionization yields. An insight on these issues would immensely widen the horizon of applications of this technique in assessing more complex and quantitative information on the samples that would have important medical and pharmaceutical implications.

10. Future Prospects and Outlook

The development of electrospray ionization mass spectrometry has enabled us to solve a wide range of biochemical and mechanistic problems as discussed above. Perhaps it is one of the instruments that has continuously evolved over the last three decades both in application and modification of instrument design. For example, the transition of source design from microspray to nanospray has improved the sensitivity. Introduction of high-resolution mass analyzer (FT-ICR) amended the accuracy and redefined the applicability.

Yet in several aspects, the understanding of the ESI-MS remains vague to date. One of the major issues is the exact mechanism of ion formation. How is the gas-phase ion formation guided mechanistically? Is the perception of CRM and IEM enough to interpret the ion formation? It is still unclear about the analyte behavior (structure and dynamics) inside the charged droplets produced by ES process. If the proteins change their structure inside the charged droplet, then the charge state distribution (CSD) should reflect the instantaneous conformation of the protein inside the charged droplet not in the original solution. Again, as discussed in the present review, the CSD is not influenced by a specific parameter but governed by multiple parameters. So the precise understanding of the cumulative effect of those different parameters on the protein CSD would no doubt help to earn more quantitative insights about structural and chemical behavior of the proteins.

So far a little attention was invoked on the detector characterization and improvement compared to that of the ion source and mass analyzer. So, the characterization and improvement of the detector would likely be the next step in the development of mass spectrometry which would enable us to give the answer whether the ion signal intensity is governed by the molecular conformation of the analyte or not.

The application of the ESI-MS got a noteworthy dimension in biochemistry laboratories after its success in protein sequencing. The sequencing is based on the idea and information of gas-phase ion fragmentation chemistry of the peptides in the tandem mass spectrometry. But the tandem mass spectra of the peptides are mostly dominated by anomalous fragment ions than the "sequence informative ions." So the characterization of those anomalous fragments and accordingly the modification of sequencing algorithms would facilitate the sequencing of proteins/peptides more precisely and rapidly using mass-spectrometric technique.

Acknowledgments

Authors are thankful to Mr. Abhishek Acharya, Max Planck Institute for Biophysics, Germany, for his help. The financial support for this endeavor was provided by Tata Institute of Fundamental Research, Department of Atomic Energy, Government of India.

References

[1] J. J. Thomson, *Rays of Positive Electricity and Their Applications to Chemical Analysis*, Longmans Green, London, UK, 1913.

[2] M. Grayson, "John Bennett Fenn: a curious road to the prize," *Journal of The American Society for Mass Spectrometry*, vol. 22, no. 8, pp. 1301–1308, 2011.

[3] M. Barber, R. S. Bordoli, R. D. Sedgwick, and A. N. Tyler, "Fast atom bombardment of solids (F.A.B.): a new ion source for mass spectrometry," *Journal of the Chemical Society, Chemical Communications*, no. 7, pp. 325–327, 1981.

[4] J. B. Fenn, M. Mann, C. K. Meng, S. F. Wong, and C. M. Whitehouse, "Electrospray ionization for mass spectrometry of large biomolecules," *Science*, vol. 246, no. 4926, pp. 64–71, 1989.

[5] M. Przybylski and M. O. Glocker, "Electrospray mass spectrometry of biomacromolecular complexes with noncovalent interactions—New analytical perspectives for supramolecular chemistry and molecular recognition processes," *Angewandte Chemie*, vol. 35, no. 8, pp. 807–826, 1996.

[6] B. Domon and R. Aebersold, "Mass spectrometry and protein analysis," *Science*, vol. 312, no. 5771, pp. 212–217, 2006.

[7] J. B. Fenn, "Electrospray wings for molecular elephants (Nobel lecture)," *Angewandte Chemie*, vol. 42, no. 33, pp. 3871–3894, 2003.

[8] Z. Ouyang, Z. Takáts, T. A. Blake et al., "Preparing protein microarrays by soft-landing of mass-selected ions," *Science*, vol. 301, no. 5638, pp. 1351–1354, 2003.

[9] B. Gologan, Z. Takáts, J. Alvarez et al., "Ion soft-landing into liquids: protein identification, separation, and purification with retention of biological activity," *Journal of the American Society for Mass Spectrometry*, vol. 15, no. 12, pp. 1874–1884, 2004.

[10] S. R. Wilson and Y. Wu, "Applications of electrospray ionization mass spectrometry to neutral organic molecules including fullerenes," *Journal of the American Society for Mass Spectrometry*, vol. 4, no. 7, pp. 596–603, 1993.

[11] C. E. C. A. Hop and R. Bakhtiar, "Electrospray ionization mass spectrometry. Part III: applications in inorganic chemistry and synthetic polymer chemistry," *Journal of Chemical Education*, vol. 73, no. 8, pp. A162–A169, 1996.

[12] M. J. Keith-Roach, "A review of recent trends in electrospray ionisation-mass spectrometry for the analysis of metal-organic ligand complexes," *Analytica Chimica Acta*, vol. 678, no. 2, pp. 140–148, 2010.

[13] I. Leito, K. Herodes, M. Huopolainen et al., "Towards the electrospray ionization mass spectrometry ionization efficiency scale of organic compounds," *Rapid Communications in Mass Spectrometry*, vol. 22, no. 3, pp. 379–384, 2008.

[14] M. Oss, A. Kruve, K. Herodes, and I. Leito, "Electrospray ionization efficiency scale of organic compound," *Analytical Chemistry*, vol. 82, no. 7, pp. 2865–2872, 2010.

[15] A. G. Baily, *Electrostatic Spraying of Liquids*, John Wiley & Sons, New York, NY, USA, 1988.

[16] M. Dole, L. L. Mack, R. L. Hines et al., "Molecular beams of macroions," *The Journal of Chemical Physics*, vol. 49, no. 5, pp. 2240–2249, 1968.

[17] L. L. Mack, P. Kralik, A. Rheude, and M. Dole, "Molecular beams of macroions. II," *The Journal of Chemical Physics*, vol. 52, no. 10, pp. 4977–4986, 1970.

[18] M. Yamashita and J. B. Fenn, "Electrospray ion source. Another variation on the free-jet theme," *Journal of Physical Chemistry*, vol. 88, no. 20, pp. 4451–4459, 1984.

[19] M. Yamashita and J. B. Fenn, "Negative ion production with the electrospray ion source," *Journal of Physical Chemistry*, vol. 88, no. 20, pp. 4671–4675, 1984.

[20] C. M. Whitehouse, R. N. Dreyer, M. Yamashita, and J. B. Fenn, "Electrospray interface for liquid chromatographs and mass spectrometers," *Analytical Chemistry*, vol. 57, no. 3, pp. 675–679, 1985.

[21] A. P. Bruins, "Mass spectrometry with ion sources operating at atmospheric pressure," *Mass Spectrometry Reviews*, vol. 10, no. 1, pp. 53–77, 1991.

[22] P. Kebarle and L. Tang, "From ions in solution to ions in the gas phase: the mechanism of electrospray mass spectrometry," *Analytical Chemistry*, vol. 65, no. 22, pp. 972A–986A, 1993.

[23] M. Karas, U. Bahr, and T. Dülcks, "Nano-electrospray ionization mass spectrometry: addressing analytical problems beyond routine," *Fresenius' Journal of Analytical Chemistry*, vol. 366, no. 6-7, pp. 669–676, 2000.

[24] J. F. Anacleto, S. Pleasance, and R. K. Boyd, "Calibration of ion spray mass spectra using cluster ions," *Organic Mass Spectrometry*, vol. 27, pp. 660–666, 1992.

[25] C. E. C. A. Hop, "Generation of high molecular weight cluster ions by electrospray ionization; implications for mass calibration," *Journal of Mass Spectrometry*, vol. 31, pp. 1314–1316, 1996.

[26] A. P. Bruins, T. R. Covey, and J. D. Henion, "Ion spray interface for combined liquid chromatography/atmospheric pressure ionization mass spectrometry," *Analytical Chemistry*, vol. 59, no. 22, pp. 2642–2646, 1987.

[27] T. R. Covey, A. P. Bruins, and J. D. Henion, "Comparison of thermospray and ion spray mass spectrometry in an atmospheric pressure ion source," *Organic Mass Spectrometry*, vol. 23, pp. 178–186, 1988.

[28] M. G. Ikonomou, A. T. Blades, and P. Kebarle, "Electrospray-ion spray: a comparison of mechanisms and performance," *Analytical Chemistry*, vol. 63, no. 18, pp. 1989–1998, 1991.

[29] J. F. Banks, J. P. Quinn, and C. M. Whilehouse, "LC/ESI-MS determination of proteins using conventional liquid chromatography and ultrasonically assisted electrospray," *Analytical Chemistry*, vol. 66, no. 21, pp. 3688–3695, 1994.

[30] J. F. Banks, S. Shen, C. M. Whitehouse, and J. B. Fenn, "Ultrasonically assisted electrospray ionization for LC/MS determination of nucleosides from a transfer RNA digest," *Analytical Chemistry*, vol. 66, no. 3, pp. 406–414, 1994.

[31] Z. Takáts, J. M. Wiseman, B. Gologan, and R. G. Cooks, "Electrosonic spray ionization. A gentle technique for generating folded proteins and protein complexes in the gas phase and for studying ion-molecule reactions at atmospheric pressure," *Analytical Chemistry*, vol. 76, no. 14, pp. 4050–4058, 2004.

[32] M. Wilm and M. Mann, "Analytical properties of the nanoelectrospray ion source," *Analytical Chemistry*, vol. 68, no. 1, pp. 1–8, 1996.

[33] E. D. Hoffmann and V. Stroobant, *Mass Spectrometry Principles and Applications*, John Wiley & Sons, Chichester, UK, 2nd edition, 2001.

[34] J. H. Gross, *Mass Spectrometry*, Springer, Heidelberg, Germany, 2004.

[35] R. E. March, "An introduction to quadrupole ion trap mass spectrometry," *Journal of Mass Spectrometry*, vol. 32, no. 4, pp. 351–369, 1997.

[36] J. S. Allen, "An improved electron multiplier particle counter," *Review of Scientific Instruments*, vol. 18, no. 10, pp. 739–749, 1947.

[37] H. E. Stanton, W. A. Chupka, and M. G. Inghram, "Electron multipliers in mass spectrometry; effect of molecular structure," *Review of Scientific Instruments*, vol. 27, no. 2, p. 109, 1956.

[38] I. J. Amster, "Fourier transform mass spectrometry," *Journal of Mass Spectrometry*, vol. 31, no. 12, pp. 1325–1337, 1996.

[39] J. B. Fenn, "Ion formation from charged droplets: roles of geometry, energy, and time," *Journal of the American Society for Mass Spectrometry*, vol. 4, no. 7, pp. 524–535, 1993.

[40] J. F. J. Todd, "Recommendations for nomenclature and symbolism for mass spectroscopy," *Pure and Applied Chemistry*, vol. 63, pp. 1541–1566, 1991.

[41] T. R. Covey, R. F. Bonner, B. I. Shushan, and J. Henion, "The determination of protein, oligonucleotide and peptide molecular weights by ion-spray mass spectrometry," *Rapid Communications in Mass Spectrometry*, vol. 2, no. 11, pp. 249–256, 1988.

[42] J. R. Chapman, R. T. Gallagher, E. C. Barton, J. M. Curtis, and P. J. Derrick, "Advantages of high-resolution and high-mass range magnetic-sector mass spectrometry for electrospray ionization," *Organic Mass Spectrometry*, vol. 27, pp. 195–203, 1992.

[43] M. Mann, C. K. Meng, and J. B. Fenn, "Interpreting mass spectra of multiply charged ions," *Analytical Chemistry*, vol. 61, no. 15, pp. 1702–1708, 1989.

[44] M. Labowsky, C. Whitehouse, and J. B. Fenn, "Three-dimensional deconvolution of multiply charged spectra," *Rapid Communications in Mass Spectrometry*, vol. 7, pp. 71–84, 1993.

[45] G. Wang and R. B. Cole, "Effect of solution intic strength on analyte charge state distributions in positive and negative ion electrospray mass spectrometry," *Analytical Chemistry*, vol. 66, no. 21, pp. 3702–3708, 1994.

[46] R. Guevremont, K. W. M. Siu, J. C. Y. Le Blanc, and S. S. Berman, "Are the electrospray mass spectra of proteins related to their aqueous solution chemistry?" *Journal of the American Society for Mass Spectrometry*, vol. 3, no. 3, pp. 216–224, 1992.

[47] M. A. Kelly, M. M. Vestling, C. C. Fenselau, and P. B. Smith, "Electrospray analysis of proteins: a comparison of positive-ion and negative-ion mass spectra at high and low pH," *Organic Mass Spectrometry*, vol. 27, pp. 1143–1147, 1992.

[48] G. Wang and R. B. Cole, "Disparity between solution-phase equilibria and charge state distributions in positive-ion electrospray mass spectrometry," *Organic Mass Spectrometry*, vol. 29, pp. 419–427, 1994.

[49] J. C. Y. Le Blanc, J. Wang, R. Guevremont, and K. W. M. Siu, "Electrospray mass spectra of protein cations formed in basic solutions," *Organic Mass Spectrometry*, vol. 29, pp. 587–593, 1994.

[50] J. A. Loo, C. G. Edmonds, H. R. Udseth, and R. D. Smith, "Effect of reducing disulfide-containing proteins on electrospray ionization mass spectra," *Analytical Chemistry*, vol. 62, no. 7, pp. 693–698, 1990.

[51] R. D. Smith, J. A. Loo, R. R. Ogorzalek Loo, M. Busman, and H. R. Udseth, "Principles and practice of electrospray ionization-mass spectrometry for large polypeptides and proteins," *Mass Spectrometry Reviews*, vol. 10, no. 5, pp. 359–451, 1991.

[52] M. Cloupeau and B. Prunet-Foch, "Electrostatic spraying of liquids: main functioning," *Journal of Electrostatics*, vol. 25, pp. 165–184, 1990.

[53] R. B. Cole, "Some tenets pertaining to electrospray ionization mass spectrometry," *Journal of Mass Spectrometry*, vol. 35, no. 7, pp. 763–772, 2000.

[54] P. Kebarle and M. Peschke, "On the mechanisms by which the charged droplets produced by electrospray lead to gas phase ions," *Analytica Chimica Acta*, vol. 406, no. 1, pp. 11–35, 2000.

[55] N. B. Cech and C. G. Enke, "Practical implications of some recent studies in electrospray ionization fundamentals," *Mass Spectrometry Reviews*, vol. 20, no. 6, pp. 362–387, 2001.

[56] A. T. Blades, M. G. Ikonomou, and P. Kebarle, "Mechanism of electrospray mass spectrometry. Electrospray as an electrolysis cell," *Analytical Chemistry*, vol. 63, no. 19, pp. 2109–2114, 1991.

[57] G. Diehl and U. Karst, "On-line electrochemistry—MS and related techniques," *Analytical and Bioanalytical Chemistry*, vol. 373, no. 6, pp. 390–398, 2002.

[58] R. B. Cole, *Electrospray and MALDI Mass Spectrometry*, John Wiley & Sons, New Jersey, NJ, USA, 2010.

[59] G. Taylor, "Disintegration of water drops in an electric field," *Proceedings of the Royal Society of London A*, vol. 280, pp. 383–397, 1964.

[60] M. S. Wilm and M. Mann, "Electrospray and Taylor-Cone theory, Dole's beam of macromolecules at last?" *International Journal of Mass Spectrometry and Ion Processes*, vol. 136, no. 2-3, pp. 167–180, 1994.

[61] J. Fernandez De La Mora, "Electrospray ionization of large multiply charged species proceeds via Dole's charged residue mechanism," *Analytica Chimica Acta*, vol. 406, no. 1, pp. 93–104, 2000.

[62] L. Rayleigh, "On the equilibrium of liquid conducting masses charged with electricity," *Philosophical Magazine*, pp. 184–186, 1882.

[63] J. V. Iribarne and B. A. Thomson, "On the evaporation of small ions from charged droplets," *The Journal of Chemical Physics*, vol. 64, no. 6, pp. 2287–2294, 1976.

[64] A. Gomez and K. Tang, "Charge and fission of droplets in electrostatic sprays," *Physics of Fluids*, vol. 6, no. 1, pp. 404–414, 1994.

[65] D. B. Hager, N. J. Dovichi, J. Klassen, and P. Kebarle, "Droplet electrospray mass spectrometry," *Analytical Chemistry*, vol. 66, no. 22, pp. 3944–3949, 1994.

[66] D. Duft, T. Achtzehn, R. Müller, B. A. Huber, and T. Leisner, "Coulomb fission: rayleigh jets from levitated microdroplets," *Nature*, vol. 421, no. 6919, p. 128, 2003.

[67] M. G. Ikonomou, A. T. Blades, and P. Kebarle, "Investigations of the electrospray interface for liquid chromatography/mass spectrometry," *Analytical Chemistry*, vol. 62, no. 9, pp. 957–967, 1990.

[68] G. Schmelzeisen-Redeker, L. Bütfering, and F. W. Röllgen, "Desolvation of ions and molecules in thermospray mass spectrometry," *International Journal of Mass Spectrometry and Ion Processes*, vol. 90, no. 2, pp. 139–150, 1989.

[69] S. Nguyen and J. B. Fenn, "Gas-phase ions of solute species from charged droplets of solutions," *Proceedings of the National Academy of Sciences of the United States of America*, vol. 104, no. 4, pp. 1111–1117, 2007.

[70] B. E. Winger, K. J. Light-Wahl, R. R. Ogorzalek Loo, H. R. Udseth, and R. D. Smith, "Observation and implications of high mass-to-charge ratio ions from electrospray ionization mass spectrometry," *Journal of the American Society for Mass Spectrometry*, vol. 4, no. 7, pp. 536–545, 1993.

[71] L. P. Tolić, G. A. Anderson, R. D. Smith, H. M. Brothers, R. Spindler, and D. A. Tomalia, "Electrospray ionization Fourier transform ion cyclotron resonance mass spectrometric characterization of high molecular mass Starburst dendrimers," *International Journal of Mass Spectrometry and Ion Processes*, vol. 165-166, pp. 405–418, 1997.

[72] S. J. Valentine, J. G. Anderson, A. D. Ellington, and D. E. Clemmer, "Disulfide-intact and-reduced lysozyme in the gas phase: conformations and pathways of folding and unfolding," *Journal of Physical Chemistry B*, vol. 101, no. 19, pp. 3891–3900, 1997.

[73] B. A. Thomson and J. V. Iribarne, "Field induced ion evaporation from liquid surfaces at atmospheric pressure," *The Journal of Chemical Physics*, vol. 71, no. 11, pp. 4451–4463, 1979.

[74] P. Kebarle, "A brief overview of the present status of the mechanisms involved in electrospray mass spectrometry," *Journal of Mass Spectrometry*, vol. 35, no. 7, pp. 804–817, 2000.

[75] Z. Olumee, J. H. Callahan, and A. Vertes, "Droplet dynamics changes in electrostatic sprays of methanol—Water mixtures," *Journal of Physical Chemistry A*, vol. 102, no. 46, pp. 9154–9160, 1998.

[76] I. A. Kaltashov and R. R. Abzalimov, "Do ionic charges in ESI MS provide useful information on macromolecular structure?" *Journal of the American Society for Mass Spectrometry*, vol. 19, no. 9, pp. 1239–1246, 2008.

[77] I. A. Kaltashov and A. Mohimen, "Estimates of protein surface areas in solution by electrospray ionization mass spectrometry," *Analytical Chemistry*, vol. 77, no. 16, pp. 5370–5379, 2005.

[78] H. Prakash and S. Mazumdar, "Direct correlation of the crystal structure of proteins with the maximum positive and negative charge states of gaseous protein ions produced by electrospray ionization," *Journal of the American Society for Mass Spectrometry*, vol. 16, no. 9, pp. 1409–1421, 2005.

[79] R. Grandori, "Origin of the conformation dependence of protein charge-state distributions in electrospray ionization mass spectrometry," *Journal of Mass Spectrometry*, vol. 38, no. 1, pp. 11–15, 2003.

[80] L. Konermann and D. J. Douglas, "Acid-induced unfolding of cytochrome c at different methanol concentrations: electrospray ionization mass spectrometry specifically monitors changes in the tertiary structure," *Biochemistry*, vol. 36, no. 40, pp. 12296–12302, 1997.

[81] L. Konermann and D. J. Douglas, "Unfolding of proteins monitored by electrospray ionization mass spectrometry: a comparison of positive and negative ion modes," *Journal of the American Society for Mass Spectrometry*, vol. 9, no. 12, pp. 1248–1254, 1998.

[82] A. K. Frimpong, R. R. Abzalimov, S. J. Eyles, and I. A. Kaltashov, "Gas-phase interference-free analysis of protein ion charge-state distributions: detection of small-scale conformational transitions accompanying pepsin inactivation," *Analytical Chemistry*, vol. 79, no. 11, pp. 4154–4161, 2007.

[83] V. J. Nesatyy, "On the conformation-dependent neutralization theory and charging of individual proteins and their non-covalent complexes in the gas phase," *Journal of Mass Spectrometry*, vol. 39, no. 1, pp. 93–97, 2004.

[84] R. Grandori, "Origin of the conformation dependence of protein charge-state distributions in electrospray ionization mass spectrometry," *Journal of Mass Spectrometry*, vol. 38, no. 1, pp. 11–15, 2003.

[85] U. H. Verkerk, M. Peschke, and P. Kebarle, "Effect of buffer cations and of H_3O^+ on the charge states of native proteins. Significance to determinations of stability constants of protein complexes," *Journal of Mass Spectrometry*, vol. 38, no. 6, pp. 618–631, 2003.

[86] R. R. Julian, R. Hodyss, and J. L. Beauchamp, "Salt bridge stabilization of charged zwitterionic arginine aggregates in the gas phase," *Journal of the American Chemical Society*, vol. 123, no. 15, pp. 3577–3583, 2001.

[87] P. D. Schnier, D. S. Gross, and E. R. Williams, "On the maximum charge state and proton transfer reactivity of peptide and protein ions formed by electrospray ionization," *Journal of the American Society for Mass Spectrometry*, vol. 6, no. 11, pp. 1086–1097, 1995.

[88] J. R. Chapman, *Mass Spectrometry of Protein and Peptides*, Humana Press, New Jersey, NJ, USA, 2000.

[89] T. B. McMahon and G. Ohanessian, "An experimental and ab initio study of the nature of the binding in gas-phase complexes of sodium ions," *Chemistry A*, vol. 6, no. 16, pp. 2931–2941, 2000.

[90] S. Hoyau, K. Norrman, T. B. McMahon, and G. Ohanessian, "A quantitative basis for a scale of Na^+ affinities of organic and small biological molecules in the gas phase," *Journal of the American Chemical Society*, vol. 121, no. 38, pp. 8864–8875, 1999.

[91] U. H. Verkerk and P. Kebarle, "Ion-ion and ion-molecule reactions at the surface of proteins produced by nanospray. Information on the number of acidic residues and control of the number of ionized acidic and basic residues," *Journal of the American Society for Mass Spectrometry*, vol. 16, no. 8, pp. 1325–1341, 2005.

[92] H. Prakash, B. T. Kansara, and S. Mazumdar, "Effects of salts on the charge-state distribution and the structural basis of the most-intense charge-state of the gaseous protein ions produced by electrospray ionization," *International Journal of Mass Spectrometry*, vol. 289, no. 2-3, pp. 84–91, 2010.

[93] N. Felitsyn, M. Peschke, and P. Kebarle, "Origin and number of charges observed on multiply-protonated native proteins produced by ESI," *International Journal of Mass Spectrometry*, vol. 219, no. 1, pp. 39–62, 2002.

[94] M. Peschke, A. Blades, and P. Kebarle, "Charged states of proteins. Reactions of doubly protonated alkyldiamines with NH_3: solvation or deprotonation. Extension of two proton cases to multiply protonated globular proteins observed in the gas phase," *Journal of the American Chemical Society*, vol. 124, no. 38, pp. 11519–11530, 2002.

[95] W. Z. Shou and W. Naidong, "Simple means to alleviate sensitivity loss by trifluoroacetic acid (TFA) mobile phases in the hydrophilic interaction chromatography-electrospray tandem mass spectrometric (HILIC-ESI/MS/MS) bioanalysis of basic compounds," *Journal of Chromatography B*, vol. 825, no. 2, pp. 186–192, 2005.

[96] A. T. Iavarone, J. C. Jurchen, and E. R. Williams, "Effects of solvent on the maximum charge state and charge state distribution of protein ions produced by electrospray ionization," *Journal of the American Society for Mass Spectrometry*, vol. 11, no. 11, pp. 976–985, 2000.

[97] R. R. Ogorzalek Loo, B. E. Winger, and R. D. Smith, "Proton transfer reaction studies of multiply charged proteins in a high mass-to-charge ratio quadrupole mass spectrometer," *Journal of the American Society for Mass Spectrometry*, vol. 5, no. 12, pp. 1064–1071, 1994.

[98] B. E. Winger, K. J. Light-Wahl, and S. Richard, "Gas-phase proton transfer reactions involving multiply charged cytochrome c ions and water under thermal conditions," *Journal of the American Society for Mass Spectrometry*, vol. 3, no. 6, pp. 624–630, 1992.

[99] J. D. Carbeck, J. C. Severs, J. Gao, Q. Wu, R. D. Smith, and G. M. Whitesides, "Correlation between the charge of proteins in solution and in the gas phase investigated by protein charge ladders, capillary electrophoresis, and electrospray ionization mass spectrometry," *Journal of Physical Chemistry B*, vol. 102, no. 51, pp. 10596–10601, 1998.

[100] A. T. Iavarone and E. R. Williams, "Mechanism of charging and supercharging molecules in electrospray ionization," *Journal of the American Chemical Society*, vol. 125, no. 8, pp. 2319–2327, 2003.

[101] A. T. Iavarone, J. C. Jurchen, and E. R. Williams, "Supercharged protein and peptide ions formed by electrospray ionization," *Analytical Chemistry*, vol. 73, no. 7, pp. 1455–1460, 2001.

[102] J. A. Loo, R. R. Loo, H. R. Udseth, C. G. Edmonds, and R. D. Smith, "Solvent-induced conformational changes of polypeptides probed by electrospray-ionization mass spectrometry," *Rapid Communications in Mass Spectrometry*, vol. 5, no. 3, pp. 101–105, 1991.

[103] R. B. Cole and A. K. Harrata, "Solvent effect on analyte charge state, signal intensity, and stability in negative ion electrospray mass spectrometry; implications for the mechanism of negative ion formation," *Journal of the American Society for Mass Spectrometry*, vol. 4, pp. 546–556, 1993.

[104] H. J. Sterling and E. R. Williams, "Origin of supercharging in electrospray ionization of noncovalent complexes from aqueous solution," *Journal of the American Society for Mass Spectrometry*, vol. 20, no. 10, pp. 1933–1943, 2009.

[105] H. J. Sterling, M. P. Daly, G. K. Feld et al., "Effects of supercharging reagents on noncovalent complex structure in electrospray ionization from aqueous solutions," *Journal of the American Society for Mass Spectrometry*, vol. 21, no. 10, pp. 1762–1774, 2010.

[106] S. H. Lomeli, I. X. Peng, S. Yin, R. R. Ogorzalek Loo, and J. A. Loo, "New reagents for increasing ESI multiple charging of proteins and protein complexes," *Journal of the American Society for Mass Spectrometry*, vol. 21, no. 1, pp. 127–131, 2010.

[107] D. F. Hunt, J. R. Yates III, and J. Shabanowitz, "Protein sequencing by tandem mass spectrometry," *Proceedings of the National Academy of Sciences of the United States of America*, vol. 83, no. 17, pp. 6233–6237, 1986.

[108] J. A. Loo, J. P. Quinn, S. I. Ryu, K. D. Henry, M. W. Senko, and F. W. McLafferty, "High-resolution tandem mass spectrometry of large biomolecules," *Proceedings of the National Academy of Sciences of the United States of America*, vol. 89, no. 1, pp. 286–289, 1992.

[109] A. T. Iavarone and E. R. Williams, "Collisionally activated dissociation of supercharged proteins formed by electrospray ionization," *Analytical Chemistry*, vol. 75, no. 17, pp. 4525–4533, 2003.

[110] X. Han, M. Jin, K. Breuker, and F. W. McLafferty, "Extending top-down mass spectrometry to proteins with masses great than 200 kilodaltons," *Science*, vol. 314, no. 5796, pp. 109–112, 2006.

[111] S. A. McLuckey and J. L. Stephenson, "Ion/ion chemistry of high-mass multiply charged ions," *Mass Spectrometry Reviews*, vol. 17, no. 6, pp. 369–407, 1998.

[112] S. J. Pitteri and S. A. McLuckey, "Recent developments in the ion/ion chemistry of high-mass multiply charged ions," *Mass Spectrometry Reviews*, vol. 24, no. 6, pp. 931–958, 2005.

[113] E. R. Williams, "Proton transfer reactivity of large multiply charged ions," *Journal of Mass Spectrometry*, vol. 31, no. 8, pp. 831–842, 1996.

[114] A. Kharlamova and S. A. McLuckey, "Negative electrospray droplet exposure to gaseous bases for the manipulation of protein charge state distributions," *Analytical Chemistry*, vol. 83, no. 1, pp. 431–437, 2011.

[115] A. Kharlamova, B. M. Prentice, T. Y. Huang, and S. A. McLuckey, "Electrospray droplet exposure to gaseous acids for the manipulation of protein charge state distributions," *Analytical Chemistry*, vol. 82, no. 17, pp. 7422–7429, 2010.

[116] C. J. Krusemark, B. L. Frey, P. J. Belshaw, and L. M. Smith, "Modifying the charge state distribution of proteins in electrospray ionization mass spectrometry by chemical derivatization," *Journal of the American Society for Mass Spectrometry*, vol. 20, no. 9, pp. 1617–1625, 2009.

[117] B. L. Frey, C. J. Krusemark, A. R. Ledvina, J. J. Coon, P. J. Belshaw, and L. M. Smith, "Ion-ion reactions with fixed-charge modified proteins to produce ions in a single, very high charge state," *International Journal of Mass Spectrometry*, vol. 276, no. 2-3, pp. 136–143, 2008.

[118] L. Tang and P. Kebarle, "Dependence of ion intensity in electrospray mass spectrometry on the concentration of the analytes in the electrosprayed solution," *Analytical Chemistry*, vol. 65, no. 24, pp. 3654–3667, 1993.

[119] G. Wang and R. B. Cole, "Mechanistic interpretation of the dependence of charge state distributions on analyte concentrations in electrospray ionization mass spectrometry," *Analytical Chemistry*, vol. 67, no. 17, pp. 2892–2900, 1995.

[120] L. Tang and P. Kebarle, "Effect of the conductivity of the electrosprayed solution on the electrospray current. Factors determining analyte sensitivity in electrospray mass spectrometry," *Analytical Chemistry*, vol. 63, no. 23, pp. 2709–2715, 1991.

[121] C. G. Enke, "A predictive model for matrix and analyte effects in electrospray ionization of singly-charged ionic analytes," *Analytical Chemistry*, vol. 69, no. 23, pp. 4885–4893, 1997.

[122] T. L. Constantopoulos, G. S. Jackson, and C. G. Enke, "Effects of salt concentration on analyte response using electrospray ionization mass spectrometry," *Journal of the American Society for Mass Spectrometry*, vol. 10, no. 7, pp. 625–634, 1999.

[123] Y. Li and R. B. Cole, "Shifts in peptide and protein charge state distributions with varying spray tip orifice diameter in nanoelectrospray fourier transform ion cycltron resonance mass spectrometry," *Analytical Chemistry*, vol. 75, no. 21, pp. 5739–5746, 2003.

[124] W. M. A. Niessen, "State-of-the-art in liquid chromatography–mass spectrometry," *Journal of Chromatography A*, vol. 856, pp. 179–197, 1999.

[125] R. D. Voyksner and H. Lee, "Improvements in LC/electrospray ion trap mass spectrometry performance using an off-axis nebulizer," *Analytical Chemistry*, vol. 71, no. 7, pp. 1441–1447, 1999.

[126] V. Gabelica, E. De Pauw, and M. Karas, "Influence of the capillary temperature and the source pressure on the internal energy distribution of electrosprayed ions," *International Journal of Mass Spectrometry*, vol. 231, no. 2-3, pp. 189–195, 2004.

[127] M. Busman, A. L. Rockwood, and R. D. Smith, "Activation energies for gas-phase dissociations of multiply charged ions from electrospray ionization mass spectrometry," *Journal of Physical Chemistry*, vol. 96, no. 6, pp. 2397–2400, 1992.
[128] M. J. Van Stipdonk, M. P. Ince, B. A. Perera, and J. A. Martin, "Cluster ions derived from sodium and potassium tetrafluoroborate and their collision induced dissociation in an ion trap mass spectrometer," *Rapid Communications in Mass Spectrometry*, vol. 16, no. 5, pp. 355–363, 2002.
[129] R. D. Smith, J. A. Loo, C. J. Barinaga, C. G. Edmonds, and H. R. Udseth, "Collisional activation and collision-activated dissociation of large multiply charged polypeptides and proteins produced by electrospray ionization," *Journal of the American Society for Mass Spectrometry*, vol. 1, no. 1, pp. 53–65, 1990.
[130] B. A. Thomson, "Declustering and fragmentation of protein ions from an electrospray ion source," *Journal of the American Society for Mass Spectrometry*, vol. 8, no. 10, pp. 1053–1058, 1997.
[131] D. S. Ashton, C. R. Beddell, D. J. Cooper, B. N. Green, and R. W. A. Oliver, "Mechanism of production of ions in electrospray mass spectrometry," *Organic Mass Spectrometry*, vol. 28, pp. 721–728, 1993.
[132] R. D. Smith, K. J. Light-Wahl, B. E. Winger, and J. A. Loo, "Preservation of non-covalent associations in electrospray ionization mass spectrometry: multiply charged polypeptide and protein dimers," *Organic Mass Spectrometry*, vol. 27, pp. 811–821, 1992.
[133] G. J. Van Berkel, F. Zhou, and J. T. Aronson, "Changes in bulk solution pH caused by the inherent controlled-current electrolytic process of an electrospray ion source," *International Journal of Mass Spectrometry and Ion Processes*, vol. 162, no. 1–3, pp. 55–67, 1997.
[134] S. Zhou, A. G. Edwards, K. D. Cook, and G. J. Van Berkel, "Investigation of the electrospray plume by laser-induced fluorescence spectroscopy," *Analytical Chemistry*, vol. 71, no. 4, pp. 769–776, 1999.
[135] C. L. Gatlin and F. Tureček, "Acidity determination in droplets formed by electrospraying methanol-water solutions," *Analytical Chemistry*, vol. 66, no. 5, pp. 712–718, 1994.
[136] S. E. Rodriguez-Cruz, J. T. Khoury, and J. H. Parks, "Protein fluorescence measurements within electrospray droplets," *Journal of the American Society for Mass Spectrometry*, vol. 12, no. 6, pp. 716–725, 2001.
[137] K. Breuker and F. W. McLafferty, "Stepwise evolution of protein native structure with electrospray into the gas phase, 10^{-12} to 10^{2} s," *Proceedings of the National Academy of Sciences of the United States of America*, vol. 105, no. 47, pp. 18145–18152, 2008.
[138] T. D. Wood, R. A. Chorush, F. M. Wampler, D. P. Little, P. B. O'Connor, and F. W. McLafferty, "Gas-phase folding and unfolding of cytochrome c cations," *Proceedings of the National Academy of Sciences of the United States of America*, vol. 92, no. 7, pp. 2451–2454, 1995.
[139] S. Banerjee, H. Prakash, and S. Mazumdar, "Evidence of molecular fragmentation inside the charged droplets produced by electrospray process," *Journal of the American Society for Mass Spectrometry*, vol. 22, no. 10, pp. 1707–1717, 2011.
[140] A. M. Gañán-Calvo, "The surface charge in electrospraying: its nature and its universal scaling laws," *Journal of Aerosol Science*, vol. 30, no. 7, pp. 863–872, 1999.
[141] *Oxford English Dictionary*, Oxford University Press, 2nd edition, 1989.
[142] F. W. McLafferty, *Tandem Mass Spectrometry*, John Wiley & Sons, New York, NY, USA, 1983.
[143] S. A. McLuckey, "Principles of collisional activation in analytical mass spectrometry," *Journal of the American Society for Mass Spectrometry*, vol. 3, no. 6, pp. 599–614, 1992.
[144] A. L. McCormack, J. L. Jones, and V. H. Wysocki, "Surface-induced dissociation of multiply protonated peptides," *Journal of the American Society for Mass Spectrometry*, vol. 3, no. 8, pp. 859–862, 1992.
[145] R. A. Chorush, D. P. Little, S. C. Beu, T. D. Wood, and F. W. McLafferty, "Surface-induced dissociation of multiply-protonated proteins," *Analytical Chemistry*, vol. 67, no. 6, pp. 1042–1046, 1995.
[146] D. P. Little, J. P. Speir, M. W. Senko, P. B. O'Connor, and F. W. McLafferty, "Infrared multiphoton dissociation of large multiply charged ions for biomolecule sequencing," *Analytical Chemistry*, vol. 66, no. 18, pp. 2809–2815, 1994.
[147] W. D. Price, P. D. Schnier, and E. R. Williams, "Tandem mass spectrometry of large biomolecule ions by blackbody infrared radiative dissociation," *Analytical Chemistry*, vol. 68, no. 5, pp. 859–866, 1996.
[148] R. C. Dunbar and T. B. McMahon, "Activation of unimolecular reactions by ambient blackbody radiation," *Science*, vol. 279, no. 5348, pp. 194–197, 1998.
[149] M. S. Thompson, W. Cui, and J. P. Reilly, "Fragmentation of singly charged peptide ions by photodissociation at λ = 157 nm," *Angewandte Chemie*, vol. 43, no. 36, pp. 4791–4794, 2004.
[150] Z. Guan, N. L. Kelleher, P. B. O'Connor, D. J. Aaserud, D. P. Little, and F. W. McLafferty, "193 nm photodissociation of larger multiply-charged biomolecules," *International Journal of Mass Spectrometry and Ion Processes*, vol. 157-158, pp. 357–364, 1996.
[151] R. A. Zubarev, N. L. Kelleher, and F. W. McLafferty, "Electron capture dissociation of multiply charged protein cations. A nonergodic process," *Journal of the American Chemical Society*, vol. 120, no. 13, pp. 3265–3266, 1998.
[152] J. E. P. Syka, J. J. Coon, M. J. Schroeder, J. Shabanowitz, and D. F. Hunt, "Peptide and protein sequence analysis by electron transfer dissociation mass spectrometry," *Proceedings of the National Academy of Sciences of the United States of America*, vol. 101, no. 26, pp. 9528–9533, 2004.
[153] B. A. Budnik, K. F. Haselmann, and R. A. Zubarev, "Electron detachment dissociation of peptide di-anions: an electron-hole recombination phenomenon," *Chemical Physics Letters*, vol. 342, no. 3-4, pp. 299–302, 2001.
[154] R. B. Cody, "Electron impact excitation of ions from organics: an alternative to collision induced dissociation," *Analytical Chemistry*, vol. 51, no. 4, pp. 547–551, 1979.
[155] R. W. Vachet and G. L. Glish, "Effects of heavy gases on the tandem mass spectra of peptide ions in the quadrupole ion trap," *Journal of the American Society for Mass Spectrometry*, vol. 7, no. 12, pp. 1194–1202, 1996.
[156] R. Boyd and A. Somogyi, "The mobile proton hypothesis in fragmentation of protonated peptides: a perspective," *Journal of the American Society for Mass Spectrometry*, vol. 21, pp. 1275–1278, 2010.
[157] X. J. Tang, P. Thibault, and R. K. Boyd, "Fragmentation reactions of multiply-protonated peptides and implications for sequencing by tandem mass spectrometry with low-energy collision-induced dissociation," *Analytical Chemistry*, vol. 65, no. 20, pp. 2824–2834, 1993.

[158] R. S. Bordoli and R. H. Bateman, "The effect of collision energy, target gas and target gas purity on the high energy collision induced product ion spectrum of renin substrate," *International Journal of Mass Spectrometry and Ion Processes*, vol. 122, pp. 243–254, 1992.

[159] W. J. Griffiths, A. P. Jonsson, S. Liu, D. K. Rai, and Y. Wang, "Electrospray and tandem mass spectrometry in biochemistry," *Biochemical Journal*, vol. 355, no. 3, pp. 545–561, 2001.

[160] M. Kinter and N. E. Sherman, *Protein Sequencing and Identification Using Tandem Mass Spectrometry*, Wiley-Interscience, New York, NY, USA, 2000.

[161] V. H. Wysocki, G. Cheng, Q. Zhang, K. A. Herrmann, R. L. Beardsley, and A. E. Hilderbrand, in *Principles of Mass Spectrometry Applied to Biomolecules*, pp. 277–300, John Wiley & Sons, 2006.

[162] M. L. Gross, "Charge-remote fragmentations: method, mechanism and applications," *International Journal of Mass Spectrometry and Ion Processes*, vol. 118-119, pp. 137–165, 1992.

[163] K. Biemann and A. M. James, in *Methods in Enzymology*, vol. 193, pp. 455–479, Academic Press, 1990.

[164] B. Paizs and S. Suhai, "Towards understanding the tandem mass spectra of protonated oligopeptides. 1: mechanism of amide bond cleavage," *Journal of the American Society for Mass Spectrometry*, vol. 15, no. 1, pp. 103–113, 2004.

[165] H. H. Hill, C. H. Hill, G. R. Asbury, C. Wu, L. M. Matz, and T. Ichiye, "Charge location on gas phase peptides," *International Journal of Mass Spectrometry*, vol. 219, no. 1, pp. 23–37, 2002.

[166] C. Afonso, F. Modeste, P. Breton, F. Fournier, and J. C. Tabet, "Proton affinities of the commonly occuring L-amino acids by using electrospray ionization-ion trap mass spectrometry," *European Journal of Mass Spectrometry*, vol. 6, no. 5, pp. 443–449, 2000.

[167] C. Bleiholder, S. Suhai, and B. Paizs, "Revising the proton affinity scale of the naturally occurring α-amino acids," *Journal of the American Society for Mass Spectrometry*, vol. 17, no. 9, pp. 1275–1281, 2006.

[168] V. H. Wysocki, G. Tsaprailis, L. L. Smith, and L. A. Breci, "Mobile and localized protons: a framework for understanding peptide dissociation," *Journal of Mass Spectrometry*, vol. 35, no. 12, pp. 1399–1406, 2000.

[169] A. R. Dongre, J. L. Jones, A. Somogyi, and V. H. Wysocki, "Influence of peptide composition, gas-phase basicity, and chemical modification on fragmentation efficiency: evidence for the mobile proton model," *Journal of the American Chemical Society*, vol. 118, no. 35, pp. 8365–8374, 1996.

[170] A. G. Harrison and T. Yalcin, "Proton mobility in protonated amino acids and peptides," *International Journal of Mass Spectrometry and Ion Processes*, vol. 165-166, pp. 339–347, 1997.

[171] I. P. Csonka, B. Paizs, G. Lendvay, and S. Suhai, "Proton mobility in protonated peptides: a joint molecular orbital and RRKM study," *Rapid Communications in Mass Spectrometry*, vol. 14, no. 6, pp. 417–431, 2000.

[172] B. Paizs and S. Suhai, "Theoretical study of the main fragmentation pathways for protonated glycylglycine," *Rapid Communications in Mass Spectrometry*, vol. 15, no. 8, pp. 651–663, 2001.

[173] B. Palzs and S. Suhal, "Fragmentation pathways of protonated peptides," *Mass Spectrometry Reviews*, vol. 24, no. 4, pp. 508–548, 2005.

[174] P. Roepstorff and J. Fohlman, "Proposal for a common nomenclature for sequence ions in mass spectra of peptides," *Biomedical Mass Spectrometry*, vol. 11, no. 11, p. 601, 1984.

[175] K. Biemann, "Contributions of mass spectrometry to peptide and protein structure," *Biomedical and Environmental Mass Spectrometry*, vol. 16, no. 1–12, pp. 99–111, 1988.

[176] K. D. Ballard and S. J. Gaskell, "Sequential mass spectrometry applied to the study of the formation of "internal" fragment ions of protonated peptides," *International Journal of Mass Spectrometry and Ion Processes*, vol. 111, no. C, pp. 173–189, 1991.

[177] K. Ambihapathy, T. Yalcin, H. W. Leung, and A. G. Harrison, "Pathways to immonium ions in the fragmentation of protonated peptides," *Journal of Mass Spectrometry*, vol. 32, no. 2, pp. 209–215, 1997.

[178] S. Sun, C. Yu, Y. Qiao et al., "Deriving the probabilities of water loss and ammonia loss for amino acids from tandem mass spectra," *Journal of Proteome Research*, vol. 7, no. 1, pp. 202–208, 2008.

[179] J. E. McClellan, S. T. Quarmby, and R. A. Yost, "Parent and neutral loss monitoring on a quadrupole ion trap mass spectrometer: screening of acylcarnitines in complex mixtures," *Analytical Chemistry*, vol. 74, no. 22, pp. 5799–5806, 2002.

[180] A. G. Harrison, A. B. Young, C. Bleiholder, S. Suhai, and B. Paizs, "Scrambling of sequence information in collision-induced dissociation of peptides," *Journal of the American Chemical Society*, vol. 128, no. 32, pp. 10364–10365, 2006.

[181] A. G. Harrison and A. B. Young, "Fragmentation of protonated oligoalanines: amide bond cleavage and beyond," *Journal of the American Society for Mass Spectrometry*, vol. 15, no. 12, pp. 1810–1819, 2004.

[182] T. Yalcin, C. Khouw, I. G. Csizmadia, M. R. Peterson, and A. G. Harrison, "Why are B ions stable species in peptide spectra?" *Journal of the American Society for Mass Spectrometry*, vol. 6, no. 12, pp. 1165–1174, 1995.

[183] H. Nair, A. Somogyi, and V. H. Wysocki, "Effect of alkyl substitution at the amide nitrogen on amide bond cleavage: electrospray ionization/surface-induced dissociation fragmentation of substance P and two alkylated analogs," *Journal of Mass Spectrometry*, vol. 31, no. 10, pp. 1141–1148, 1996.

[184] R. W. Vachet, B. M. Bishop, B. W. Erickson, and G. L. Glish, "Novel peptide dissociation: gas-phase intramolecular rearrangement of internal amino acid residues," *Journal of the American Chemical Society*, vol. 119, no. 24, pp. 5481–5488, 1997.

[185] J. V. Olsen and M. Mann, "Improved peptide identification in proteomics by two consecutive stages of mass spectrometric fragmentation," *Proceedings of the National Academy of Sciences of the United States of America*, vol. 101, no. 37, pp. 13417–13422, 2004.

[186] Tang Xue Jun and R. K. Boyd, "Rearrangements of doubly charged acylium ions from lysyl and ornithyl peptides," *Rapid Communications in Mass Spectrometry*, vol. 8, no. 9, pp. 678–686, 1994.

[187] J. Yaqüe, A. Paradela, M. Ramos et al., "Peptide rearrangement during quadrupole ion trap fragmentation: added complexity to MS/MS spectra," *Analytical Chemistry*, vol. 75, no. 6, pp. 1524–1535, 2003.

[188] C. Bleiholder, S. Osburn, T. D. Williams et al., "Sequence-scrambling fragmentation pathways of protonated peptides," *Journal of the American Chemical Society*, vol. 130, no. 52, pp. 17774–17789, 2008.

[189] C. Jia, W. Qi, and Z. He, "Cyclization reaction of peptide fragment Ions during multistage collisionally activated decomposition: an inducement to lose internal amino-acid residues," *Journal of the American Society for Mass Spectrometry*, vol. 18, no. 4, pp. 663–678, 2007.

[190] A. G. Harrison, "Peptide sequence scrambling through cyclization of b5 Ions," *Journal of the American Society for Mass Spectrometry*, vol. 19, no. 12, pp. 1776–1780, 2008.

[191] S. Molesworth, S. Osburn, and M. Van Stipdonk, "Influence of amino acid side chains on apparent selective opening of cyclic b5 ions," *Journal of the American Society for Mass Spectrometry*, vol. 21, no. 6, pp. 1028–1036, 2010.

[192] L. Mouls, J. L. Aubagnac, J. Martinez, and C. Enjalbal, "Low energy peptide fragmentations in an ESI-Q-Tof type mass spectrometer," *Journal of Proteome Research*, vol. 6, no. 4, pp. 1378–1391, 2007.

[193] L. Yu, Y. Tan, Y. Tsai, D. R. Goodlett, and N. C. Polfer, "On the relevance of peptide sequence permutations in shotgun proteomics studies," *Journal of Proteome Research*, vol. 10, no. 5, pp. 2409–2416, 2011.

[194] A. A. Goloborodko, M. V. Gorshkov, D. M. Good, and R. A. Zubarev, "Sequence scrambling in shotgun proteomics is negligible," *Journal of the American Society for Mass Spectrometry*, vol. 22, no. 7, pp. 1121–1124, 2011.

[195] D. Pu and C. J. Cassady, "Negative ion dissociation of peptides containing hydroxyl side chains," *Rapid Communications in Mass Spectrometry*, vol. 22, no. 2, pp. 91–100, 2008.

[196] J. H. Bowie, C. S. Brinkworth, and S. Dua, "Collision-induced fragmentations of the (M-H)-parent anions of underivatized peptides: an aid to structure determination and some unusual negative ion cleavages," *Mass Spectrometry Reviews*, vol. 21, no. 2, pp. 87–107, 2002.

[197] A. G. Harrison and A. B. Young, "Fragmentation of deprotonated N-benzoylpeptides: formation of deprotonated oxazolones," *Journal of the American Society for Mass Spectrometry*, vol. 15, no. 4, pp. 446–456, 2004.

[198] N. L. Clipston, J. Jai-nhuknan, and C. J. Cassady, "A comparison of negative and positive ion time-of-flight post-source decay mass spectrometry for peptides containing basic residues," *International Journal of Mass Spectrometry*, vol. 222, no. 1–3, pp. 363–381, 2003.

[199] A. L. McCormack, A. Somogyi, A. R. Dongre, and V. H. Wysocki, "Fragmentation of protonated peptides: surface-induced dissociation in conjunction with a quantum mechanical approach," *Analytical Chemistry*, vol. 65, no. 20, pp. 2859–2872, 1993.

[200] G. E. Reid, R. J. Simpson, and R. A. J. O'Hair, "Leaving group and gas phase neighboring group effects in the side chain losses from protonated serine and its derivatives," *Journal of the American Society for Mass Spectrometry*, vol. 11, no. 12, pp. 1047–1060, 2000.

[201] R. J. Ferrier, *Carbohydrate Chemistry*, vol. 33, Royal Society of Chemistry, Cambridge, UK, 2002.

[202] B. Domon and C. E. Costello, "A systematic nomenclature for carbohydrate fragmentations in FAB-MS/MS spectra of glycoconjugates," *Glycoconjugate Journal*, vol. 5, no. 4, pp. 397–409, 1988.

[203] J. Lemoine, B. Fournet, D. Despeyroux, K. R. Jennings, R. Rosenberg, and E. de Hoffmann, "Collision-induced dissociation of alkali metal cationized and permethylated oligosaccharides: influence of the collision energy and of the collision gas for the assignment of linkage position," *Journal of the American Society for Mass Spectrometry*, vol. 4, no. 3, pp. 197–203, 1993.

[204] D. J. Harvey, "Collision-induced fragmentation of underivatized N-linked carbohydrates ionized by electrospray," *Journal of Mass Spectrometry*, vol. 35, no. 10, pp. 1178–1190, 2000.

[205] A. Pfenninger, M. Karas, B. Finke, and B. Stahl, "Structural analysis of underivatized neutral human milk oligosaccharides in the negative ion mode by nano-electrospray M^{Sn} (Part 1: methodology)," *Journal of the American Society for Mass Spectrometry*, vol. 13, no. 11, pp. 1331–1340, 2002.

[206] W. Chai, V. Piskarev, and A. M. Lawson, "Negative-ion electrospray mass spectrometry of neutral underivatized oligosaccharides," *Analytical Chemistry*, vol. 73, no. 3, pp. 651–657, 2001.

[207] D. Garozzo, M. Giuffrida, G. Impallomeni, A. Ballistreri, and G. Mon taudo, "Determination of linkage position and identification of the reducing end in linear oligosaccharides by negative ion fast atom bombardment mass spectrometry," *Analytical Chemistry*, vol. 62, no. 3, pp. 279–286, 1990.

[208] S. A. McLuckey, G. J. Van Berker, and G. L. Glish, "Tandem mass spectrometry of small, multiply charged oligonucleotides," *Journal of the American Society for Mass Spectrometry*, vol. 3, no. 1, pp. 60–70, 1992.

[209] E. Nordhoff, M. Karas, R. Cramer et al., "Direct mass spectrometric sequencing of low-picomole amounts of oligodeoxynucleotides with up to 21 bases by matrix-assisted laser desorption/ionization mass spectrometry," *Journal of Mass Spectrometry*, vol. 30, no. 1, pp. 99–112, 1995.

[210] J. P. Barry, P. Vouros, A. Van Schepdael, and S. J. Law, "Mass and sequence verification of modified oligonucleotides using electrospray tandem mass spectrometry," *Journal of Mass Spectrometry*, vol. 30, no. 7, pp. 993–1006, 1995.

[211] M. G. Bartlett, J. A. McCloskey, S. Manalili, and R. H. Griffey, "The effect of backbone charge on the collision-induced dissociation of oligonucleotides," *Journal of Mass Spectrometry*, vol. 31, no. 11, pp. 1277–1283, 1996.

[212] S. A. McLuckey and S. Habibi-Goudarzi, "Decompositions of multiply charged oligonucleotide anions," *Journal of the American Chemical Society*, vol. 115, no. 25, pp. 12085–12095, 1993.

[213] R. L. Hettich and E. A. Stemmler, "Investigation of oligonucleotide fragmentation with matrix-assisted laser desorption/ionization fourier-transform mass spectrometry and sustained off-resonance irradiation," *Rapid Communications in Mass Spectrometry*, vol. 10, no. 3, pp. 321–327, 1996.

[214] W. J. Griffiths, Y. Yang, J. Å. Lindgren, and J. Sjövall, "Charge remote fragmentation of fatty acid anions in 400 eV collisions with xenon atoms," *Rapid Communications in Mass Spectrometry*, vol. 10, no. 1, pp. 21–28, 1996.

[215] E. Fahy, S. Subramaniam, H. A. Brown et al., "A comprehensive classification system for lipids," *Journal of Lipid Research*, vol. 46, no. 5, pp. 839–861, 2005.

[216] K. B. Tomer, F. W. Crow, and M. L. Gross, "Location of double bond position in unsaturated fatty acids by negative ion MS/MS," *Journal of the American Chemical Society*, vol. 105, no. 16, pp. 5487–5488, 1983.

[217] W. J. Griffiths, "Tandem mass spectrometry in the study of fatty acids, bile acids, and steroids," *Mass Spectrometry Reviews*, vol. 22, no. 2, pp. 81–152, 2003.

[218] V. H. Wysocki and M. M. Ross, "Charge-remote fragmentation of gas-phase ions: mechanistic and energetic considerations in the dissociation of long-chain functionalized alkanes and alkenes," *International Journal of Mass Spectrometry and Ion Processes*, vol. 104, no. 3, pp. 179–211, 1991.

[219] W. J. Griffiths, A. Brown, R. Reimendal, Y. Yang, J. Zhang, and J. Sjövall, "A comparison of fast-atom bombardment and electrospray as methods of ionization in the study of

sulphated- and sulphonated-lipids by tandem mass spectrometry," *Rapid Communications in Mass Spectrometry*, vol. 10, no. 10, pp. 1169–1174, 1996.

[220] P. Roepstorff, "Mass spectrometry in protein studies from genome to function," *Current Opinion in Biotechnology*, vol. 8, no. 1, pp. 6–13, 1997.

[221] J. R. Yates, "Mass spectrometry and the age of the proteome," *Journal of Mass Spectrometry*, vol. 33, no. 1, pp. 1–19, 1998.

[222] D. J. C. Pappin, P. Hojrup, and A. J. Bleasby, "Rapid identification of proteins by peptide-mass fingerprinting," *Current Biology*, vol. 3, no. 6, pp. 327–332, 1993.

[223] B. M. Mayr, O. Kohlbacher, K. Reinert et al., "Absolute myoglobin quantitation in serum by combining two-dimensional liquid chromatography-electrospray ionization mass spectrometry and novel data analysis algorithms," *Journal of Proteome Research*, vol. 5, no. 2, pp. 414–421, 2006.

[224] Y. Oda, K. Huang, F. R. Cross, D. Cowburn, and B. T. Chait, "Accurate quantitation of protein expression and site-specific phosphorylation," *Proceedings of the National Academy of Sciences of the United States of America*, vol. 96, no. 12, pp. 6591–6596, 1999.

[225] J. A. Loo, "Electrospray ionization mass spectrometry: a technology for studying noncovalent macromolecular complexes," *International Journal of Mass Spectrometry*, vol. 200, no. 1–3, pp. 175–186, 2000.

[226] E. T. J. Van den Bremer, W. Jiskoot, R. James et al., "Probing metal ion binding and conformational properties of the colicin E9 endonuclease by electrospray ionization time-of-flight mass spectrometry," *Protein Science*, vol. 11, no. 7, pp. 1738–1752, 2002.

[227] S. K. Chowdhury, V. Katta, and B. T. Chait, "Probing conformational changes in proteins by mass spectrometry," *Journal of the American Chemical Society*, vol. 112, no. 24, pp. 9012–9013, 1990.

[228] J. S. Andersen, B. Svensson, and P. Roepstorff, "Electrospray ionization and matrix assisted laser desorption/ionization mass spectrometry: powerful analytical tools in recombinant protein chemistry," *Nature Biotechnology*, vol. 14, no. 4, pp. 449–457, 1996.

[229] H. D. Niall and S. N. T. C. H. W. Hirs, in *Methods in Enzymology*, vol. 27, pp. 942–1010, Academic Press, 1973.

[230] J. A. Loo, C. G. Edmonds, and R. D. Smith, "Primary sequence information from intact proteins by electrospray ionization tandem mass spectrometry," *Science*, vol. 248, no. 4952, pp. 201–204, 1990.

[231] K. Biemann, "Sequencing of peptides by tandem mass spectrometry and high-energy collision-induced dissociation," *Methods in Enzymology*, vol. 193, pp. 455–479, 1990.

[232] J. A. Taylor and R. S. Johnson, "Sequence database searches via de Novo peptide sequencing by tandem mass spectrometry," *Rapid Communications in Mass Spectrometry*, vol. 11, no. 9, pp. 1067–1075, 1997.

[233] A. Armirotti, "Bottom-up proteomics," *Current Analytical Chemistry*, vol. 5, no. 2, pp. 116–130, 2009.

[234] H. Zhong, Y. Zhang, Z. Wen, and L. Li, "Protein sequencing by mass analysis of polypeptide ladders after controlled protein hydrolysis," *Nature Biotechnology*, vol. 22, no. 10, pp. 1291–1296, 2004.

[235] G. E. Reid and S. A. McLuckey, ""Top down" protein characterization via tandem mass spectrometry," *Journal of Mass Spectrometry*, vol. 37, no. 7, pp. 663–675, 2002.

[236] M. Mann and O. N. Jensen, "Proteomic analysis of post-translational modifications," *Nature Biotechnology*, vol. 21, no. 3, pp. 255–261, 2003.

[237] M. A. Freitas, A. R. Sklenar, and M. R. Parthun, "Application of mass spectrometry to the identification and quantification of histone post-translational modifications," *Journal of Cellular Biochemistry*, vol. 92, no. 4, pp. 691–700, 2004.

[238] H. Oberacher, C. G. Huber, and P. J. Oefner, "Mutation scanning by ion-pair reversed-phase high-performance liquid chromatography-electrospray ionization mass spectrometry (ICEMS)," *Human Mutation*, vol. 21, no. 1, pp. 86–95, 2003.

[239] L. Zhang, E. E. Eugeni, M. R. Parthun, and M. A. Freitas, "Identification of novel histone post-translational modifications by peptide mass fingerprinting," *Chromosoma*, vol. 112, no. 2, pp. 77–86, 2003.

[240] H. Prakash and S. Mazumdar, "Succinylation of cytochrome c investigated by electrospray ionization mass spectrometry: reactive lysine residues," *International Journal of Mass Spectrometry*, vol. 281, no. 1-2, pp. 55–62, 2009.

[241] S. Goyal, M. S. Deshpande, and S. Mazumdar, "Structural design of the active site for covalent attachment of the heme to the protein matrix: atudies on a thermostable cytochrome P450," *Biochemistry*, vol. 50, no. 6, pp. 1042–1052, 2011.

[242] T. Y. Yen, H. Yan, and B. A. Macher, "Characterizing closely spaced, complex disulfide bond patterns in peptides and proteins by liquid chromatography/electrospray ionization tandem mass spectrometry," *Journal of Mass Spectrometry*, vol. 37, no. 1, pp. 15–30, 2002.

[243] D. L. Smith, Z. Zhou, and A. M. James, in *Methods in Enzymology*, vol. 193, pp. 1296–1291, Academic Press, 1990.

[244] M. Scigelova, P. S. Green, A. E. Giannakopulos et al., "A practical protocol for the reduction of disulfide bonds in proteins prior to analysis by mass spectrometry," *European Journal of Mass Spectrometry*, vol. 7, pp. 29–34, 2001.

[245] S. Y. Gauthier, C. M. Kay, B. D. Sykes, V. K. Walker, and P. L. Davies, "Disulfide bond mapping and structural characterization of spruce budworm antifreeze protein," *European Journal of Biochemistry*, vol. 258, no. 2, pp. 445–453, 1998.

[246] M. Zhang and I. A. Kaltashov, "Mapping of protein disulfide bonds using negative ion fragmentation with a broadband precursor selection," *Analytical Chemistry*, vol. 78, no. 14, pp. 4820–4829, 2006.

[247] J. Wu, "Disulfide bond mapping by cyanylation-induced cleavage and mass spectrometry," *Methods in Molecular Biology*, vol. 446, pp. 1–20, 2008.

[248] R. D. Smith and K. J. Light-Wahl, "The observation of non-covalent interactions in solution by electrospray ionization mass spectrometry: promise, pitfalls and prognosis," *Biological Mass Spectrometry*, vol. 22, no. 9, pp. 493–501, 1993.

[249] T. D. Veenstra, "Electrospray ionization mass spectrometry in the study of biomolecular non-covalent interactions," *Biophysical Chemistry*, vol. 79, no. 2, pp. 63–79, 1999.

[250] B. N. Pramanik, P. L. Bartner, U. A. Mirza, Y. H. Liu, and A. K. Ganguly, "Electrospray ionization mass spectrometry for the study of non-covalent complexes: an emerging technology," *Journal of Mass Spectrometry*, vol. 33, no. 10, pp. 911–920, 1998.

[251] S. Y. Sheu, D. Y. Yang, H. L. Selzle, and E. W. Schlag, "Energetics of hydrogen bonds in peptides," *Proceedings of the National Academy of Sciences of the United States of America*, vol. 100, no. 22, pp. 12683–12687, 2003.

[252] M. F. Jarrold, "Peptides and proteins in the vapor phase," *Annual Review of Physical Chemistry*, vol. 51, pp. 179–207, 2000.

[253] S. Banerjee and S. Mazumdarc, "Non-covalent dimers of the lysine containing protonated peptide ions in gaseous state: electrospray ionizationmass spectrometric study," *Journal of Mass Spectrometry*, vol. 45, no. 10, pp. 1212–1219, 2010.

[254] E. N. Kitova, M. Seo, P. N. Roy, and J. S. Klassen, "Elucidating the intermolecular interactions within a desolvated protein-ligand complex. An experimental and computational study," *Journal of the American Chemical Society*, vol. 130, no. 4, pp. 1214–1226, 2008.
[255] B. Qiu, J. Liu, Z. Qin, G. Wang, and H. Luo, "Quintets of uracil and thymine: a novel structure of nucleobase self-assembly studied by electrospray ionization mass spectrometry," *Chemical Communications*, no. 20, pp. 2863–2865, 2009.
[256] J. A. Loo, "Studying noncovalent protein complexes by electrospray ionization mass spectrometry," *Mass Spectrometry Reviews*, vol. 16, no. 1, pp. 1–23, 1997.
[257] D. Lafitte, A. J. R. Heck, T. J. Hill, K. Jumel, S. E. Harding, and P. J. Derrick, "Evidence of noncovalent dimerization of calmodulin," *European Journal of Biochemistry*, vol. 261, no. 1, pp. 337–344, 1999.
[258] J. A. Loo, "Probing protein-metal ion interactions by electrospray ionization mass spectrometry: enolase and nucleocapsid protein," *International Journal of Mass Spectrometry*, vol. 204, no. 1–3, pp. 113–123, 2001.
[259] K. Benkestock, P. O. Edlund, and J. Roeraade, "Electrospray ionization mass spectrometry as a tool for determination of drug binding sites to human serum albumin by noncovalent interaction," *Rapid Communications in Mass Spectrometry*, vol. 19, no. 12, pp. 1637–1643, 2005.
[260] H. Steen and N. Jensen, "Analysis of protein-nucleic acid interactions by photochemical cross-linking and mass spectrometry," *Mass Spectrometry Reviews*, vol. 21, no. 3, pp. 163–182, 2002.
[261] A. Tjernberg, S. Carnö, F. Oliv et al., "Determination of dissociation constants for protein-ligand complexes by electrospray ionization mass spectrometry," *Analytical Chemistry*, vol. 76, no. 15, pp. 4325–4331, 2004.
[262] L. Liu, D. Bagal, E. N. Kitova, P. D. Schnier, and J. S. Klassen, "Hydrophobic protein-ligand interactions preserved in the gas phase," *Journal of the American Chemical Society*, vol. 131, no. 44, pp. 15980–15981, 2009.
[263] N. P. Barrera, N. Di Bartolo, P. J. Booth, and C. V. Robinson, "Micelles protect membrane complexes from solution to vacuum," *Science*, vol. 321, no. 5886, pp. 243–246, 2008.
[264] M. C. Crowe and J. S. Brodbelt, "Evaluation of noncovalent interactions between peptides and polyether compounds via energy-variable collisionally activated dissociation," *Journal of the American Society for Mass Spectrometry*, vol. 14, no. 10, pp. 1148–1157, 2003.
[265] W. M. David and J. S. Brodbelt, "Threshold dissociation energies of protonated amine/polyether complexes in a quadrupole ion trap," *Journal of the American Society for Mass Spectrometry*, vol. 14, no. 4, pp. 383–392, 2003.
[266] C. S. Ho, C. W. K. Lam, M. H. M. Chan et al., "Electrospray ionisation mass spectrometry: principles and clinical applications," *Clinical Biochemistry Review*, vol. 24, pp. 3–12, 2003.
[267] M. S. Rashed, M. P. Bucknall, D. Little et al., "Screening blood spots for inborn errors of metabolism by electrospray tandem mass spectrometry with a microplate batch process and a computer algorithm for automated flagging of abnormal profiles," *Clinical Chemistry*, vol. 43, no. 7, pp. 1129–1141, 1997.
[268] D. H. Chace, J. E. Sherwin, S. L. Hillman, F. Lorey, and G. C. Cunningham, "Use of phenylalanine-to-tyrosine ratio determined by tandem mass spectrometry to improve newborn screening for phenylketonuria of early discharge specimens collected in the first 24 hours," *Clinical Chemistry*, vol. 44, no. 12, pp. 2405–2409, 1998.
[269] D. H. Chace, J. C. DiPerna, B. L. Mitchell, B. Sgroi, L. F. Hofman, and E. W. Naylor, "Electrospray tandem mass spectrometry for analysis of acylcarnitines in dried postmortem blood specimens collected at autopsy from infants with unexplained cause of death," *Clinical Chemistry*, vol. 47, no. 7, pp. 1166–1182, 2001.
[270] T. Ito, A. B. P. Van Kuilenburg, A. H. Bootsma et al., "Rapid screening of high-risk patients for disorders of purine and pyrimidine metabolism using HPLC-electrospray tandem mass spectrometry of liquid urine or urine-soaked filter paper strips," *Clinical Chemistry*, vol. 46, no. 4, pp. 445–452, 2000.
[271] U. G. Jensen, N. J. Brandt, E. Christensen, F. Skovby, B. Nørgaard-Pedersen, and H. Simonsen, "Neonatal screening for galactosemia by quantitative analysis of hexose monophosphates using tandem mass spectrometry: a retrospective study," *Clinical Chemistry*, vol. 47, no. 8, pp. 1364–1372, 2001.
[272] D. W. Johnson, "A rapid screening procedure for the diagnosis of peroxisomal disorders: quantification of very long-chain fatty acids, as dimethylaminoethyl esters, in plasma and blood spots, by electrospray tandem mass spectrometry," *Journal of Inherited Metabolic Disease*, vol. 23, no. 5, pp. 475–486, 2000.
[273] A. H. Bootsma, H. Overmars, A. Van Rooij et al., "Rapid analysis of conjugated bile acids in plasma using electrospray tandem mass spectrometry: application for selective screening of peroxisomal disorders," *Journal of Inherited Metabolic Disease*, vol. 22, no. 3, pp. 307–310, 1999.
[274] B. J. Wild, B. N. Green, E. K. Cooper et al., "Rapid identification of hemoglobin variants by electrospray ionization mass spectrometry," *Blood Cells, Molecules, and Diseases*, vol. 27, no. 3, pp. 691–704, 2001.
[275] W. Hoelzel and K. Miedema, "Development of a reference system for the international standardization of HbA1c/glycohemoglobin determinations," *Journal of the International Federation of Clinical Chemistry*, vol. 9, no. 2, pp. 62–67, 1996.
[276] U. Krishnamurti and M. W. Steffes, "Glycohemoglobin: a primary predictor of the development or reversal of complications of diabetes mellitus," *Clinical Chemistry*, vol. 47, no. 7, pp. 1157–1165, 2001.
[277] S. K. Manna, A. D. Patterson, Q. Yang et al., "UPLC-MS-based urine metabolomics reveals indole-3-lactic acid and phenyllactic acid as conserved biomarkers for alcohol-induced liver disease in the Ppara-null mouse model," *Journal of Proteome Research*, vol. 10, no. 9, pp. 4120–4133, 2011.
[278] S. K. Manna, A. D. Patterson, Q. Yang et al., "Identification of noninvasive biomarkers for alcohol-induced liver disease using urinary metabolomics and the ppara-null mouse," *Journal of Proteome Research*, vol. 9, no. 8, pp. 4176–4188, 2010.
[279] C. Lifshitz and J. Laskin, *Principles of Mass Spectrometry Applied to Biomolecules*, John Wiley & Sons, 2006.
[280] E. Kalenius, D. Moiani, E. Dalcanale, and P. Vainiotalo, "Measuring H-bonding in supramolecular complexes by gas phase ion-molecule reactions," *Chemical Communications*, no. 37, pp. 3865–3867, 2007.
[281] D. P. Weimann, H. D. F. Winkler, J. A. Falenski, B. Koksch, and C. A. Schalley, "Highly dynamic motion of crown ethers along oligolysine peptide chains," *Nature Chemistry*, vol. 1, no. 7, pp. 573–577, 2009.
[282] T. E. Wales and J. R. Engen, "Hydrogen exchange mass spectrometry for the analysis of protein dynamics," *Mass Spectrometry Reviews*, vol. 25, no. 1, pp. 158–170, 2006.

[283] Y. Hamuro, S. J. Coales, M. R. Southern, J. F. Nemeth-Cawley, D. D. Stranz, and P. R. Griffin, "Rapid analysis of protein structure and dynamics by hydrogen/deuterium exchange mass spectrometry," *Journal of Biomolecular Techniques*, vol. 14, no. 3, pp. 171–182, 2003.

[284] B. E. Winger, K. J. Light-Wahl, A. L. Rockwood, and R. D. Smith, "Probing qualitative conformation differences of multiply protonated gas-phase proteins via hydrogen/deuterium isotopic exchange with water-d2," *Journal of the American Chemical Society*, vol. 114, no. 14, pp. 5897–5898, 1992.

[285] F. Wang, M. A. Freitas, A. G. Marshall, and B. D. Sykes, "Gas-phase memory of solution-phase protein conformation: H/D exchange and Fourier transform ion cyclotron resonance mass spectrometry of the N-terminal domain of cardiac troponin C," *International Journal of Mass Spectrometry*, vol. 192, no. 1–3, pp. 319–325, 1999.

[286] H. Han and S. A. McLuckey, "Selective covalent bond formation in polypeptide ions via gas-phase ion/ion reaction chemistry," *Journal of the American Chemical Society*, vol. 131, no. 36, pp. 12884–12885, 2009.

[287] R. Qian, J. Zhou, S. Yao, H. Wang, and Y. Guo, in *Reactive Intermediates*, pp. 113–131, Wiley-VCH, 2010.

[288] V. M. Williams, J. R. Kong, B. J. Ko et al., "ESI-MS, DFT, and synthetic studies on the H_2-mediated coupling of acetylene: insertion of C=X bonds into rhodacyclopentadienes and Brønsted acid cocatalyzed hydrogenolysis of organorhodium intermediates," *Journal of the American Chemical Society*, vol. 131, no. 44, pp. 16054–16062, 2009.

[289] A. O. Aliprantis and J. W. Canary, "Observation of catalytic intermediates in the Suzuki reaction by electrospray mass spectrometry," *Journal of the American Chemical Society*, vol. 116, no. 15, pp. 6985–6986, 1994.

[290] L. S. Santos, C. H. Pavam, W. P. Almeida, F. Coelho, and M. N. Eberlin, "Probing the mechanism of the Baylis-Hillman reaction by electrospray ionization mass and tandem mass spectrometry," *Angewandte Chemie*, vol. 43, no. 33, pp. 4330–4333, 2004.

[291] A. A. Sabino, A. H. L. Machado, C. R. D. Correia, and M. N. Eberlin, "Probing the mechanism of the Heck reaction with arene diazonium salts by electrospray mass and tandem mass spectrometry," *Angewandte Chemie*, vol. 43, no. 19, pp. 2514–2518, 2004.

[292] H. Guo, R. Qian, Y. Liao, S. Ma, and Y. Guo, "ESI-MS studies on the mechanism of Pd(0)-catalyzed three-component tandem double addition-cyclization reaction," *Journal of the American Chemical Society*, vol. 127, no. 37, pp. 13060–13064, 2005.

[293] S. R. Wilson and Y. Wu, "A study of nickel-catalyzed coupling reactions by electrospray ionization mass spectrometry," *Organometallics*, vol. 12, no. 4, pp. 1478–1480, 1993.

[294] Y. Xie, L. F. He, S. C. Lin et al., "Desorption electrospray ionization mass spectrometry for monitoring the kinetics of baeyer-villiger solid-state organic reactions," *Journal of the American Society for Mass Spectrometry*, vol. 20, no. 11, pp. 2087–2092, 2009.

[295] D. G. Harman and S. J. Blanksby, "Trapping of a tert-adamantyl peroxyl radical in the gas phase," *Chemical Communications*, no. 8, pp. 859–861, 2006.

[296] M. N. Eberlin, "Electrospray ionization mass spectrometry: a major tool to investigate reaction mechanisms in both solution and the gas phase," *European Journal of Mass Spectrometry*, vol. 13, no. 1, pp. 19–28, 2007.

[297] V. Kertesz and G. J. V. Berkel, "Chemical imaging with desorption electrospray ionization mass spectrometry," *Methods in Molecular Biology*, vol. 656, pp. 231–241, 2010.

[298] J. M. Wiseman, D. R. Ifa, Y. Zhu et al., "Desorption electrospray ionization mass spectrometry: imaging drugs and metabolites in tissues," *Proceedings of the National Academy of Sciences of the United States of America*, vol. 105, no. 47, pp. 18120–18125, 2008.

[299] D. R. Ifa, J. M. Wiseman, Q. Song, and R. G. Cooks, "Development of capabilities for imaging mass spectrometry under ambient conditions with desorption electrospray ionization (DESI)," *International Journal of Mass Spectrometry*, vol. 259, no. 1–3, pp. 8–15, 2007.

[300] J. M. Wiseman and B. C. Laughlin, "Desorption electrospray ionization (DESI) mass spectrometry: a brief introduction and overview," *Current Separations and Drug Development*, vol. 22, pp. 11–14, 2007.

8

Determination of Tetracycline in Pharmaceutical Preparation by Molecular and Atomic Absorption Spectrophotometry and High Performance Liquid Chromatography via Complex Formation with Au(III) and Hg(II) Ions in Solutions

Ahlam Jameel Abdulghani,[1] Hadi Hassan Jasim,[2] and Abbas Shebeeb Hassan[2]

[1] *Department of Chemistry, College of Science, Baghdad University, Jaderiya, Baghdad, Iraq*
[2] *Department of Chemistry, College of Science, Al-Mustansiriyah University, Baghdad, Iraq*

Correspondence should be addressed to Ahlam Jameel Abdulghani; prophahlam@yahoo.com

Academic Editor: Peter A. Tanner

UV-visible and atomic spectrophotometry and HPLC techniques were applied for the determination of tetracycline (TC) in pharmaceutical preparations via complexation of the drug with Au(III) and Hg(II) ions in solutions. The mole ratio of TC to metal ions was 1 : 1. Maximum peak absorption at λ 425 and 320 nm for the two ions, respectively, was optimized at heating temperature 75°C for 15 minutes at pH = 4 followed by the extraction with ethyl acetate. The percentage of extraction and stability constants for the two complexes was 95.247, 95.335% and 2.518×10^4, $1.162 \times 10^5\ M^{-1}$, respectively. HPLC method was applied without extraction process. The analytical data obtained from direct calibration curves of UV-visible absorption, FAAS, and HPLC for Au(III) complexes were recovery (100.78, 104.85, and 101.777%, resp.); detection limits (0.7403, 0.0997, and 2.647 μg/ml, resp.); linearity (5–70, 5–30, and 10–150 μg/ml, resp.), and correlation coefficient (0.9991, 0.9967, and 0.9986, resp.). The analytical data obtained from direct calibration curves for Hg(II) complexes by UV-visible spectrophotometry and HPLC were recovery (100.95 and 102.000%, resp.); detection limits (0.5867 and 2.532 μg/ml, resp.); linearity (5–70 and 10–150 μg/ml, resp.); and correlation coefficients (0.9989 and 0.9997, resp.).

1. Introduction

Tetracyclines possess a wide range of antimicrobial activity against Gram-positive and Gram-negative bacteria. They have been used not only in human medicine for the treatment of infectious diseases but also as additives in animal feed to promote growth. Beside the pharmacological importance of tetracycline (TC), this molecule possesses many potential metal-binding sites. Therefore a wide range of research work has been reported on chelation of TCs with various metal ions which have been utilized successfully in pharmaceutical analysis [1–15].

Depending on the chosen experimental conditions such as solvent medium, pH, metal ion type, and ligand: metal ratio, the O(10), O(11), and O(12) on the BCD-ring and O(1), O(3), and N(4) in ring A and at the carboxamide group on the A-ring are the main coordination sites that were frequently proposed in the literature [1–6, 16]. Pharmacokinetics and bioavailability of TC are affected by its coordination with metal ions such as calcium in blood plasma and magnesium ion in the intracellular medium [4]. By combining with the copper, zinc, iron, and other trace metal elements in enzymes such as collagenase, tetracyclines inhibit the enzymatic destruction of tissues [5].

Complexes of (TC) with Co(II), Ni(II), and Fe(III) ions showed higher activity against *Bacillus subtilis, Serratia* species, and *Escherichia coli* than that the original tetracycline hydrochloride [6]. Platinum(II) complexes of tetracycline

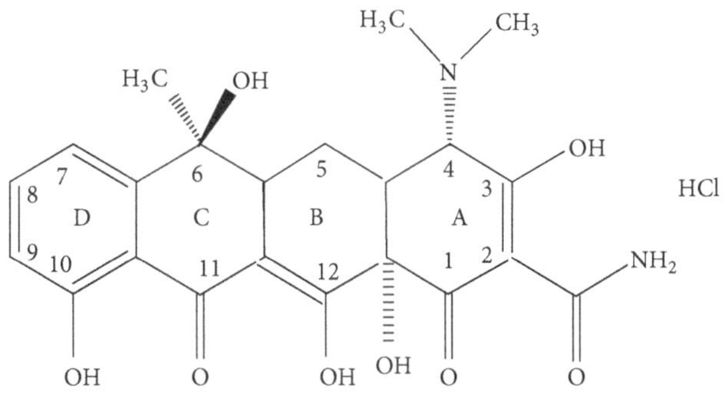

FIGURE 1: TC. HCl.

and doxycycline were reported to bind to DNA and inhibit tumoral cell growth [15]. Although the Au(III) is as important as Pt(II) ions in this aspect, little attention has been paid to Au(III) complexes of TC. Environmental studies on TCs showed that complexation of these drugs with heavy metal cations in natural waters and soils may induce chronic toxic effects in aquatic plants [7, 8, 17].

In this work we study the determination of TC (see Figure 1) in pure and dosage form via complexation with gold(III) and mercury(II) ions in solution using UV-visible and atomic absorption spectrophotometry and HPLC techniques.

2. Experimental

UV-visible spectra were recorded on Varian Gary 100 UV-visible spectrophotometer supplied with UV-probe software. FTIR spectra were recorded on Shimadzu FT-IR 8400S Fourier transforms infra-red spectrophotometer. Flame atomic absorption spectrophotometric studies (FAAS) on Au(III) ion and its complexes were carried out using GBC 933 plus flame atomic absorption spectrophotometer supplied with hollow cathode lamp, D_2 Lamp (GBC), flame autosampler, and air-acetylene flame. High performance liquid chromatography was performed on HPLC, Shimadzu, and Shimadzu LC 2010A, supplied with UV-visible detector. HPLC conditions for determination of Tc-Au(III) complexes were : stationary phase or column 150 × 4.65 mm supelcosil LC-18, mobile phase acetonitrile : acetate buffer solution (40 : 60), flow rate (1 mL/min), injection volume 100 μL, detection wavelength (220 nm). HPLC conditions for determination of Hg(II)-TC complexes were : mobile phase acetonitrile : acetate buffer solution (20 : 80), stationary phase or column 150 × 4.65 mm ODS C-18, flow rate (2 mL/min), injection volume 200 μL, detection wavelength (276 nm). pH of solutions were measured by using HANA, HI 98150 GLP PH/ORP-meter calibrated with reference pH solutions at 4.0 and 9.0.

All chemicals used were of analytical reagent grade, and tetracycline hydrochloride was provided from state company for drug industries, and medical appliance (SDI), Samarra Iraq.

2.1. Antibiotic Standard Solutions. A stock solution of tetracycline hydrochloride (TC. HCl) (1000 μg/mL) was prepared by dissolving 0.100 g of TC. HCl standard powder in 100 mL distilled deionized water (DDW). Working standard solutions of TC. HCl (100–150 μg/mL) were prepared by diluting 10–15 mL of stock solution to 100 mL with DDW in 100 mL volumetric flask.

2.2. Metal Ions Standard Solutions (100 μg/mL). A stock solution of Hg(II) ions (1000 μg/mL) was prepared by dissolving 0.1350 g of $HgCl_2$ in a 100 mL (DDW) in 100 mL volumetric flask. Working solutions (100 μg/mL) were prepared by diluting 10 mL of the stock solution to 100 mL with DDW in a 100 mL volumetric flask. Working solutions of Au(III) ions (100 μg/mL) were prepared from 1000 $\mu g \cdot mL^{-1}$ Au(III) solution provided from suppliers for atomic absorption spectrophotometric analysis.

2.3. UV-Visible Spectrophotometry Complex Formation. To 1 mL aliquots of metal ion solutions (5.08×10^{-3} M) in a series of 10 mL volumetric flasks were added 0.25, 0.5, 0.75, 1.0, 2.0, 3.0, and 4.0 mL of TC. HCl solution (5.08×10^{-3} M) and volumes were completed to the marks. All experimental conditions were set to their optimum values, and resulting complexes were extracted from aqueous solutions with ethyl acetate (5 : 1 v/v aq : EA) before measuring the absorbance.

2.3.1. Determination of Antibiotic Complexes by Direct Method. Solution mixtures of 5–70 μg/mL TC. HCl and 10 μg/mL Au(III) ion or 15 μg/mL of Hg(II) standard solutions were prepared by adding different volumes of antibiotics' standard solutions (100 μg/mL) to 1 ml or 1.5 mL, respectively, of metal ion standard solutions (100 μg/mL) in 10 mL volumetric flasks. The volumes were completed to the mark, and the experimental conditions were adjusted to the respective optimum values of concentration, pH and temperature, and liquid extraction with ethyl acetate (EA) (5 : 1 v/v aq : EA). The absorbance of complexes in each case was recorded at the recommended λ_{max} and plotted against the concentration of the cited antibiotic.

2.3.2. Determination of Antibiotic Complexes in Dosage Form by Direct Method. 0.1 g of TC powder obtained from 20 capsules of samacycline 250 mg was weighted and dissolved in 100 mL distilled water in volumetric flask. Ten mL of the resulted solution was diluted to 100 mL by DDW in volumetric flasks. Then 1, 2, and 3 mL of the end solution were transferred to 10 mL volumetric flasks containing 1 mL of Au(III) or 1.5 mL of Hg(II) standard solutions (100 μg/mL), and the volumes were completed to the marks with DDW. After adjusting to the optimum conditions, the absorbance of these solutions was measured against blank solution, and the concentration of the studied analyst was calculated depending upon the respective standard direct calibration curve.

2.3.3. Determination of Antibiotic Complexes in Dosage Form by Standard Addition Method. To a series of solutions containing 1 mL of Au(III) or 1.5 mL of Hg(II) standard solutions

(100 μg/mL) and various amounts 0.5–6.0 mL or 0.5–5.5 mL, respectively, of TC. HCl standard solutions (100 $\mu g \cdot mL^{-1}$) was added 1 mL of dosage antibiotic (samacycline) solutions (100 $\mu g \cdot mL^{-1}$) in 10 mL volumetric flasks. The volumes were completed to the marks with DDW. After adjusting the optimum conditions in each case, the absorbance was measured, and the relationship between absorbance and concentration was plotted to construct standard addition curves.

2.4. Flame Atomic Absorption Spectrometry FAAS for Au(III) Complexes. This method was worked out for the determination of dosage TC at the same optimum conditions fixed in the UV-visible method except that the suitable concentration of Au(III) ion was 6 μg/mL.

2.4.1. Determination of TC-Au(III) Complexes by Direct Method. Solution mixtures containing 1–30 μg/mL of TC. HCl and 6 $\mu g \cdot mL^{-1}$ of Au(III) ion were prepared by adding different volumes of antibiotics' standard solutions (100 μg/mL) to 0.6 mL of Au(III) ion standard solution (100 μg/mL) in 10 mL volumetric flasks. The volumes were completed to the mark and the experimental conditions were adjusted. The absorbance was measured, and the relationship between absorbance and concentration was plotted to produce standard direct calibration curve

2.4.2. Determination of TC-Au(III) Complexes in Dosage Form by Direct Method. From a prepared solution of dosage TC (100 μg/mL) 0.6, 0.8, and 1.0 mL were transferred to 10 mL volumetric flasks, containing 0.6 mL of Au(III) standard solution (100 μg/mL). The volumes were completed to the marks, and all optimum conditions were adjusted. The absorbance of these solutions against blank solution was measured, and the concentrations of solutions were calculated depending on standard calibration curve.

2.4.3. Determination of TC-Au(III) Complexes in Dosage Form by Standard Addition Method. In this method 0.6 mL of dosage TC (100 μg/mL) solutions was added to aqueous solutions containing a mixture of 0.6 mL Au(III) standard solutions (100 μg/mL) and different volumes of TC. HCl standard solutions to form 1–25 μg/mL solutions. After all optimum conditions have been adjusted, the relationship between absorbance and concentration was plotted to produce standard addition curves for determination of TC in dosage form.

2.5. High Performance Liquid Chromatography (HPLC). Solutions were prepared as was mentioned in the direct UV-spectrophotometric method but with different concentration ranges of standard TC. HCl (10-150 μg/mL) and without solvent extraction. The peak area of complexes was plotted against antibiotic concentration to construct the direct calibration curves.

2.5.1. Determination of Dosage TC Au(III) and Hg(II) Complexes by Direct HPLC Method. A powder of 20 capsules (0.1 g) was weighted and dissolved in (100 mL) DDW in a volumetric flask. Twenty-five mL of this solution was diluted to 100 mL with DDW in a 100 mL volumetric flask. Then 1.2, 2.0, and 4.0 mL of the end solution were transferred to (10 mL) volumetric flasks containing 1.0 mL of Au(III) or 1.5 mL of Hg(II) standard solutions (100 μg/mL). The volumes were completed to the marks, and all optimum conditions were adjusted. Peak area of these solutions was measured, and the concentration of solutions was calculated depending on standard direct calibration curve.

3. Result and Discussion

3.1. UV-Visible Spectrophotometry. The UV-visible spectra of TC. HCl, metal ions, and their complexes (50 $\mu g \cdot mL^{-1}$) in aqueous solutions are shown in Figure 2. The spectrum of the drug exhibited a multiplet with maximum absorption peaks at λ 235, 270, and 370 nm corresponding mainly to $\pi \rightarrow \pi^*$ transitions [18]. The spectrum of the yellow gold(III) aqueous solution exhibited two absorption bands, a doublet appeared at λ 240 and 295 nm and a low intensity band at 385 nm and were assigned to ligand to metal charge transfer and $^1A_1g \rightarrow {}^1Eg$ transitions of square planar tetrachloroaurate(III) anion $[AuCl_4]^-$ [19]. The spectrum of the Hg(II) ion exhibited a single high intensity band at λ 285 nm and was assigned to ligand $\rightarrow$ metal charge transfer [19]. The spectrum of TC. HCl with the Au(III) ion exhibited hypsochromic shift of the $\pi \rightarrow \pi^*$transitions band and the appearance of two new absorption bands at λ_{max} 350 and 425 nm assigned to $^1A_1g \rightarrow {}^1Eg$ and $^1A_1g \rightarrow {}^1B_2g$ transitions of square planar gold(III) complexes, respectively [16, 19, 20]. The spectrum of the Hg(II) complex solution exhibited shifts of ligand bands to shorter wavelengths and the appearance of additional band at λ_{max} 320 nm attributed to ligand to metal charge transfer transition [16, 19, 20].

3.1.1. Optimisation of the Experimental Conditions. Figure 3 shows the effect of metal ion concentration, pH of solution, temperature, and heating time on the absorbance of the TC-Au(III) and TC-Hg(II) complex solutions at λ_{max} 425 and 320 nm, respectively, using 50 μg/mL of drug solutions. The optimum concentration of the ions that gave maximum absorbance was 10 μg/mL of Au(III) ions and 15 μg/mL of Hg(II) ions. The best values of pH recorded for the highest absorbance values were 2–4. At high pH values formation of metal hydroxides took place [21, 22]. It is also important to mention that tetracycline is a very adaptive molecule, capable of easily modifying itself through tautomerism in response to various chemical environments [10, 23–25]. Therefore protonation and deprotonation of Tcs in acidic and basic aqueous solutions have important effects on their coordination behavior and their absorption spectra [23, 24]. However the hypsochromic shift of the $\pi \rightarrow \pi^*$ transitions band in this study supports the suggested structures of Au(III) and Hg(II) complexes [16]. The extraction efficiency and the choice of extracting solvent have also an important effect on absorbance values [26, 27]. Optimum experimental conditions for TC. HCl complexes with Au(III) and Hg(II) ions are described in Table 1.

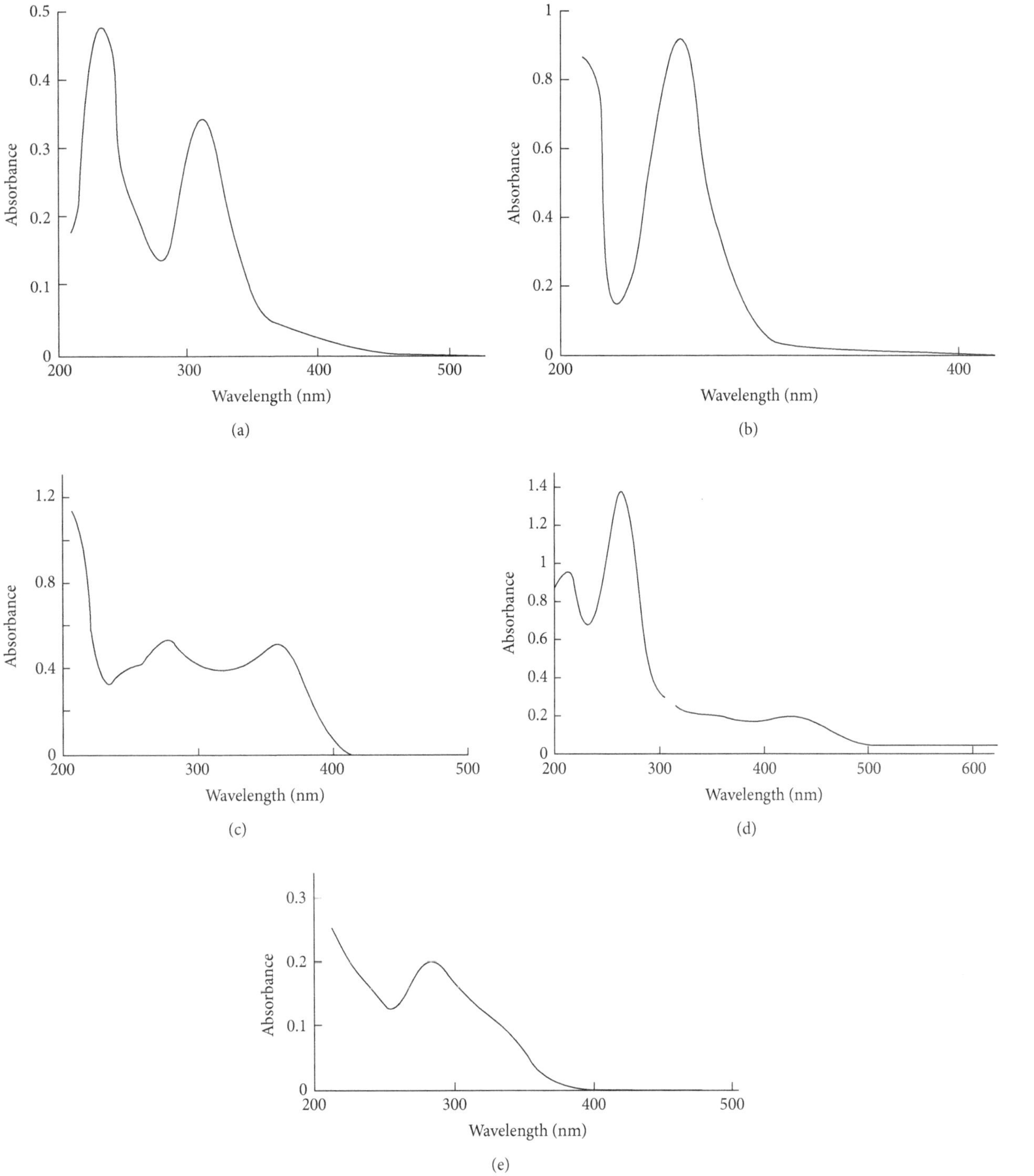

FIGURE 2: UV-visible spectra of (a) gold(III) ion, (b) mercury(II) ion, (c) TC. HCl, (d) TC-Au(III), and (e) TC-Hg(II) complexes in aqueous solutions (50 μg/mL each).

TABLE 1: Optimum experimental conditions for TC. HCl complexes with Au(III) and Hg(II) ions.

Complex	Ion conc. (μg/mL)	pH	Heating temp. (°C)	Heating time (min.)	Phase ratio (EA : aq)	Extraction time (min.)	Extraction% (E%)	Distribution ratio
TC-Au	10	4	75	15	1 : 5	1	95.242	100.196
TC-Hg	15	2–4	75	15	1 : 5	1-2	95.335	102.181

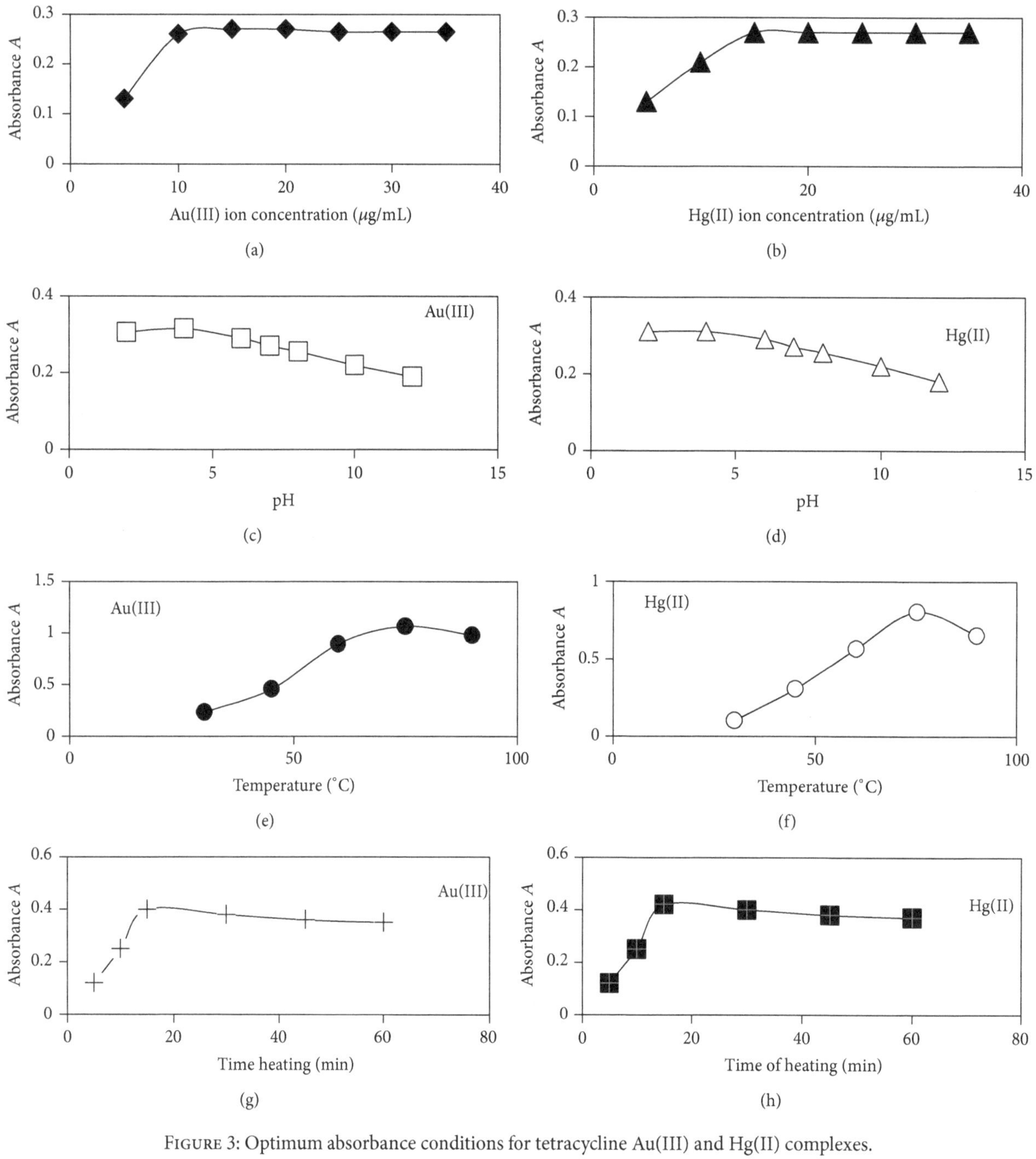

FIGURE 3: Optimum absorbance conditions for tetracycline Au(III) and Hg(II) complexes.

TABLE 2: Stability constant (k) and molar absorptivity for TC-Au(III) and TC-Hg(II) complexes in solutions.

Complex	ε_{max} ($L \cdot mol^{-1} \cdot cm^{-1}$)	Stability constant (k) (M^{-1})	λ_{max} (nm)
TC-Au(III)	3.4820×10^3	2.5187×10^4	425
TC-Hg(II)	6.9656×10^3	1.6122×10^5	320

Figure 4 shows the effect of volume ratio of complex aqueous solution to ethyl acetate (aq : EA or aq : org.), extraction percentage, and extraction time on the absorbance of complexes. The highest absorbance value was achieved when the ratio of aqu : EA was 5 : 1. The percentage of extraction (%E) and distribution ratio (D) were obtained from following equations [27]:

$$\%E = \frac{\text{Initial Conc. (org.)} - \text{Final Conc. (aq)}}{\text{Initial Conc. (org.)}} \times 100,$$

$$\%E = \frac{100D}{D + \left(V_{aq}/V_o\right)}, \tag{1}$$

where V_{aq} and V_o are volumes of aqueous and organic layers, respectively.

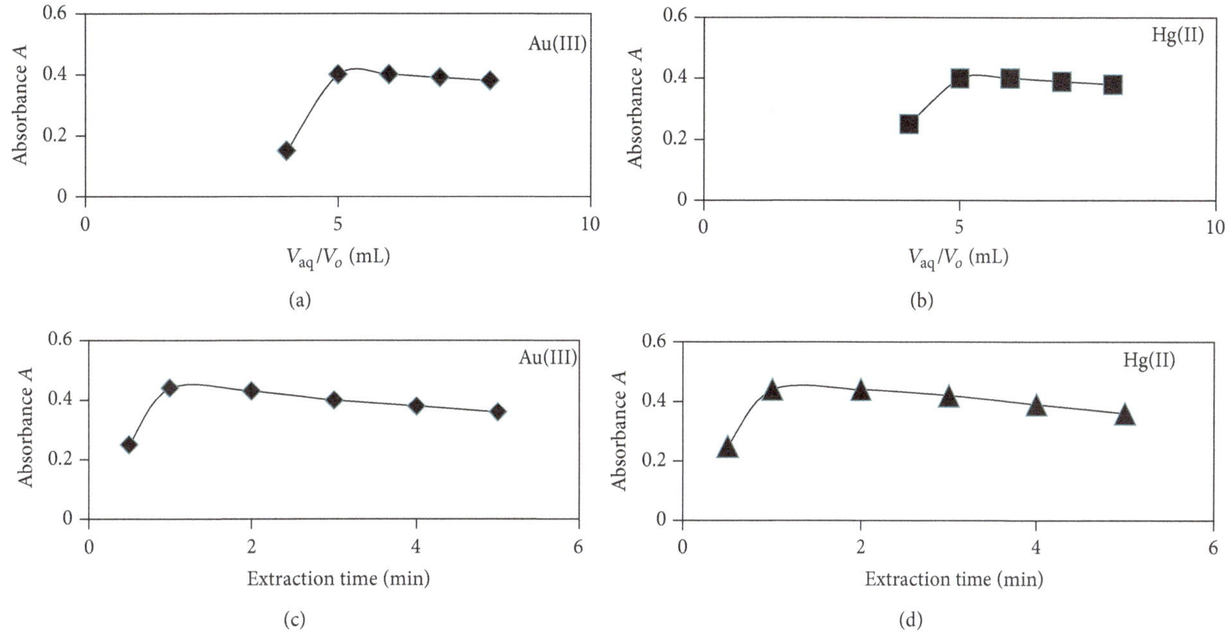

FIGURE 4: Phase ratio of aq : EA (V_{aq}/V_o) extraction effect and extraction time effects on absorbance of TC-Au(III) and TC-Hg(II) complexes.

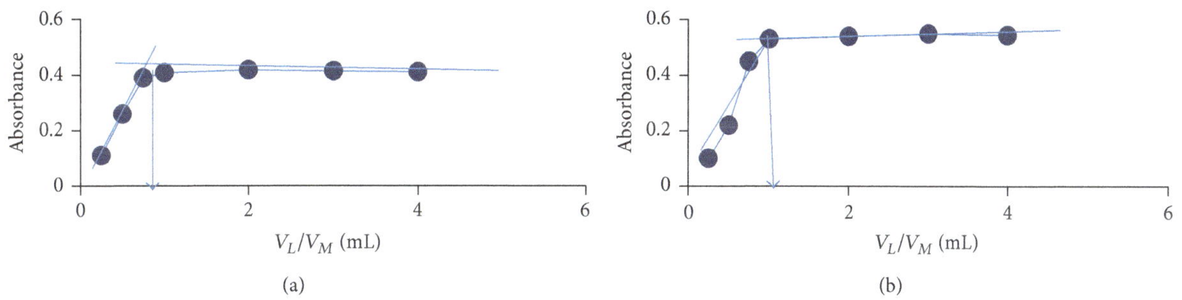

FIGURE 5: The mole ratio curves for the formation of (a) TC-Au(III) and (b) TC-Hg(II) complexes in solutions at optimum conditions.

TABLE 3: Regression equation, correlation coefficient, t-test concentration ranges, detection limits, and RSD% direct and standard addition calibration curves of TC-Au complex by using uv-visible spectrophotometry.

Method	Regr. eq. $Y = Bx + A$	Correlation coefficient (r)	Line range (μg/mL)	D.L. (μg/mL)	t-test statistic (calcd)	Tabulated t-test two tailed 95% C.I.	RSD% ($n = 4$)	Mean Rec.%
TC-Au(III) complex								
Direct	$Y = 0.0093x - 0.0080$	0.9991	5–70	0.7403	0.3994	2.365	0.9617	100.780
Standard addition	$Y = 0.0115x + 0.1133$	0.9989	5–70		0.8320	2.365	1.2855	103.970
TC-Hg(II) complex								
Direct	$Y = 0.0100x - 0.037$	0.9989	5–70	0.5867	0.4791	2.365	0.9617	100.950
Standard addition	$Y = 0.0102x + 0.0963$	0.9986	5–70		1.2126	2.365	1.2854	103.103

TABLE 4: Percentage relative error for determination of TC-Au(III) and TC-Hg(II) complexes by direct and standard addition UV-vis methods.

Complex	State of drug	Stated concentration (mg/unit)	Found direct calb. (mg/unit)	% Erel	Found st. add. calb. (mg/unit)	% Erel
TC-Au(III)	Capsule	250	251.950	+0.780	259.925	+3.970
TC-Hg(II)	Capsule	250	252.375	+0.950	257.757	+3.103

TABLE 5: Regression equation, correlation coefficient, t-test concentration ranges, detection limits, RSD%, and recovery (rec.%) for the determination of TC-Au(III) complex by FAAS.

Regr. eq. $Y = Bx + A$	Corr. coefficient (r)	Linear range (μg/mL)	D.L. (μg/mL)	t-test statistic (calcd)	Tabulated t-test two tailed 95% C.I.	RSD% ($n = 4$)	Rec.%
Direct method							
$Y = 0.0520x + 0.0088$	0.9967	5–30	0.0995	1.7603	2.571	1.054	104.753
Standard addition method							
$Y = 00549x + 0.2791$	0.9988	5–28		0.6569	2.571	3.253	95.639

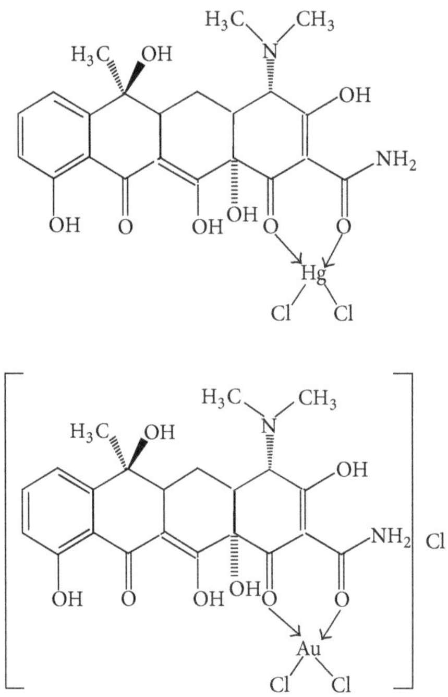

FIGURE 6: The suggested structures of TC-Au(III) and TC-Hg(II) complexes.

3.1.2. Complex Formation by Mole Ratio Method and Calculation of Stability Constant (k) of Complexes. Figure 5 shows the variation of absorbance of TC-Au(III) and TC-Hg(II) complexes against mole ratio of TC : M (5.8×10^{-3} M each). The ratios for both complexes were 1 : 1. The stability constant values for both complexes (Table 2) were calculated depending on mole ratio curves according to the equation

$$k = \frac{(A_1 - A_3)(A_2 - A_3)}{(A_2 - A_1)^2 C}, \tag{2}$$

where k is the formation constant, C is the molar concentration, A_1 is the absorbance which represents two tangents intercept, A_2 is the absorbance which represents the highest absorbance, A_3 is the absorbance of first point.

3.1.3. Suggested Structures. The results of UV-visible and IR spectra, conductivity measurements, and metal content obtained from the synthesis of the two complexes in the solid state [28] led to the suggested structure illustrated in Figure 6 and supported the results of complex formation in solution. The IR spectrum of the free TC. HCl showed two absorption bands at 3280–3480 cm^{-1} attributed to O-H stretching vibrations that undergo intermolecular hydrogen bonding [6, 15]. This band was shifted to higher frequencies at 3379–3439 cm^{-1} and to 3340–3440 cm^{-1} in the spectra of Au(III) and Hg(III) complexes, respectively, which rules out the coordination of the hydroxyl group with the two metal ions. The band observed at 3122 cm^{-1} was attributed to νNH of amide group [3]. The bands observed at 1672, 1618.2, and 1583 cm^{-1} in the spectrum of TC. HCl are characteristic of amide carbonyl, carbonyl groups of rings A and C, respectively [17]. Complexation of TC with Au(III) and Hg(II) ions caused shifts of the amide C=O and ring A C=O to lower frequencies at 1639.55, 1608, and 1637.6, 1599 cm^{-1}, respectively. This provides a strong evidence on bonding of the two carbonyls with the two ions [16]. Stretching modes of M-O bonds related to the two carbonyls were observed at 555, 525 and 550, 520 for the two complexes respectively. The new bands appeared at 324, 315 cm^{-1} and 317, 312 cm^{-1} may be assigned to stretching vibrations of Au-Cl and Hg-Cl, respectively [6].

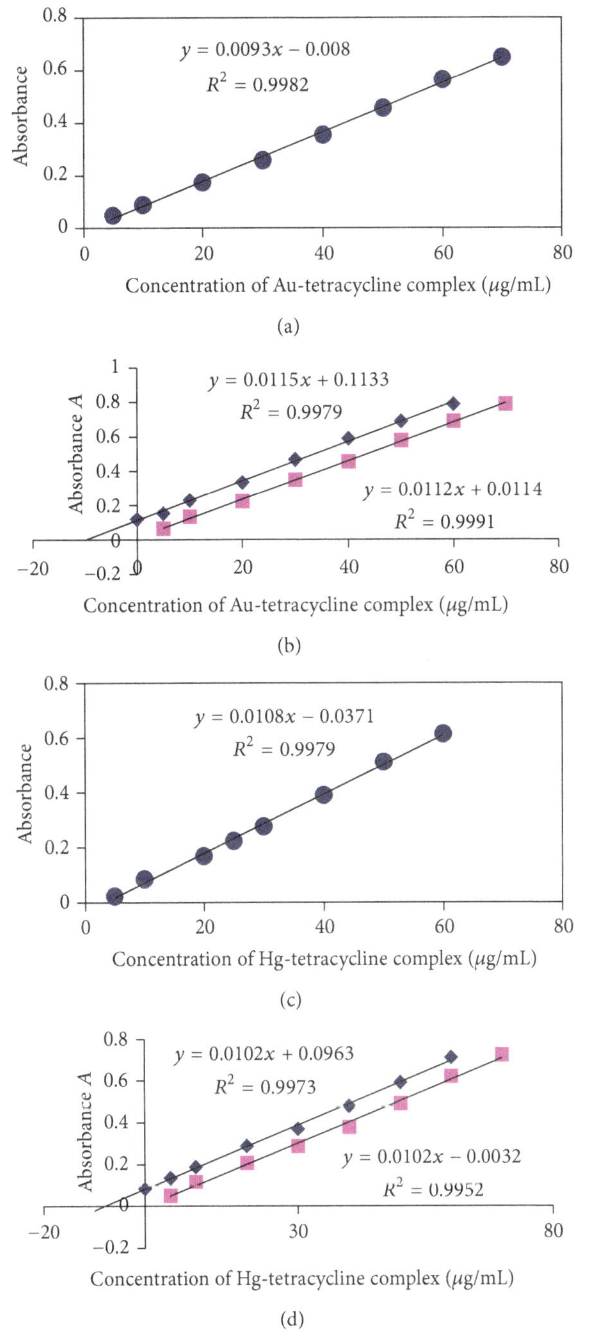

FIGURE 7: Direct and standard addition calibration curves for determination of TC-Au(III) and TC-Hg(II) complexes by spectrophotometric methods at λ_{max} 425 and 320 nm, respectively.

3.1.4. Determination of TC-Au(III) and TC-Hg(II) Complexes by UV-Vis Method. Figure 7 shows the direct and standard addition calibration curves of absorbance against concentration for determination TC-Au(III) and TC-Hg(II) complexes. The analytical parameters described in Tables 3 and 4 were calculated depending on these curves. Table 3 show that t-tabulated is more than t-calculated which indicates that the results of the slope of standard addition curve were parallel to slope of direct method which means that there was no matrix interference of additives on absorbance of complexes [29]. The correlation coefficient values are within the accepted ranges. The high range of concentration and low detection limits show that the adopted method is applicable and reliable. The RSD% and percentage relative errors (Table 4) for the determination of the two complexes by direct method were less than those of standard addition methods.

TABLE 6: Relative error percentage for determination of TC-Au complex by direct and standard addition FAAS method.

Method	State of drug	Stated concentration (mg per unit)	Found (mg per unit)	Erel.%
Direct	Capsule	250	261.883	+4.753
Standard addition	Capsule	250	239.098	−4.361

3.2. Determination of Drug-Au Complexes by Direct FAAS Method. The flame atomic absorption spectroscopy was applied for determination of drugs indirectly by determination of gold(III) ions in the extracted complexes depending on direct and standard addition curves of absorbance of Au(III) complex against TC concentrations (Figure 8). Tables 5 and 6 describe the calculated analytical data obtained from the two curves. Compared with the UV-visible method, this method showed similar percentage recovery, lower concentration range, and higher percentage error (Table 6). However, the percentage recovery was comparable to that obtained for the determination of other drugs by Au(III) ion [30]. On the other hand this method showed very low detection limit which makes it more sensitive than the UV-visible method.

The t-test calculated values are much lower than t-tabulated, and the slopes of standard addition curves were also parallel to slopes of direct calibration curve, which indicates that the results of the applied method are acceptable with no matrix interference effect observed [26, 29].

3.3. Determination of TC Complexes by (HPLC) Method. This method was applied for the determination of TC and its complexes directly and simultaneously. Figure 9 shows the HPLC chromatograms in which the retention time for the parent ligand and metal complexes was obtained simultaneously at detection wavelength without solvent extraction because HPLC is a separation technique. Tables 7 and 8 describe the results obtained from direct calibration curves by plotting peak area of complexes against concentration. In spite of higher detection limits the t-test, linearity, and correlation coefficients indicate that the applied method is acceptable besides being faster and easier than the other two previous methods. The detection limits and sensitivity may be improved by finding a better separation conditions [29].

4. Conclusions

Tetracycline is a very adaptive molecule, capable of easily modifying itself (chemical bonds as well as geometry)

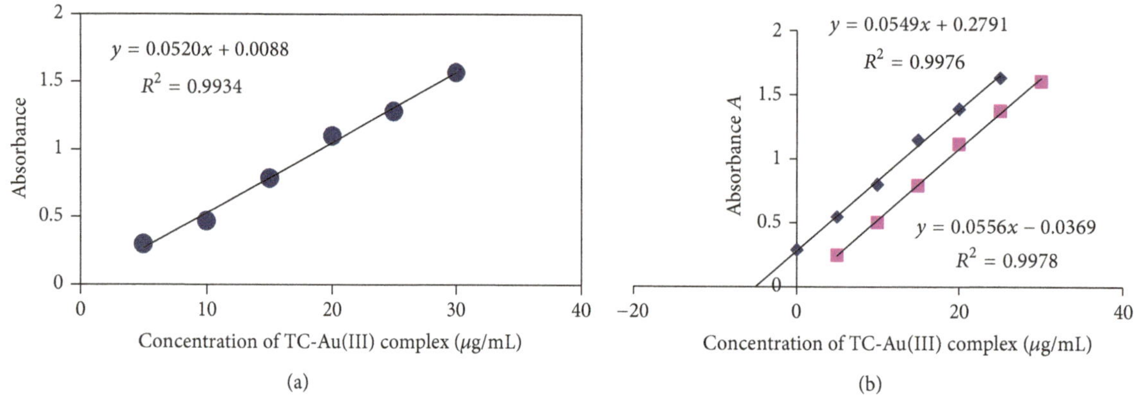

FIGURE 8: Direct and standard addition calibration curves for TC-Au(III) complex by FAAS.

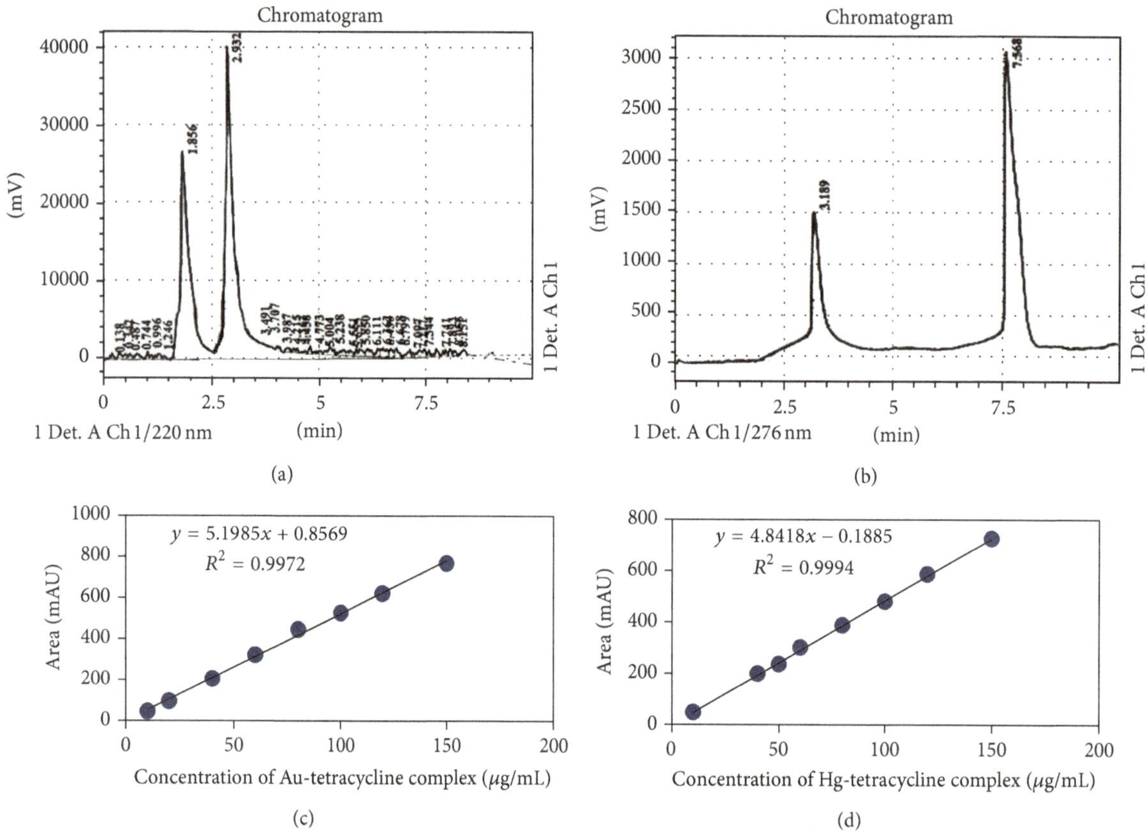

FIGURE 9: Upper: chromatograms of TC and its (a) Au(III) and (b) Hg(II) complexes with retention times (1.865, 2.932 min) and (3.189, 7.568 min), respectively, at detection λ_{max} 220 and 276 nm, respectively. Lower: direct calibration curves for the determination of TC-Au(III) and TC-Hg(II) complexes by HPLC method.

TABLE 7: Regression equation, correlation coefficient, t-test concentration ranges, detection limits, and RSD% for TC-Au(III) and TC-Hg(II) complexes by HPLC.

Regr. eq. $Y = Bx + A$	Correlation coefficient (r)	Linear range (μg/mL)	D.L. (μg/mL)	t-tests statistic (calculated)	Tabulated t-test two tailed 95% C.I.	RSD% ($n = 4$)	Rec.%
TC-Au(III) complex							
$Y = 5.1985x + 0.8569$	0.9986	10–150	2.647	1.2864	2.365	1.2339	101.777
TC-Hg(II) complex							
$Y = 4.8418x - 0.1885$	0.9997	10–150	2.532	1.4720	2.365	1.0733	102.000

TABLE 8: Relative percentage error for determination of TC-Au(III) and TC-Hg(II) complexes by HPLC direct method.

Complex	State of drug	Stated concentration (mg/unit)	Found (mg/unit)	% Erel.
TC-Au(III)	Capsule	250	254.4425	+1.777
TC-Hg(II)	Capsule	250	255.000	+2.000

through tautomerism in response to various chemical environments which have a strong effect on its coordination behavior with metal ion and consequently on the determination method. UV-visible absorption spectrophotometry, flame atomic absorption spectrophotometry FAAS, and high performance liquid chromatography HPLC have been applied successfully to determine the antibiotic (tetracycline hydrochloride) in pharmaceutical preparations by studying, for the first time, the complexation behavior of this drug with gold(III) and mercury(II) ions in solutions. The results of mole-ratio in solution at pH 4 agreed with the suggested square planar and tetrahedral structural formula for the solid Au(III) and Hg(II) chelate complexes, respectively. The FTIR spectra showed that TC coordinates with the metal ions through amide C=O and ring A C=O showing bidentate behavior. The results of comparison between FAAS, UV-visible spectrophotometry, and HPLC methods for the studied chelate complexes showed that the FAAS method, although of higher cost, showed higher sensitivity, lower detection limits, and linear ranges than UV-visible spectrophotometry and HPLC techniques. The HPLC method showed high detection limits but also high linear range and less time consumption for analysis by omitting the extraction step compared with the other two methods.

References

[1] D. M. Janrao, R. P. Shimpi, and R. B. Fadat, "Potentiometric studies of ternary complexes of zinc with vitamins and medicinally important drugs," *Journal of Chemical and Pharmaceutical Research*, vol. 4, no. 4, pp. 1965–1968, 2012.

[2] G. J. Palm, T. Lederer, P. Orth et al., "Specific binding of divalent metal ions to tetracycline and to the Tet repressor/tetracycline complex," *Journal of Biological Inorganic Chemistry*, vol. 13, no. 7, pp. 1097–1110, 2008.

[3] G. Karthikeyan, K. Mohanraj, K. P. Elango, and K. Girishkumar, "Synthesis, spectroscopic characterization and antibacterial activity of lanthanide-tetracycline complexes," *Transition Metal Chemistry*, vol. 29, no. 1, pp. 86–90, 2004.

[4] P. Mishra, "Biocoordination and computational modeling of novel ligands with Bi (V)," *International Journal of ChemTech Research*, vol. 1, no. 3, pp. 401–419, 2009.

[5] B. S. Sekhon, "Metalloantibiotics and antibiotic mimics—an overview," *Journal of Pharmaceutical Education and Research*, vol. 1, no. 1, pp. 1–20, 2010.

[6] J. A. Obaleye and A. Lawal, "A synthesis and antibacterial activity of some tetracycline metal complexes," *Centrepoint*, vol. 14, no. 1-2, pp. 1–6, 2007.

[7] W. R. Chen and C. H. Huang, "Transformation of tetracyclines mediated by Mn(II) and Cu(II) ions in the presence of oxygen," *Environmental Science & Technology*, vol. 43, pp. 401–407, 2009.

[8] Y. Zhao, J. Geng, X. Wang, X. Gu, and S. Gao, "Tetracycline adsorption on kaolinite: PH, metal cations and humic acid effects," *Ecotoxicology*, vol. 20, no. 5, pp. 1141–1147, 2011.

[9] S. M. Sultan, I. Z. Alzamil, and N. A. Alarfaj, "Complexometric-spectrophotometric assay of tetracyclines in drug formulations," *Talanta*, vol. 35, no. 5, pp. 375–378, 1988.

[10] S. Schneider, M. O. Schmitt, G. Brehm, M. Reiher, P. Matousek, and M. Towrie, "Fluorescence kinetics of aqueous solutions of tetracycline and its complexes with Mg^{2+} and Ca^{2+}," *Photochemical and Photobiological Sciences*, vol. 2, no. 11, pp. 1107–1117, 2003.

[11] C. Jiang and L. Luo, "Spectrofluorimetric determination of human serum albumin using a tetracycline-europium complex," *Analytical Letters*, vol. 37, no. 6, pp. 1129–1137, 2004.

[12] H. Li, J. Zhang, X. He, and J. Li, "Spectrofluorimetric determination of human serum albumin using a tetracycline-europium complex," *Chinese Journal of Chemistry*, vol. 22, pp. 177–183, 2004.

[13] S. Liawruangrath, B. Liawruangrath, S. Watanesk, and W. Ruengsitagoon, "Flow injection spectrophotometric determination of tetracycline in a pharmaceutical preparation by complexation with aluminium(III)," *Analytical Sciences*, vol. 22, no. 1, pp. 15–19, 2006.

[14] A. El Said Mahgoub, E. M. Khairy, and A. Kasem, "Complex formation of uranyl acetate with tetracycline and its utilization for their microdetermination," *Journal of Pharmaceutical Sciences*, vol. 63, no. 9, pp. 1451–1455, 1974.

[15] P. P. Silva, F. C. S. De Paula, W. Guerra et al., "Platinum(II) compounds of tetracyclines as potential anticancer agents: cytotoxicity, uptake and interactions with DNA," *Journal of the Brazilian Chemical Society*, vol. 21, no. 7, pp. 1237–1246, 2010.

[16] B. L. Marcial, L. A. S. Costa, W. B. De Almeida, and H. F. Dos Santos, "Structure and properties of the new complexes of platinum (II) with the chemically modified tetracycline CMT-3: a theoretical DFT study," *Journal of Molecular Structure*, vol. 916, no. 1–3, pp. 94–104, 2009.

[17] C. Gu and K. G. Karthikeyan, "Interaction of tetracycline with aluminum and iron hydrous oxides," *Environmental Science and Technology*, vol. 39, no. 8, pp. 2660–2667, 2005.

[18] R. M. Silverstein and F. X. Webster, *Spectrometric Identification of Organic Compounds*, John Wiley and Sons, New York, NY, USA, 6th edition, 1997.

[19] E. Solomon and A. Lever, *Inorganic Electronic Structure and Spectroscopy*, John Wiley and Sons, Toronto, Canada, 1999.

[20] B. Figgis, *Introduction to Ligand Fields*, Interscience, New York, NY, USA, 1996.

[21] M. Sneed, J. Maynard, and R. Brasted, *Comprehensive Inorganic Chemistry*, vol. 2, D. Van Nostard Company, 1965.

[22] M. Sneed and R. Brasted, *Comprehensive Inorganic Chemistry*, vol. 4, D. Van Nostard Company, 1965.

[23] P. Pospisil, P. Ballmer, L. Scapozza, and G. Folkers, "Tautomerism in computer-aided drug design," *Journalof Receptors and Signal Transduction*, vol. 23, no. 4, pp. 361–371, 2003.

[24] J. Kang, H. Liu, Y.-M. Zheng, J. Qu, and J. P. Chen, "Application of nuclear magnetic resonance spectroscopy, Fourier transform infrared spectroscopy, UV-Visible spectroscopy and kinetic modeling for elucidation of adsorption chemistry in uptake of tetracycline by zeolite beta," *Journal of Colloid and Interface Science*, vol. 354, no. 1, pp. 261–267, 2011.

[25] H. A. Duarte, S. Carvalho, E. B. Paniago, and A. M. Simas, "Importance of tautomers in the chemical behavior of tetracyclines," *Journal of Pharmaceutical Sciences*, vol. 88, no. 1, pp. 111–120, 1999.

[26] D. Harvey, *Modern Analytical Chemistry*, McGraw-Hill Higher Education a Division of the McGraw-Hill, New York, NY, USA, 2000.

[27] G. H. Morrison, *Solvent Extraction in Analytical Chemistry*, Butter Worth, 1972.

[28] A. S. H. Al-Kadhumi, *Atomic absorption, HPLC and spectrophotometric determination of some antibiotics in pharmaceutical preparations via complexes formation with Au(III) and Hg(II) ions [Ph.D. thesis]*, Department of Chemistry, College of Science, Baghdad University, Baghdad, Iraq, 2011.

[29] D. Skoog, D. West, F. Holler, and S. Cronch, *Fundamentals of Analytical Chemistry*, International Student Edition (ISE), 8th edition, 2004.

[30] M. S. El-Shahawi, A. S. Bashammakh, and S. O. Bahaffi, "Chemical speciation and recovery of gold(I, III) from wastewater and silver by liquid-liquid extraction with the ion-pair reagent amiloride mono hydrochloride and AAS determination," *Talanta*, vol. 72, no. 4, pp. 1494–1499, 2007.

Monitoring of Heavy Metals Content in Soil Collected from City Centre and Industrial Areas of Misurata, Libya

M. A. Elbagermi,[1] H. G. M. Edwards,[2] and A. I. Alajtal[1]

[1] *Department of Chemistry, Faculty of Science, University of Misurata, P.O. Box 1338, Misurata, Libya*
[2] *Raman Spectroscopy Group, University Analytical Centre, Division of Chemical and Forensic Sciences, University of Bradford, West Yorkshire, BD7 1DP, UK*

Correspondence should be addressed to M. A. Elbagermi; m.elbagermi@yahoo.co.uk

Academic Editor: Jan Åke Jönsson

The present paper deals with the assessment of heavy metals in soil and roadside dust around Misurata City Centre and industrial areas/roads in the period of October 2011–May 2012. The levels of Pb, Fe, Zn, Ni, Cd, Cr, and Cu in settled dust samples collected near small streets, playgrounds, gas stations and main streets in the Misurata Area have been determined by inductively coupled plasma atomic emission spectroscopy (ICP-AES). Also, the levels of same heavy metals in industrial areas have been determined. Metal concentration trend variation was also discussed in relation with traffic density and other sources of fugitive emission around different sites on each road/area. The overall mean concentration for main streets was significantly higher ($P < 0.05$) than for other small streets, where Misurata has been the centre of fierce fighting and is located in a frontline battle zone in the Libyan war; therefore most of metal concentrations in surface soil in the fighting area Tripoli Street and Benghazi Street were higher than those from the other sites (outside fighting area).

1. Introduction

Global industrialization and human social and agricultural activities have an effect on environmental pollution and the global ecosystem. The pollution of soil by heavy metals from automobile sources is a serious environmental issue. These metals are released during different operations of the road transport such as combustion, component wear, fluid leakage, and corrosion of metals. Lead, cadmium, copper, and zinc are the major metal pollutants of the roadside environments and are released from fuel burning, wear out of tires, leakage of oils, and corrosion of batteries, and metallic parts such as radiators [1]. Intake of heavy metals. In urban area, heavy metals in urban soil and urban road dusts can be accumulated in human body via direct inhalation, ingestion, and dermal contact absorption.

The most important sources of heavy metals in the environment are the anthropogenic activities such as mining, smelting procedures, steel and iron industry, chemical industry, traffic, and agriculture as well as domestic activities [2–12]. Chemical and metallurgical industries are the most important sources of heavy metals in soil [13–15]. Tracing an life and the environment [16].

The problem of environmental pollution due to toxic metals has begun to cause concern now in most of the major cities. Pollution of the environment with toxic metals has increased dramatically since the onset of the industrial revolution [17]. Soil pollution by heavy metals, such as cadmium, lead, chromium, and copper, and iron, is a problem of concern. Although heavy metals are naturally present in soil, contamination comes from local sources: mostly industry, agriculture, waste incineration, combustion of fossil fuels, and road traffic. Long-range transport of atmospheric pollutants adds to the metals in the natural environment. In recent years, it has been shown that lead levels in soil and vegetation have increased considerably due to traffic pollution, especially from the usage of leaded petrol and exhaust combustion [18].

Biomonitoring of heavy metal is still new in Libya and has its own advantages such as sample is available throughout

Table 1: Heavy metal concentration $\mu g\,g^{-1}$ and pH in roadside dust.

Soil location	pH	Pb	Fe	Zn	Ni	Cd	Cr	Cu
Benghazi Street	6.67	2.1–5.6	42–48	65–96	32.5–39.1	39.1–49.5	36.7–46.8	41–60
Mean		3.85	45	80.5	34.6	46.9	45.3	53.6
Tripoli Street	6.25	2.25–2.92	30–46	73–107	33.5–42.8	39.8–51.2	45.3–52.4	35–59
Mean		2.585	38	90	38.7	48.8	51.3	48.6
Airport Street	6.55	2.8–3.7	20–29	48–116	26.4–30.8	25.1–33.8	30.4–39.8	21–45
Mean		3.25	24.5	82	29.8	27.5	34.5	31.9
Sadon Swihli Street	6.89	2.5–6.5	45–53	65–136	36.4–45.2	40.1–46.7	33.8–43.9	42–66
Mean		4.5	49	100.5	42.5	45.7	37.7	55.9
Qasr Ahmed Street	6.90	4.03–6.65	48–72	55–146	34.5–42.8	38.5–47.8	29.7–36.8	38–55
Mean		5.34	60	100.5	37.2	44.1	31.3	51.5
Aljazera Street	7.30	1.2–3.3	22–28	42–93	13.8–28.4	12.5–34.8	16.7–25.9	23–41
Mean		2.25	25	67.5	22.5	29.1	19.7	32.1

the year, easy to identify the potential biomonitoring agent and to sample, has a wide geographical distribution, and can be relatively tolerant to pollutants [19].

The aim of the present study was to demonstrate the factors affecting human health through the analysis of soil samples collected from industrial areas, nonindustrial areas, and areas of heavy traffic in the City Centre of Misurata during the period of October 2011–May 2012.

2. Materials and Methods

All chemicals and reagents were of analytical grade and were purchased from: (*Sigma-Aldrich, UK*). All glassware was rinsed successively with detergent and distilled water three times prior to use.

2.1. Data Collection and Analysis. Samples were collected during October 2011 and May 2012. 15 Soil samples (three replicates) were collected at surface level (0–10 cm in depth) from various locations to cover industrial, commercial, and residential areas. The procedure of metals determination in soil and dust samples was followed according to the reported method [20]. The soil dust samples were grounded and sieved. The sieved samples were dried at 70°C/24 hrs. One gram of soil was treated with 10 mL concentrated nitric acid heated up to dryness and then cooled. This procedure was repeated with another 10 mL concentrated nitric acid followed by 10 mL of 12 N HCl. The digested soil and dust samples were then warmed in 20 mL of 2 N HCl to re-dissolve the metal salts. Extracts were filtered using Whatman filter paper no. 40, and the volume was then adjusted to 25 mL with 1.5% HNO_3. Heavy metal concentrations of each fraction were inductively coupled plasma atomic emission spectroscopy (ICP-AES) assurance that was guaranteed through double determinations and use of blanks for correction of background and other sources of error. Soil pH was determined with a glass electrode with water-soil slurry (1 : 10).

3. Results and Discussion

Levels of heavy metals in soil and dust samples supplied from different areas of Misurata City are given in Table 1. The results of heavy metal analysis are given below with the subheadings.

The limit of detection (LOD) of the analytical method for each metal was calculated as triple the standard deviation of a series of measurements of a solution, the concentration of which is distinctly detectable above. These values were 0.001, 0.002, 0.001, 0.001, 0.003, 0.001, and 0.002 mg/kg for Pb, Fe, Cd, Zn, Cu, Cr, and Ni, respectively. Also the limit of quantification (LOQ) of the element was determined; these were 0.003, 0.003, 0.003, 0.003, 0.01, 0.003, and 0.007 mg/kg for Pb, Fe, Cd, Zn, Cu, Cr, and Ni, respectively.

Lead. In the present study, the lead content of the roadside dust ranged from 1.2 to 6.65 $\mu g\,g^{-1}$ (Table 1 and Figure 1). Maximum lead content was measured as 6.65 $\mu g\,g^{-1}$ in roadside dust of Qasr Ahmed street this area contains (Libyan Iron and Steel Company (LISCO) Port of Misurata and central petroleum station). The lead level in the Sadon Swihli Street ranges from 2.5 to 6.5 $\mu g\,g^{-1}$. The reason for the high Pb content at Taorghae Street is the heavy traffic in the area and industrial area. The lead content in the Aljazera Street was lower than in other areas.

Iron. The highest Fe value as a mean was at Qasr Ahmed Street and Sadon Swihli Street samples 60 $\mu g\,g^{-1}$ and 49 $\mu g\,g^{-1}$, respectively whereas the lowest Fe concentration was at Airport Street 24.5 $\mu g\,g^{-1}$, Figure 2.

Zinc. The amount of zinc in the roadside dust ranged from 42 to 146 $\mu g\,g^{-1}$ with the mean value of 94 $\mu g\,g^{-1}$. The highest Zn value was at Qasr ahmed street (146 $\mu g\,g^{-1}$), while the lowest value was at Aljazera Street (42 $\mu g\,g^{-1}$). The highest value as a mean was at Qasr ahmed street and Sadon Swihli Street samples (100.5 $\mu g\,g^{-1}$). Alloway [21] and Mcgrath and Loveland [22] reported the mean zinc concentration of

FIGURE 1: Concentration of pb, in soil.

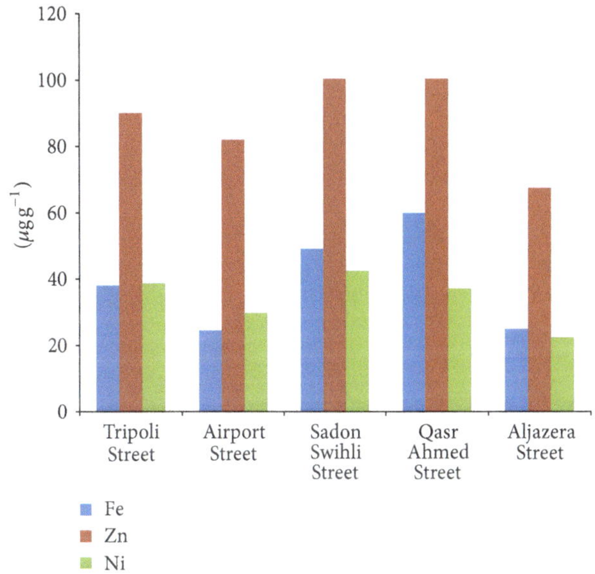

FIGURE 2: Concentration of Fe, Zn, and Ni in soil.

410 $\mu g\,g^{-1}$ in soil collected from the urban roadside soil in Bradford. The industrial area showed the highest mean value of 146 $\mu g\,g^{-1}$, and the seaside had the lowest concentration of 42 $\mu g\,g^{-1}$.

The mobility of the metal depends on the soil pH and also depends on the organic matter and granulometric composition of the soil. Acidic pH makes easier the solubilisation of the Zn compounds.

Nickel. The nickel content for the Sadon Swihli Street (36.4–45.2 $\mu g\,g^{-1}$) was the highest of all the sampling places, whereas the lowest value was at Aljazera street (13.8 $\mu g\,g^{-1}$). The concentration of nickel at Qasr Ahmed Street, Tripoli Street, and Benghazi Street were 34.5–42.8, 33.5–42.8, and 32.5–39.1 $\mu g\,g^{-1}$, respectively. Also, the mobility of the metal depends on soil pH and also depends on the organic matter and granulometric.

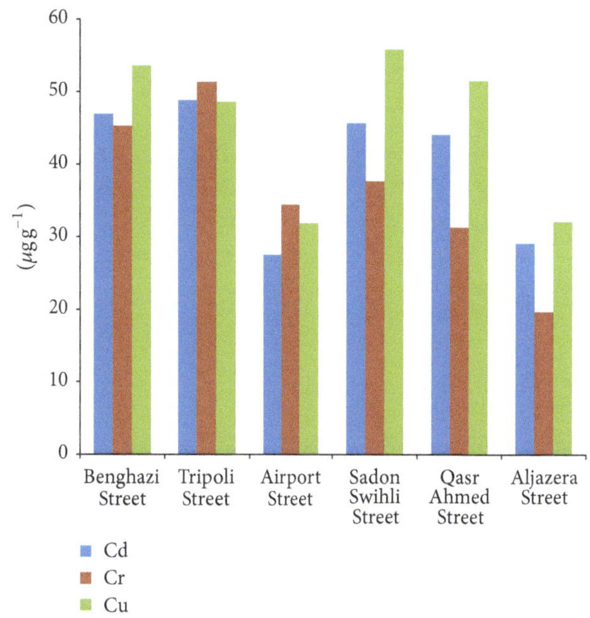

FIGURE 3: Concentration of Cd, Cr, and Cu in soil.

Cadmium. In the present study, the cadmium content of the roadside dust ranged from 27.5 to 48.8 $\mu g\,g^{-1}$ (Table 1 and Figure 3). Maximum cadmium content was measured as 51.2 $\mu g\,g^{-1}$ in roadside dust Tripoli street, and the lowest concentration was 12.5 8 $\mu g\,g^{-1}$ at Aljazera Street. Cadmium levels in roadside soil decrease as distancing from the main road. Also, as previously observed for Pb, the Cd levels in the heavy traffic areas (Tripoli Street, Benghazi Street, and Sadon Swihli Street) were greater than Cd levels along the residential street (Aljazera Street and Airport Street). This feature is attributed to the wear and tear of tires and the greater traffic density on the busy road compared to the residential street.

Chromium. The results of average Cr levels in roadside soil samples from different sites collected from Misurata Area are represented in Table 1. It is observed that the overall level of Cd lies between 45.3 and 52.4 $\mu g\,g^{-1}$ for Tripoli street, 36.7–46.8 $\mu g\,g^{-1}$ for Benghazi street, 30.4–39.8 $\mu g\,g^{-1}$ for Airport street, 33.8–43.9 $\mu g\,g^{-1}$ for Sadon swihli street, 29.7–36.8 $\mu g\,g^{-1}$ for Qasr ahmed street, and 16.7–25.9 $\mu g\,g^{-1}$ for Aljazera street. These results indicate that Cr levels among the sites of each road are significantly different. This indicates that the existence of Cr in roadside soil may be due to the tire erosion. The Cr levels in the heavy traffic areas and industrial areas were greater than Cr levels along the residential street.

Copper. The copper content in the roadside soil ranged from 21 to 60 $\mu g\,g^{-1}$ with the mean values of 31.9–55.9 $\mu g\,g^{-1}$ (Table 1 and Figure 3). Copper is usually present in soil within the range of 0–250 $\mu g\,g^{-1}$ [23].

Muller [24] reported the range of 1.2–1507.7 $mg\,kg^{-1}$ for copper in the soil of England and Wales with a median value of 18.1 $mg\,kg^{-1}$. Total copper content in most of the

TABLE 2: Geoaccumulation index of heavy metals in soil in the areas of Misurata.

Soil location	Pb	Fe	Zn	Ni	Cd	Cr	Cu
Benghazi Street	1.52	1.21	0.39	0.61	2.23	0.13	1.02
Tripoli Street	1.48	0.54	3.42	0.74	2.01	0.21	1.46
Airport Street	0.5	0.45	0.41	−0.38	0.75	−0.24	0.34
Sadon Swihli Street	1.2	2.65	3.56	0.41	0.18	0.43	1.76
Qasr Ahmed Street	0.7	3.43	3.54	−0.27	0.98	−0.8	0.32
Aljazera Street	−0.56	−0.23	0.76	−0.76	−0.35	−1.2	−0.08

roadside soil was below or within the limits of the critical soil concentration of 60–125 μg g^{-1} [25].

3.1. Contamination Levels of Heavy Metals in Soil. The contamination levels of heavy metals in urban soil, urban road dusts, and agricultural soil are assessed by using geoaccumulation index (I_{geo}) introduced by Muller (1969). The method has been widely employed in European trace metal studies since the late 1960s [24]. The I_{geo} is used to assess heavy metal contamination in urban soil by comparing current and preindustrial concentrations, although it is not always easy to reach the preindustrial sediment layers. It is also employed in pollution assessment of heavy metals in urban road dust. Geoaccumulation index is computed using the following equation [24, 25]:

$$I_{geo} = \log_2 (\mathrm{Cn}/1.5\mathrm{Bn}), \quad (1)$$

where Cn is the measured concentration of the element in environment and Bn is the geochemical background value in soil. The constant 1.5 allows us to analyze natural fluctuations in the content of a given substance in the environment and to detect very small anthropogenic influences [26, 27]. According to Muller (1969) [24], the I_{geo} for each metal is calculated and classified as uncontaminated ($I_{geo} \leq 0$); uncontaminated to moderately contaminated ($0 < I_{geo} \leq 1$); moderately contaminated ($1 < I_{geo} \leq 2$); moderately to heavily contaminated ($2 < I_{geo} \leq 3$); heavily contaminated ($3 < I_{geo} \leq 4$); heavily to extremely contaminated ($4 < I_{geo} \leq 5$); and extremely contaminated ($I_{geo} \geq 5$). The I_{geo} values for the metals in urban soil, urban road dusts, and agricultural soil for each area are presented in Table 2, respectively.

In general, Cr and Ni appear to be the least contaminated elements in all the cities, while Pb, Fe, Zn, Cu, and Cd show the highest I_{geo} values for most areas (Table 2). In all the areas, ranges of I_{geo} values for the metals are very wide. The areas of Airport street, Qasr ahmed street, and Aljazera street appear to be the least contaminated areas with low I_{geo} values for Cr, Cu, Pb, Zn, Ni, and Cd, while Tripoli Street, Benghazi Street, and Sadon swihli street, three heavy industrial and traffic areas, show the highest I_{geo} values for the metals.

The highest I_{geo} values for Pb (1.52) and Cd (2.23) are in Benghazi Street. The highest I_{geo} values for Cu (1.76) and Zn (3.56) are found in Sadon Swihli Street, while the highest I_{geo} value for Fe (3.43) is found in Qasr ahmed street. This indicates that the urban soil in these areas are significantly contaminated by the corresponding metals. In general, Tripoli Street and Benghazi Street have the highest I_{geo} values of heavy metals as they are crowded areas and were the centre of fierce fighting and are located in a frontline battle zone in the Misurata-Libyan war.

4. Conclusion

Heavy metal contamination in the soil from the busy roadside verges in the study area was higher as compared to the background levels for lead, iron, zinc, nickel, cadmium, chromium, and copper in residential street. These concentrations, however, were below the critical maximum levels above which toxicity is possible. The highest concentrations were detected in the samples collected from the border zone of the verges, and there was a trend of gradual decrease in the metal contents with the increasing distance from the paved roads. Also, the heavy metals concentration in industrial area was higher as compared to the nonindustrial area.

Higher Pb concentrations were found in sites with a high traffic volume on main roads and in the entrances of petrol station. It seems reasonable to conclude that Pb and Cd in roadside soil levels are significantly higher on busy roads compared to residential roads. Pb and Cd concentration levels in roadside soil decline as distancing from the main roads. We, therefore, conclude that there is a correlation between the roadside soil concentration of heavy metal and the distance from the road. It can also be observed that the frequencies with which motor vehicles stop, start, and accelerate, especially at traffic lights, may help to explain differences in the Pb levels in roadside soil. It is clear that Pb levels vary from time to time and depend on the volume of traffic.

References

[1] L. M. J. Dolan, H. Van Bohemen, P. Whelan et al., "Towards the sustainable development of modern road ecosystem," in *The Ecology of Transportation: Managing Mobility For the Environment*, J. Davenport and J. L. Davenport, Eds., pp. 275–331, Springer, Amsterdam, The Netherlands, 2006.

[2] I. Suciu, C. Cosma, M. Todică, S. D. Bolboacă, and L. Jäntschi, "Analysis of soil heavy metal pollution and pattern in central transylvania," *International Journal of Molecular Sciences*, vol. 9, no. 4, pp. 434–453, 2008.

[3] E. I. B. Chopin and B. J. Alloway, "Distribution and mobility of trace elements in soils and vegetation around the mining and smelting areas of Tharsis, Ríotinto and Huelva, Iberian Pyrite Belt, SW Spain," *Water, Air, and Soil Pollution*, vol. 182, no. 1–4, pp. 245–261, 2007.

[4] C. Stihi, A. Bancuta, I. V. Popescu et al., "Air pollution studies using PIXE and ICP methods," *Journal of Physics*, vol. 41, no. 1, article 070, pp. 565–568, 2006.

[5] R. Garcia and E. Millán, "Assessment of Cd, Pb and Zn contamination in roadside soils and grasses from Gipuzkoa (Spain)," *Chemosphere*, vol. 37, no. 8, pp. 1615–1625, 1998.

[6] X. Li, C. S. Poon, and P. S. Liu, "Heavy metal contamination of urban soils and street dusts in Hong Kong," *Applied Geochemistry*, vol. 16, no. 11-12, pp. 1361–1368, 2001.

[7] N. Sezgin, H. K. Ozcan, G. Demir, S. Nemlioglu, and C. Bayat, "Determination of heavy metal concentrations in street dusts in

Istanbul E-5 highway," *Environment International*, vol. 29, no. 7, pp. 979–985, 2004.

[8] B. Viard, F. Pihan, S. Promeyrat, and J. C. Pihan, "Integrated assessment of heavy metal (Pb, Zn, Cd) highway pollution: bioaccumulation in soil, graminaceae and land snails," *Chemosphere*, vol. 55, no. 10, pp. 1349–1359, 2004.

[9] G. Nabulo, H. Oryem-Origa, and M. Diamond, "Assessment of lead, cadmium, and zinc contamination of roadside soils, surface films, and vegetables in Kampala City, Uganda," *Environmental Research*, vol. 101, no. 1, pp. 42–52, 2006.

[10] S. R. Oliva and A. J. F. Espinosa, "Monitoring of heavy metals in topsoils, atmospheric particles and plant leaves to identify possible contamination sources," *Microchemical Journal*, vol. 86, no. 1, pp. 131–139, 2007.

[11] M. Kampa and E. Castanas, "Human health effects of air pollution," *Environmental Pollution*, vol. 151, no. 2, pp. 362–367, 2008.

[12] L. Guo-li, L. Da-xue, and L. Quan-ming, "Heavy metals contamination characteristics in soil of different mining activity zone Trans," *Transactions of Nonferrous Metals Society of China*, vol. 18, no. 1, pp. 207–211, 2008.

[13] A. Pantelica, V. Cercasov, E. Steinnes, P. Bode, and B. Wolterbeek, in *Proceedings of the 4th National Conference of Applied Physics (NCAP '08)*, pp. 25–26, Galati, Romania, September 2008.

[14] W. de Vries, P. F. Römkens, and G. Schütze, "Critical soil concentrations of cadmium, lead, and mercury in view of health effects on humans and animals," *Reviews of Environmental Contamination and Toxicology*, vol. 191, pp. 91–130, 2007.

[15] V. Cojocaru, A. Pantelică, E. Pincovschi, and I. I. Georgescu, "EDXRF versus INAA in a pollution control of soil," *Journal of Radioanalytical and Nuclear Chemistry*, vol. 268, no. 1, pp. 71–78, 2006.

[16] Bangalore Metropolitan Rapid Transport Limited, *Environmental Impact Analysis*, 2006.

[17] J. O. Nriagu, "Global inventory of natural and anthropogenic emissions of trace metals to the atmosphere," *Nature*, vol. 279, no. 5712, pp. 409–411, 1979.

[18] F. Cabrera, L. Clemente, E. Díaz Barrientos, R. López, and J. M. Murillo, "Heavy metal pollution of soils affected by the Guadiamar toxic flood," *Science of the Total Environment*, vol. 242, no. 1–3, pp. 117–129, 1999.

[19] R. Wittig, *General Aspects of Bio-Monitoring Heavy Metals by Plants, Plants as Bio-Monitors. Indicators for Heavy Metal in Terrestrial Environment*, VCH Press, Weinheim, Germany, 1993.

[20] X. Chen, X. Xia, S. Wu, F. Wang, and X. Guo, "Mercury in urban soils with various types of land use in Beijing, China," *Environmental Pollution*, vol. 158, no. 1, pp. 48–54, 2010.

[21] B. J. Alloway, *HeavyMetals in Soils*, Chapman & Hall, London, UK, 1995.

[22] S. P. Mcgrath and P. J. Loveland, *The Soil Geochemical Atlas of England and Wales*, Blackie Academic & Professional, London, UK, 1992.

[23] ICRCL, *Interdepartmental Committee on the Redevelopment of Contaminated Land. Guidance on the Assessment and Redevelopment of Contaminated Land. Guidance Note. 59/83*, Department of Environment, London, UK, 1987.

[24] G. Muller, "Index of geo-accumulation in sediments of the Rhine River," *Geo Journal*, vol. 24, no. 2, pp. 108–118, 1969.

[25] Y. Ji, Y. Feng, J. Wu, T. Zhu, Z. Bai, and C. Dua, "Using geo-accumulation index to study source profiles of soil dust in China," *Journal of Environmental Sciences*, vol. 20, no. 5, pp. 571–578, 2008.

[26] CEPA, *(Chinese Environmental Protection Administration, Elemental Background Values of Soils in China*, Environmental Science Press of China, Beijing, China, 1990.

[27] CEPA, *(Chinese Environmental Protection Administration, Environmental Quality Standard For Soils (GB15618-1995)*, Beijing, China, 1995.

10

Low-Level Detection of Poly(amidoamine) PAMAM Dendrimers Using Immunoimaging Scanning Probe Microscopy

Chevelle A. Cason,[1] Thomas A. Fabré,[1] Andrew Buhrlage,[1] Kristi L. Haik,[2] and Heather A. Bullen[1]

[1] *Department of Chemistry, Northern Kentucky University, Highland Heights, KY 41099, USA*
[2] *Department of Biological Sciences, Northern Kentucky University, Highland Heights, KY 41099, USA*

Correspondence should be addressed to Heather A. Bullen, bullenh1@nku.edu

Academic Editor: Charles L. Wilkins

Immunoimaging scanning probe microscopy was utilized for the low-level detection and quantification of biotinylated G4 poly(amidoamine) PAMAM dendrimers. Results were compared to those of high-performance liquid chromatography (HPLC) and found to provide a vastly improved analytical method for the low-level detection of dendrimers, improving the limit of detection by a factor of 1000 (LOD = 2.5×10^{-13} moles). The biorecognition method is reproducible and shows high specificity and good accuracy. In addition, the capture assay platform shows a promising approach to patterning dendrimers for nanotechnology applications.

1. Introduction

Dendrimers are at the forefront of research in nanoscience due to the many interesting properties of these macromolecular systems including their precise architecture, highly reproducible shape, high uniformity and purity, low immunogenicity and toxicity, high loading capacity, and high shear resistance [1–5]. They have shown a great deal of versatility with applications in numerous areas such as drug delivery [6, 7], gene therapy [8, 9], chemotherapy [10], electrochemistry [11, 12], metal recovery [13], catalysis [14, 15], and sensors [16–18]. Development of new low-level detection and quantification methods is needed with the utilization of these nanomaterials. Currently, high-performance liquid chromatography (HPLC) is the predominate approach reported for dendrimer quantification [19, 20]. However, the primary focus of HPLC, along with capillary electrophoresis, has been to evaluate dendrimer purity and degree of conjugation [21–27]. Little has been reported within the literature with regard to the advancement of new quantification methods for dendrimers.

This work introduces a biorecognition readout technique that has the potential to provide low-level detection of dendrimers. Biotinylated poly(amidoamine) PAMAM dendrimers were chosen as a model target. PAMAM dendrimers, which are highly water soluble, represent the most widely studied class of dendrimers. Functionalization of PAMAM dendrimer surfaces has proven useful in their utilization for various applications including drug delivery and chemical sensing [5, 6, 16]. Biotin-labeled dendrimers have been utilized in tumor [28] and antibody [29] targeting studies and biosensor design [30]. Biotinylated PAMAM dendrimers may also have the potential for delivering therapeutic drugs to the brain [31, 32].

We report here a readout method using an immunoassay platform and scanning probe microscopy (SPM) for low-level quantification of biotinylated G4 PAMAM dendrimers. The assay takes advantage of the documented specificity of biotin-avidin. Results are correlated with HPLC analysis. In addition, we briefly highlight the potential of this capture assay platform to selectively pattern PAMAM dendrimers onto a surface. Patterning of nanoparticles is relevant to a wide variety of applications in the fields of sensing, drug delivery, or development of nanodevices [33–35]. Dendritic architectures show promise in designing and developing sensor platforms with high sensitivity and stability [16].

2. Experimental

2.1. Reagents. Poly(amidoamine) PAMAM dendrimers [core: ethylene diamine] (G = 4) *dendri*-PAMAM-$(NH_2)_{32}$ were obtained from Dendritic Nanotechnologies, Inc. (Mt. Pleasant, MI). Biotinylated PAMAM dendrimers were prepared using sulfo-NHS-LC-biotin (Pierce EZ-Link Kit) as described previously [36]. Briefly, a 3 : 1 molar ratio of biotin/PAMAM dendrimers in 0.1 M phosphate buffer saline (PBS) was allowed to react for 2 h on an orbital shaker. Excess, unreacted biotin was then removed using Microcon filters (Millipore. Bedford, MA, USA). The biotinylation of dendrimers was evaluated using NMR spectroscopy. Biotinylated dendrimers were resuspended (1.0 mg/mL) in 1.0 M PBS until used. Octadecanethiol (ODT), 3,3′-dithio-bis(propionic acid N-hydroxysuccinimide ester) (DSP), bovine serum albumin (BSA), Triton X-100, and avidin >98% were obtained from Sigma (Sigma-Aldrich, St. Louis, MO). Avidin conjugated to Alexa Fluor 488 was purchased from Invitrogen (Invitrogen, Carlsbad, CA). Poly(dimethyl siloxane) (PDMS) was obtained from Dow Corning (Midland, MI). All organic solvents used were analytical, HPLC grade, from Sigma (Sigma-Aldrich, St. Louis, MO). DI water was obtained using a Milli-Q plus water purification system (Millipore, Bedford, MA). PBS and Borate buffers were prepared from Pierce buffer packs (Pierce Protein Research Products, Rockford, IL).

2.2. Capture Substrate Preparation. A modified approach was used for preparation of the capture substrate [37–39] utilizing template-stripped gold (TSG) for SPM imaging, as shown in Figure 1. TSG was prepared by evaporating gold onto p-type silicon wafers (University Wafer) with a resistive evaporator and affixing 1 × 1 cm glass pieces (ultrasonically cleaned 30 min each in diluted Contrad 70, DI water, and methanol) using two-part epoxy (Epoxy Technology) followed by curing at 150°C for 2 h. The glass pieces were gently detached from the silicon wafer revealing a smooth gold surface atop the glass chip.

The TSG substrates were exposed for ~30 s to an ODT soaked PDMS stamp (with a 3 mm diameter hole cut in the center), rinsed with ethanol, and dried under high-purity nitrogen. The substrates were then placed in a 0.1 mM solution of DSP in ethanol overnight. The capture platform was then rinsed with ethanol and dried under N_2. This formed the DSP-based adlayer in the areas on the substrate not covered by ODT. The hydrophobic ODT localized reagents in a confined sample area (3 mm ODT spot size) for the capture assay platform.

To form the capture avidin surface, a 20 μL aliquot of avidin solution (500 μg/mL diluted in 50 mM borate buffer) was placed on top of the sample area and allowed to incubate for 6 h at room temperature in a humidity chamber. Substrates were then rinsed with 5 mL of 10 mM PBS (with 0.1% Triton X-100), and the surface area was incubated with a 20 μL solution of blocking buffer (1% (w/v) BSA in 20 mM borate buffer with 0.1% Triton X) for 5 h, followed by rinsing with 5 mL of 10 mM PBS. The capture substrate was then exposed to 20 μL aliquots of various concentrations

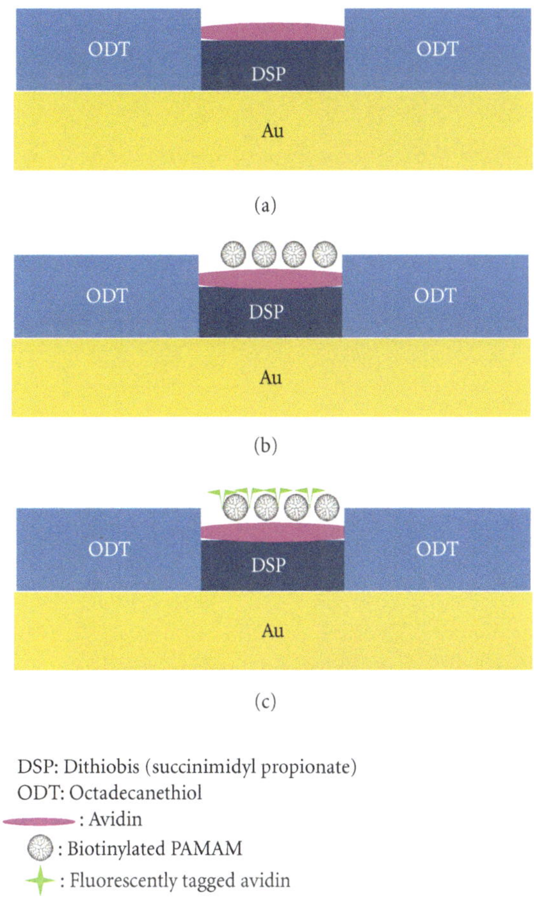

Figure 1: Schematic of capture assay platform: (a) immobilize avidin onto Au substrate with controlled surface area using DSP and (b) expose substrate to biotinylated PAMAM dendrimers. (c) The captured biotinylated PAMAM dendrimers can be fluorescently labeled for additional verification of immobilization and patterning of dendrimers.

of biotinylated G4 PAMAM dendrimers diluted in 10 mM PBS for 8 h at room temperature in a humidity chamber. Substrates were then rinsed with 5 mL of 10 mM borate buffer before drying with nitrogen. A blank sample consisted of all stages except the avidin capture surface.

To demonstrate patterning of biotinylated G4 PAMAM dendrimers, a PDMS stamp with a positive structure of 10 μm wide stripes separated by 5 μm was utilized. Preparation of the patterned substrate consisted of all stages, with the following modifications: (1) a sample volume of 200 μL was used instead of 20 μL and (2) after incubation with biotinylated G4 PAMAM dendrimers, the substrate was rinsed with 10 mM PBS and the dendrimers were fluorescently tagged by incubating the substrate with 200 μL of 250 ug/mL of Alexa Fluor conjugated avidin. Following this step the substrates were rinsed with 5 mL of 10 mM borate buffer and dried with nitrogen.

2.3. Scanning Probe Microscopy. A Dimension 3100 Digital Instruments SPM was utilized in tapping mode equipped with 125 μm n-doped silicon cantilevers with resonance frequencies between 110 and 220 kHz and typical scan rates

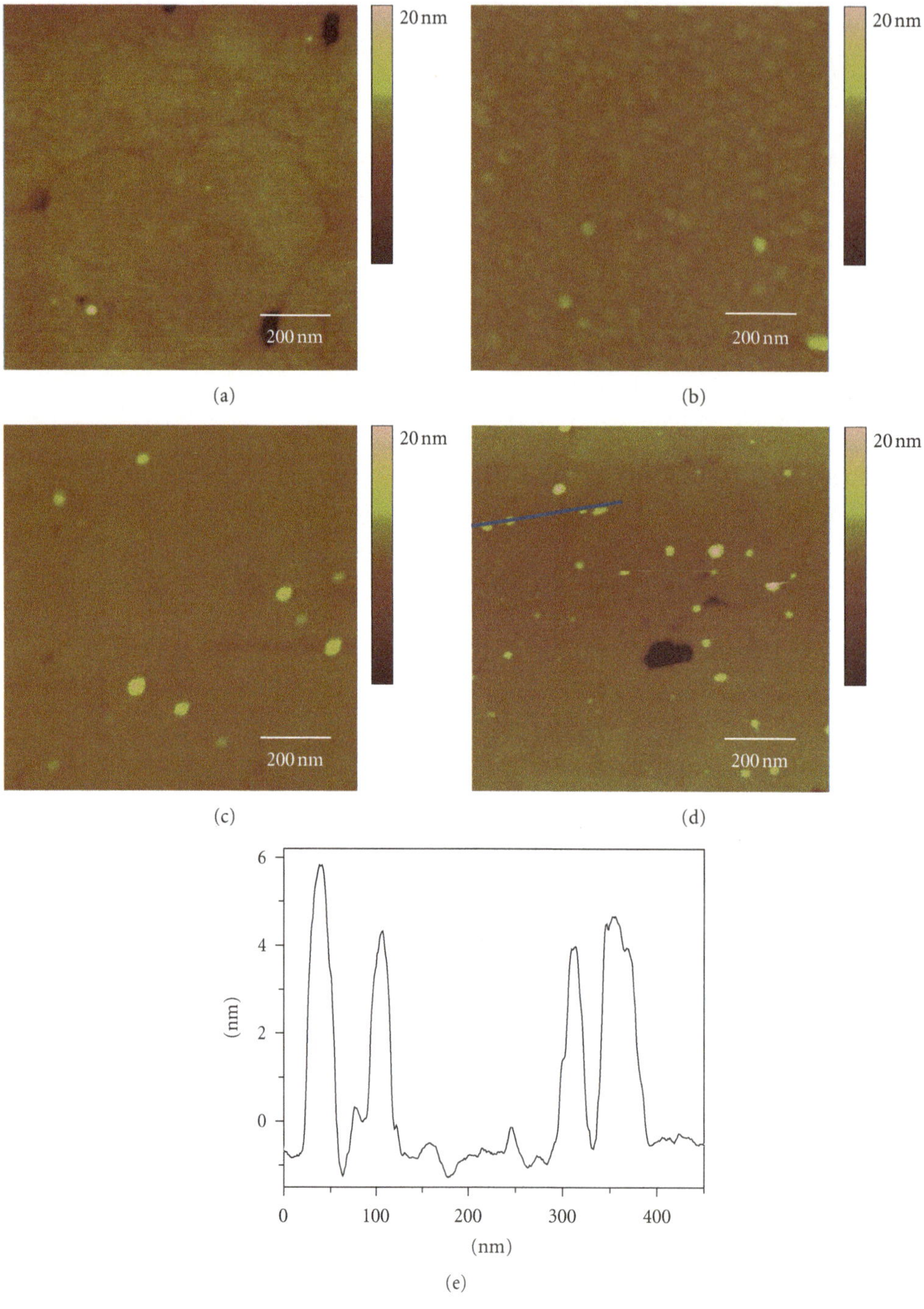

FIGURE 2: Representative SPM images of capture surface (1×1 μm) of different concentrations of bound biotinylated G4 PAMAM dendrimers to avidin capture substrate: (a) blank and (b) 7.0×10^{-3} μmol/L, (c) 7.0×10^{-2} μmol/L, and (d) 3.28×10^{-1} μmol/L; (e) a cross-sectional analysis of captured biotinylated G4 PAMAM dendrimers in (d).

of 0.75–1 Hz (NSC 14, MikroMasch). Typical scan rates were 1 Hz, and all images were acquired in air. For quantification, multiple areas of the capture surface were analyzed and the dendrimers in each 1×1 μm image were enumerated manually.

2.4. High-Performance Liquid Chromatography. HPLC analysis of G4 biotinylated PAMAM dendrimers was conducted using parameters reported previously [36]. Briefly, a reversed phase HPLC system consisting of a Waters Breeze HPLC (Waters Corporation, Milford, MA, USA), equipped with a 717plus autosampler, 2487 dual λ UV detector, 5 μm Symmetry300 C18 column (4.6 mm × 150 mm), and a Waters Sentry Symmetry C18 guard column was used. The mobile phase consisted of a linear gradient beginning with 90% water and 10% ACN (each with 0.14% TFA) at a flow rate of 1 mL/min reaching 68% water and 32% ACN over 10 minutes. The injection volume was 100 μL, and detection of eluted samples was performed at 214 nm.

2.5. Fluorescent Microscopy. A Nikon E600 fluorescent microscope with a Mercury-100 W light source was utilized to take images at 100x total magnification.

3. Results and Discussion

The principle goal of this work was the development and application of capture platform for quantification of PAMAM dendrimers. A model assay was evaluated using biotinylated G4 PAMAM dendrimers and avidin. Avidin was immobilized onto a patterned thiolate layer, DSP, via succinimidyl ester chemistry (Figure 1). In the presence of biotinylated G4 PAMAM dendrimers, the immobilized avidin layer extracts the dendrimers. SPM analysis of the capture platform revealed the presence of nanometrically sized objects, ~4.5 nm in dimension, according to cross-sectional height profiles. These are consistent in size and shape of G4 biotinylated PAMAM dendrimers, taking into account tip convolution effects (Figure 2). At this scale, the tip convolution effects distorted the resolution in the lateral dimensions, as the dimensions of the particles are smaller than the radius of curvature of the AFM probe used (radius = 20 nm) [40, 41]. The dendrimers largely appear to be captured individually on the platform surface. In some cases dendrimers were found grouped together (2-3 dendrimer clusters) but could be individually distinguished using cross-sectional analysis. SPM analysis reveals that the dendrimers do not have a tendency to cluster together in domains, indicating that sedimentation is not a factor during the incubation. Analysis of control (blank) substrates, devoid of the avidin-specific capture surface, shows limited nonspecific binding (0.67 ± 0.45 dendrimers/μm^2) and that blocking steps are effective.

The patterning of DSP provides a reproducible surface area for quantification. Figure 2 shows that the number of captured G4 biotinylated PAMAM dendrimers tracks with the increase in dendrimer solution concentration. The degree of bound dendrimers was similar for various different areas of the capture surface. A calibration plot of the amount of captured G4 biotinylated PAMAM dendrimers (determined by SPM analysis) as a function of solution concentration is shown in Figure 3(a). The number of captured dendrimers correlates with G4 biotinylated PAMAM dendrimer concentrations, showing a linear dynamic range under the concentrations range evaluated (7.0×10^{-3}–3.28×10^{-1} μmol/L). This readout method has the potential for low limits of detection. The limit of detection can be estimated as the signal three times above the measurement noise, where the noise is the sample blank response. This corresponds to a limit of detection of ~2.5×10^{-13} moles. Results were correlated with a standard HPLC analysis approach for quantification of dendrimers, as shown in Figure 3(b). HPLC quantification of dendrimers shows a linear dynamic range under the concentrations evaluated

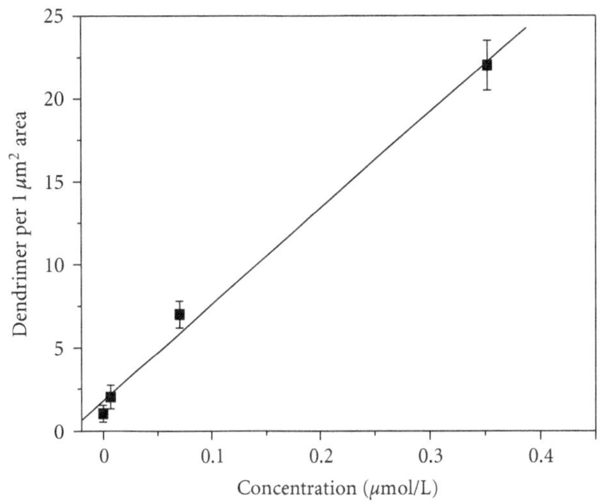

(a)

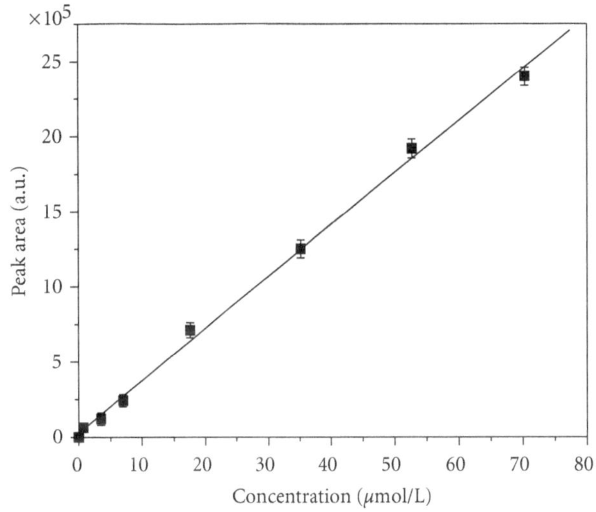

(b)

FIGURE 3: Calibration curves for (a) immunoimaging SPM assay and (b) HPLC analysis of G4 biotinylated PAMAM dendrimers. The SPM analysis represents the average number of dendrimers bound within a $1 \times 1\,\mu m$ area for different concentrations of dendrimers. Data acquired from 10 capture substrates, with 10 images per sample ($y = 58.1x + 1.76$, $n = 20$, $r^2 = 0.9964$). The HPLC analysis represents an average of 10 samples for each concentration evaluated ($y = 34620.85x + 28845.984$, $n = 10$, $r^2 = 0.9986$).

(7.0×10^{-1}–$70.35\,\mu$mol/L) with a corresponding limit of detection of 1.8×10^{-10} moles. The capture assay platform improves the limit of detection by a factor of 1000. It should also be noted that the concentrations investigated by the capture assay were not detectable by HPLC analysis.

The capture assay platform has shown the potential for low-level detection and quantification of G4 biotinylated PAMAM dendrimers. Further improvements in the performance of this immunosorbent assay approach such as the reduction of incubation times may be possible by utilizing a rotating capture substrate to increase the flux of analyte to the capture surface [38]. In addition, the limit of detection may be further improved by reducing the area dimension

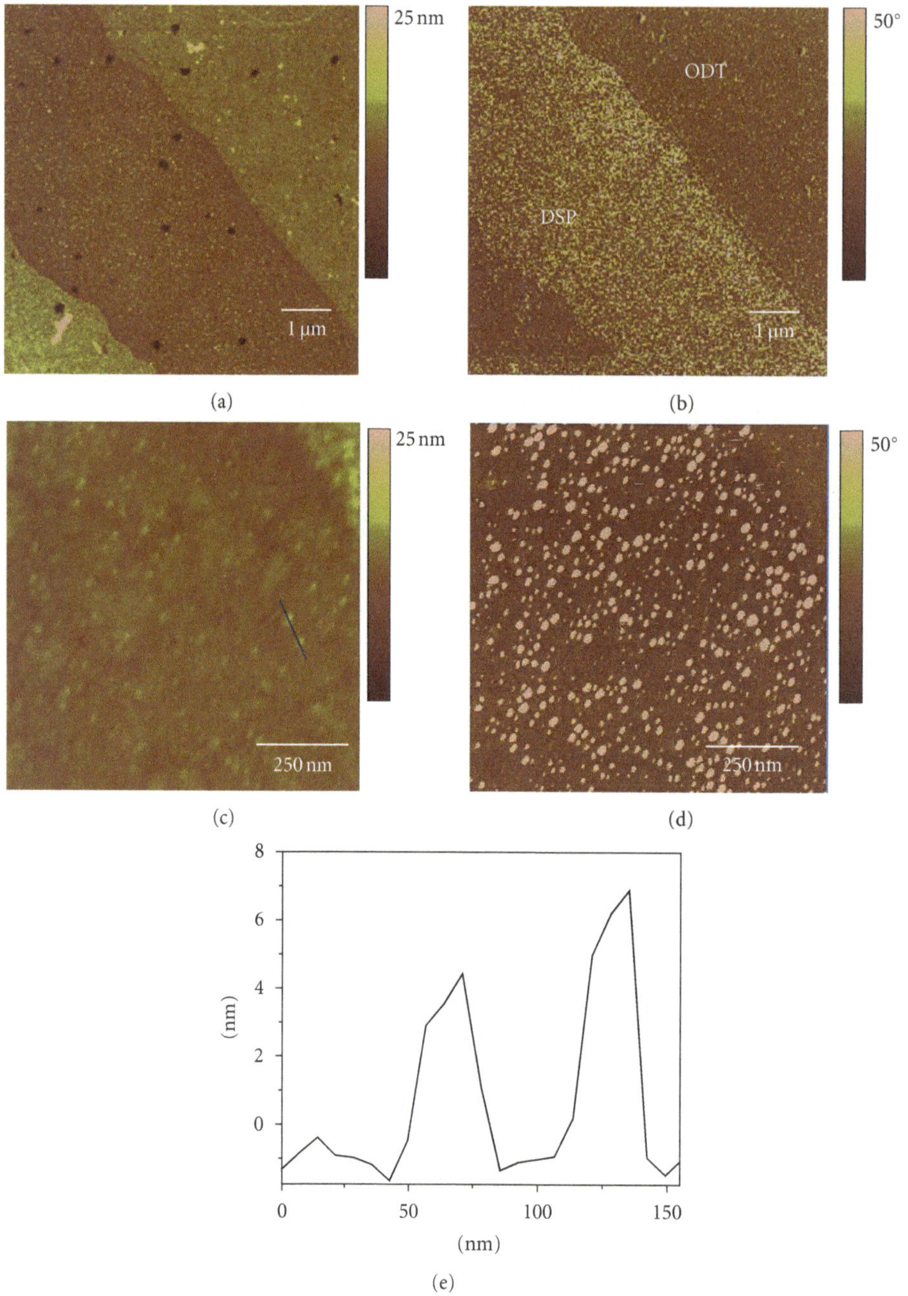

FIGURE 4: SPM image of capture surface patterning: (a) height and (b) phase ($8 \times 8\,\mu$m), zoomed in analysis; (c) height and (b) phase of capture stripe (avidin immobilized on DSP) containing bound G4 biotinylated PAMAM dendrimers. Cross-section analysis of bound G4 biotinylated PAMAM dendrimers in (d) is shown in (e). The capture platform was exposed to 200 μL of 5.0 μmol/L concentration of G4 biotinylated PAMAM dendrimer solution.

of the DSP capture platform. The range of dendrimers that can be quantified using this approach is potentially limited to the analyte capture surface and labeling strategy of the dendrimers. This quantification method can be applied to any type of immunosorbant assay approach (e.g., antibody antigen binding). In addition, the capture platform design could be utilized for multidetection capability. It is important to note, however, that quantification of unfunctionalized dendrimers would require a labeling strategy for this capture assay approach to be used.

The capture assay platform also has the ability to selectively pattern PAMAM dendrimers onto a surface, as shown in Figure 4. In this study a PDMS stamp with a positive structure of 10 μm wide stripes separated by 5 μm stripes was used to pattern G4 biotinylated PAMAM dendrimers using the assay approach. Figures 4(a) and 4(b) show the height and phase analysis of the patterned surface. The bright stripes in the height analysis (Figure 4(a)) correspond to the ODT layer (10 μm wide), and the darker stripe corresponds to the DSP layer (5 μm wide). The height difference (~2.5 nm)

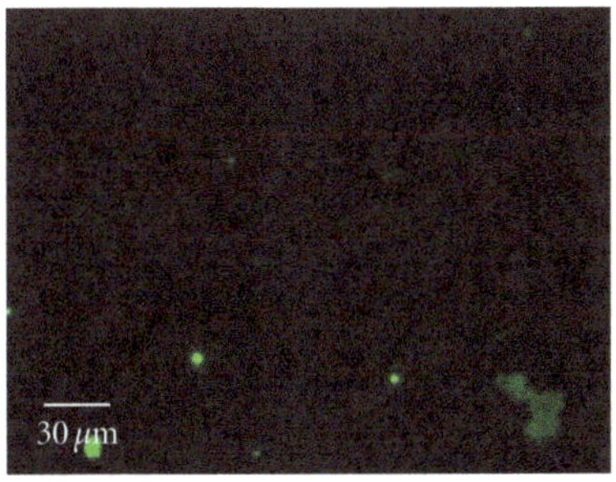

(a)

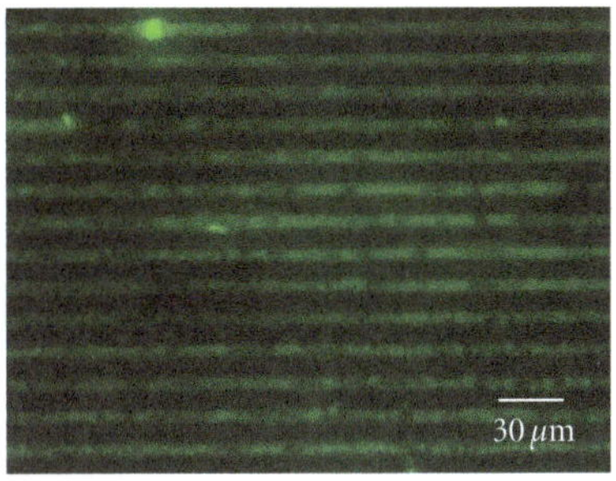

(b)

FIGURE 5: Fluorescence image of capture surface assay utilizing Alexa Fluor conjugated avidin: (a) blank and (b) G4 biotinylated PAMAM dendrimers bound to avidin capture substrate.

is consistent with the differing dimensions of ODT and DSP monolayers. It can be seen from the phase analysis that the captured dendrimers are more easily identifiable in the DSP stripe (Figure 4(b)). Figures 4(c) and 4(d) present higher-resolution height and phase analysis of the patterned G4 biotinylated PAMAM dendrimers. In these SPM images the individual dendrimers bound to the DSP stripe can be ascertained. The dendrimers are individually distributed within the DSP stripe. The dendrimers do not clump together and multilayers of dendrimers are not evident. Furthermore, the ODT layers do not show any evidence of nonspecific adsorption. Cross-sectional height profiles (~4.5 nm) are consistent with G4 dendrimers. The slight variabilities in the height may be due to variations in how the biotin on the dendrimer surface is bound to the avidin capture platform. Further validation of the patterning of the G4 biotinylated PAMAM dendrimers was conducted by tagging the bound dendrimers with Alexa Fluor conjugated avidin, as shown in Figure 5. Fluorescence imaging verifies that the dendrimers are patterned on the avidin capture surface over large surface areas. Analysis of the blank confirms that nonspecific adsorption of G4 biotinylated PAMAM dendrimers is minimal. The bound patterned dendrimers have been found to be stable on the capture platform for >1 year under ambient storage conditions. Although, methods such as direct microcontact printing [42] dip-pen lithography [43], and electron-beam patterning [44] of dendrimers have been reported previously, they do not provide the selective patterning of this approach. The dimensions and profiles of the patterned dendrimers can be modified using different PDMS platforms.

4. Conclusions

This paper has demonstrated the potential value of immunoimaging SPM to detect and quantify G4 biotinylated PAMAM dendrimers. Results were presented that showed that this new approach using an immunoassay platform provides a 1000-fold improvement in the limit of detection of dendrimers compared to current quantification methods using HPLC. The range of dendrimers that could be evaluated with this approach is potentially limited to the availability of analyte capture and labeling strategies. Findings have also shown the potential of this platform to pattern dendrimers for sensor and nanodevice applications.

Acknowledgments

This project has been funded by the Merck Institute for Science Education, Kentucky NSF EPSCoR, Northern Kentucky University, Center for Integrated Natural Science and Mathematics, and the NKU Research Foundation. This work was also supported by a grant from the Kentucky Biomedical Research Infrastructure Network (P20 RR016481-08).

References

[1] G. M. Dykes, "Dendrimers: a review of their appeal and applications," *Journal of Chemical Technology and Biotechnology*, vol. 76, no. 9, pp. 903–918, 2001.

[2] D. A. Tomalia, "Birth of a new macromolecular architecture: dendrimers as quantized building blocks for nanoscale synthetic polymer chemistry," *Progress in Polymer Science*, vol. 30, no. 3-4, pp. 294–324, 2005.

[3] D. A. Tomalia, "The emergence of a new macromolecular architecture: the dendritic state," in *Physical Properties of Polymers Handbook*, J. E. Mark, Ed., pp. 671–692, Springer, New York, NY, USA, 2007.

[4] D. A. Tomalia, S. A. Henderson, and M. S. Diallo, "Dendrimers—an enabling synthetic science to controlled organic nanostructures," in *Handbook of Nanoscience, Engineering and Technology*, W. A. I. Goddard, D. W. Brenner, S. E. Lyshevski, and G. J. Iafrate, Eds., pp. 24.1–24.47, CRC Press, Taylor and Francis, Boca Raton, Fla, USA, 2007.

[5] D. Astruc, E. Boisselier, and C. Ornelas, "Dendrimers designed for functions: from physical, photophysical, and supramolecular properties to applications in sensing, catalysis, molecular electronics, photonics, and nanomedicine," *Chemical Reviews*, vol. 110, no. 4, pp. 1857–1959, 2010.

[6] V. Gajbhiye, V. K. Palanirajan, R. K. Tekade, and N. K. Jain, "Dendrimers as therapeutic agents: a systematic review," *Journal of Pharmacy and Pharmacology*, vol. 61, no. 8, pp. 989–1003, 2009.

[7] W. Wijagkanalan, S. Kawakami, and M. Hashida, "Designing dendrimers for drug delivery and imaging: pharmacokinetic considerations," *Pharmaceutical Research*, vol. 28, no. 7, pp. 1500–1519, 2011.

[8] K. C. Petkar, S. S. Chavhan, S. Agatonovik-Kustrin, and K. K. Sawant, "Nanostructured materials in drug and gene delivery: a review of the state of the art," *Critical Reviews in Therapeutic Drug Carrier Systems*, vol. 28, no. 2, pp. 101–164, 2011.

[9] M. A. Mintzer and M. W. Grinstaff, "Biomedical applications of dendrimers: a tutorial," *Chemical Society Reviews*, vol. 40, no. 1, pp. 173–190, 2011.

[10] Y. Cheng, L. Zhao, Y. Li, and T. Xu, "Design of biocompatible dendrimers for cancer diagnosis and therapy: current status and future perspectives," *Chemical Society Reviews*, vol. 40, no. 5, pp. 2673–2703, 2011.

[11] E. Bustos, J. Manríquez, G. Orozco, and L. A. Godínez, "Preparation, characterization, and electrocatalytic activity of surface anchored, Prussian Blue containing starburst PAMAM dendrimers on gold electrodes," *Langmuir*, vol. 21, no. 7, pp. 3013–3021, 2005.

[12] N. Zhu, Y. Gu, Z. Chang, P. He, and Y. Fang, "PAMAM dendrimers-based DNA biosensors for electrochemical detection of DNA hybridization," *Electroanalysis*, vol. 18, no. 21, pp. 2107–2114, 2006.

[13] M. S. Diallo, S. Christie, P. Swaminathan, J. H. Johnson Jr., and W. A. Goddard III, "Dendrimer enhanced ultrafiltration—1. Recovery of Cu(II) from aqueous solutions using PAMAM dendrimers with ethylene diamine core and terminal NH 2 groups," *Environmental Science and Technology*, vol. 39, no. 5, pp. 1366–1377, 2005.

[14] L. J. Twyman, A. S. H. King, and I. K. Martin, "Catalysis inside dendrimers," *Chemical Society Reviews*, vol. 31, no. 2, pp. 69–82, 2002.

[15] J. N. H. Reek, D. De Groot, G. E. Oosterom, P. C. J. Kamer, and P. W. N. M. Van Leeuwen, "Core and periphery functionalized dendrimers for transition metal catalysis; a covalent and a non-covalent approach," *Reviews in Molecular Biotechnology*, vol. 90, no. 3-4, pp. 159–181, 2002.

[16] J. Satija, V. V. R. Sai, and S. Mukherji, "Dendrimers in biosensors: concept and applications," *Journal of Materials Chemistry*, vol. 21, pp. 14367–14386, 2011.

[17] J. Losada, M. Zamora, P. García Armada, I. Cuadrado, B. Alonso, and C. M. Casado, "Bienzyme sensors based on novel polymethylferrocenyl dendrimers," *Analytical and Bioanalytical Chemistry*, vol. 385, no. 7, pp. 1209–1217, 2006.

[18] H. Cavaye, P. E. Shaw, A. R.G. Smith et al., "Solid state dendrimer sensors: effect of dendrimer dimensionality on detection and sequestration of 2,4-dinitrotoluene," *Journal of Physical Chemistry C*, vol. 115, no. 37, pp. 18366–18371, 2011.

[19] J. Giri, M. S. Diallo, W. A. Goddard, N. F. Dalleska, X. Fang, and Y. Tang, "Partitioning of poly(amidoamine) dendrimers between n-octanol and water," *Environmental Science and Technology*, vol. 43, no. 13, pp. 5123–5129, 2009.

[20] D. G. Mullen, E. L. Borgmeier, A. M. Desai et al., "Isolation and characterization of dendrimers with precise numbers of functional groups," *Chemistry*, vol. 16, no. 35, pp. 10675–10678, 2010.

[21] S. Mutalik, A. K. Hewavitharana, P. N. Shaw, Y. G. Anissimov, M. S. Roberts, and H. S. Parekh, "Development and validation of a reversed-phase high-performance liquid chromatographic method for quantification of peptide dendrimers in human skin permeation experiments," *Journal of Chromatography B*, vol. 877, no. 29, pp. 3556–3562, 2009.

[22] S. Lalwani, V. J. Venditto, A. Chouai, G. E. Rivera, S. Shaunak, and E. E. Simanek, "Electrophoretic behavior of anionic triazine and pamam dendrimers: methods for improving resolution and assessing purity using capillary electrophoresis," *Macromolecules*, vol. 42, no. 8, pp. 3152–3161, 2009.

[23] M. T. Islam, I. J. Majoros, and J. R. Baker Jr., "HPLC analysis of PAMAM dendrimer based multifunctional devices," *Journal of Chromatography B*, vol. 822, no. 1-2, pp. 21–26, 2005.

[24] M. T. Islam, X. Shi, L. Balogh, and J. R. Baker, "HPLC separation of different generations of poly(amidoamine) dendrimers modified with various terminal groups," *Analytical Chemistry*, vol. 77, no. 7, pp. 2063–2070, 2005.

[25] X. Shi, I. Bányai, K. Rodriguez et al., "Electrophoretic mobility and molecular distribution studies of poly(amidoamine)dendrimers of defined charges," *Electrophoresis*, vol. 27, no. 9, pp. 1758–1767, 2006.

[26] X. Shi, I. Bányai, M. T. Islam et al., "Generational, skeletal and substitutional diversities in generation one poly(amidoamine) dendrimers," *Polymer*, vol. 46, no. 9, pp. 3022–3034, 2005.

[27] X. Shi, X. Bi, T. R. Ganser et al., "HPLC analysis of functionalized poly(amidoamine) dendrimers and the interaction between a folate-dendrimer conjugate and folate binding protein," *Analyst*, vol. 131, no. 7, pp. 842–848, 2006.

[28] N. Sato, H. Kobayashi, T. Saga et al., "Tumor targeting and imaging of intraperitoneal tumors by use of antisense oligo-DNA complexed with dendrimers and/or avidin in mice," *Clinical Cancer Research*, vol. 7, no. 11, pp. 3606–3612, 2001.

[29] D. S. Wilbur, P. M. Pathare, D. K. Hamlin, K. R. Buhler, and R. L. Vessella, "Biotin reagents for antibody pretargeting. 3. Synthesis, radioiodination, and evaluation of biotinylated starburst dendrimers," *Bioconjugate Chemistry*, vol. 9, no. 6, pp. 813–825, 1998.

[30] H. C. Yoon, M. Y. Hong, and H. S. Kim, "Affinity biosensor for avidin using a double functionalized dendrimer monolayer on a gold electrode," *Analytical Biochemistry*, vol. 282, no. 1, pp. 121–128, 2000.

[31] R. Spector and D. Mock, "Biotin transport through the blood-brain barrier," *Journal of Neurochemistry*, vol. 48, no. 2, pp. 400–404, 1987.

[32] S. Beg, A. Samad, M. I. Alam, and I. Nazish, "Dendrimers as novel systems for delivery of neuropharmaceuticals to the brain," *CNS and Neurological Disorders*, vol. 10, no. 5, pp. 576–588, 2011.

[33] N. L. Rosi and C. A. Mirkin, "Nanostructures in biodiagnostics," *Chemical Reviews*, vol. 105, no. 4, pp. 1547–1562, 2005.

[34] B. Basnar and I. Willner, "Dip-pen-nanolithographic patterning of metallic, semiconductor, and metal oxide nanostructures on surfaces," *Small*, vol. 5, no. 1, pp. 28–44, 2009.

[35] A. N. Shipway, E. Katz, and I. Willner, "Nanoparticle arrays on surfaces for electronic, optical, and sensor applications," *ChemPhysChem*, vol. 1, no. 1, pp. 18–52, 2000.

[36] C. A. Cason, S. A. Oehrle, T. A. Fabre et al., "Improved methodology for monitoring poly(amidoamine) dendrimers surface transformations and product quality by ultra performance liquid chromatography," *Journal of Nanomaterials*, vol. 2008, Article ID 456082, 7 pages, 2008.

[37] J. D. Driskell, K. M. Kwarta, R. J. Lipert, M. D. Porter, J. D. Neill, and J. F. Ridpath, "Low-level detection of viral pathogens by a surface-enhanced Raman scattering based immunoassay," *Analytical Chemistry*, vol. 77, no. 19, pp. 6147–6154, 2005.

[38] J. D. Driskell, J. M. Uhlenkamp, R. J. Lipert, and M. D. Porter, "Surface-enhanced Raman scattering immunoassays using a rotated capture substrate," *Analytical Chemistry*, vol. 79, no. 11, pp. 4141–4148, 2007.

[39] R. L. Millen, J. Nordling, H. A. Bullen, M. D. Porter, M. Tondra, and M. C. Granger, "Giant magenetoresistive sensors. 2. Detection of biorecognition events at self-referencing and magnetically tagged arrays," *Analytical Chemistry*, vol. 80, no. 21, pp. 7940–7946, 2008.

[40] D. E. Q. Yang, Y. U. Q. Xiong, Y. Guo, D. A. Da, and W. G. Lu, "Sizes correction on AFM images of nanometer spherical particles," *Journal of Materials Science*, vol. 36, no. 1, pp. 263–267, 2001.

[41] P. Markiewicz and M. C. Goh, "Simulation of atomic force microscope tip-sample/sample-tip reconstruction," *Journal of*

Vacuum Science and Technology B, vol. 13, no. 3, pp. 1115–1118, 1995.

[42] X. C. Wu, A. M. Bittner, and K. Kern, "Microcontact printing of CDS/dendrimer nanocomposite patterns on silicon wafers," *Advanced Materials*, vol. 16, no. 5, pp. 413–417, 2004.

[43] R. McKendry, W. T. S. Huck, B. Weeks, M. Fiorini, C. Abell, and T. Rayment, "Creating nanoscale patterns of dendrimers on silicon surfaces with dip-pen nanolithography," *Nano Letters*, vol. 2, no. 7, pp. 713–716, 2002.

[44] P. Bhatnagar, S. S. Mark, I. Kim et al., "Dendrimer-Scaffold-based electron-beam patterning of biomolecules," *Advanced Materials*, vol. 18, no. 3, pp. 315–319, 2006.

11

Chemical Constituents of Essential Oil from *Lippia sidoides* Cham. (Verbenaceae) Leaves Cultivated in Hidrolândia, Goiás, Brazil

Sandra Ribeiro de Morais,[1,2] Thiago Levi Silva Oliveira,[2,3] Maria Teresa Freitas Bara,[4] Edemilson Cardoso da Conceição,[4] Maria Helena Rezende,[5] Pedro Henrique Ferri,[6] and José Realino de Paula[4]

[1] *Programa de Pós-Graduação em Biologia, Instituto de Ciências Biológicas, Universidade Federal de Goiás, 74001 970 Goiânia, GO, Brazil*
[2] *Instituto de Ciências da Saúde, Universidade Paulista, Campus Flamboyant, 74845 090 Goiânia, GO, Brazil*
[3] *Programa de Pós-Graduação em Ciências Farmacêuticas, Laboratório de Pesquisa de Produtos Naturais, Faculdade de Farmácia, Universidade Federal de Goiás, 74605 220 Goiânia, GO, Brazil*
[4] *Faculdade de Farmácia, Universidade Federal de Goiás, 74605 220 Goiânia, GO, Brazil*
[5] *Instituto de Ciências Biológicas, Universidade Federal de Goiás, Campus Samambaia, 74001 970 Goiânia, GO, Brazil*
[6] *Instituto de Química, Laboratório de Bioatividade Molecular, Universidade Federal de Goiás, Campus Samambaia, 74001 970 Goiânia, GO, Brazil*

Correspondence should be addressed to Sandra Ribeiro de Morais, sandrarmorais@hotmail.com

Academic Editor: Norberto Peporine Lopes

Several studies involving the family Verbenaceae, occurring in the Brazilian Cerrado, have emphasized the popular use of many aromatic species. We highlight the use of *Lippia sidoides* Cham., known as "alecrim pimenta," native to northeastern Brazil and northern Minas Gerais. Leaves of this species were collected in antropized Brazilian Cerrado area, in Hidrolândia, Goiás, and their essential oils were extracted by hydrodistillation in a Clevenger-type apparatus and thereafter analyzed GC/MS. Among the compounds identified in this study were the most abundant oxygenated monoterpenes, followed by sesquiterpenes hydrocarbons. The oxygenated monoterpene 1,8-cineole was the major constituent followed by isoborneol and bornyl acetate. The chemical composition of essential oil described in this paper differs from that described in the literature for *L. sidoides* found in its native environment, where the major constituents are thymol and carvacrol.

1. Introduction

The knowledge of chemical constituents of essential oils is of fundamental importance to the pharmaceutical, food, and perfumery industries. As the use of aromatic compounds requires detailed chemical characterization and evaluation of possible modifications within their compositions, which are due to the different geographical origins and/or climatic conditions and various population genetics that can lead to the formation of different chemotypes [1, 2].

Various studies involving the Verbenaceae family have highlighted the importance of many species used within popular medicine by the presence of principle aromas [3–9]. It is worth noting that in this family, the species *Lippia sidoides*, popularly known as "alecrim-pimenta," native to the northeastern region of Brazil and north of the state of Minas Gerais, is an aromatic species commonly used in the form of infusions and inhalations, allergic rhinitis, and in the treatment of vaginal, mouth, and throat infections [10].

Within the chemical components described of this species, thymol and carvacrol are major constituents of the essential oil [11–13], with a remarkable inhibitory activity regarding the development of microorganisms [14–17]. Moreover, the studies show variations in the concentration

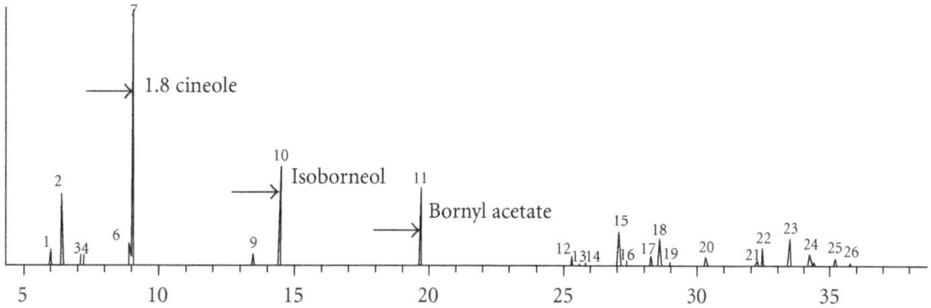

Figure 1: Total ion chromatogram (TIC) of chemical constituents of essential oil from *Lippia sidoides* Cham., (Verbenaceae) leaves, cultivated in Hidrolândia, Goiás, Brazil. (arrow: major constituents).

of thymol in different stages of the plant's development [18]. This work aims to determine the chemical composition of the essential oil of *L. sidoides* cultivated in an area of antropized cerrado in Hidrolândia, Goiás, Brazil.

2. Experimental

The leaves of *Lippia sidoides* Cham. were collected from three plants grown in the municipality of Hidrolândia, Goiás, Brazil (altitude 835 m, 16° 54′ 1.3″ south, 49° 15′ 35.2″ west) in august/2010, both northwest Minas Gerais, Brazil origin. Botanic material was identified by Dr. Marcos José da Silva, of Departamento de Biologia Geral do Instituto de Ciências Biológicas/UFG, and vouchers were deposited in the Herbarium of Universidade Federal de Goiás (UFG) under code number 45121.

Leaves were dried at room temperature and then pulverized by blade mill. Essential oil was extracted by hydrodistillation in a Clevenger-type apparatus for 2 hours from 50 g of powered leaves in 1000 mL of water. At the end of each distillation, the oils were measured in Clevenger trap, collected, dried with anhydrous Na_2SO_4, stored in hermetically sealed glass containers with rubber lids, covered with aluminum foil to protect the contents from light, and kept under refrigeration at −10°C until used. The essential oil was submitted to GC/MS analysis performed on Shimadzu QP5050A apparatus using a CBP-5 (Shimadzu) fused silica capillary column (30 m × 0.25 mm; 0.25 μm film thickness composed of 5% phenylmethylpolysiloxane) and programmed temperature as follows: 60°–240°C at 3°C/min, then to 280°C at 10°C/min, ending with 10 min at 280°C. The carrier gas was He at a flow rate of 1.0 mL/min and the split mode had a ratio of 1 : 20. Compounds were identified by computer search using digital libraries of mass spectral data [19] and by comparison of their retention indices and authentic mass spectra, relative to C8–C32 n-alkane series [20] in a temperature-programmed run.

3. Results and Discussion

The yield of *Lippia sidoides* essential oil was 0.8%. Within the identified components of the essential oil, the most abundant were oxygenated monoterpenes, followed by sesquiterpenes hydrocarbons (Table 1).

Table 1: Percentage of chemical constituents of essential oil from *Lippia sidoides* Cham. (Verbenaceae) leaves, cultivated in Hidrolândia, Goiás, Brazil.

Constituent	RI	%
Artemisia triene	929	1.71
Camphene	954	6.19
Sabinene	975	1.27
β-pinene	979	1.23
ρ-cymene	1024	0.38
Sylvestrene	1030	3.03
1,8 cineole	1031	26.67
Cis-sabinene hydrate	1070	0.50
Camphor	1146	1.60
Isoborneol	1160	14.60
Bornyl acetate	1288	10.77
α-cedrene	1411	1.75
(e)-caryophyllene	1419	1.09
Cis-thujopsene	1431	1.12
α-himachalene	1451	1.37
α-humulene	1454	5.66
Ar-curcumene	1480	1.83
β-selinene	1490	4.33
Cis-calamenene	1529	2.68
Zierone	1575	1.99
Rosifoliol	1600	4.53
Citronellyl pentanoate	1625	1.79
Alo-himachalol	1662	0.59
Oxygenated monoterpenes	—	53.64
Sesquiterpenes hydrocarbons	—	19.83
Monoterpene hydrocarbons	—	14.31
Oxygenated sesquiterpenes	—	8.90
Unidentified	—	3.32
Total identified (%)	—	96.68

RI: retention indices.

Altogether, 96.68% of the chemical constituents of the essential oil were identified. As described in Table 1 and by chromatogram showed in Figure 1, 1,8 cineole, an oxygenated monoterpene was mostly constituent (26.67%), followed by isoborneol (14.60%) and bornyl acetate (10.77%).

In *L. sidoides* cultivated in Minas Gerais, Brazil, the 1,8 cineole was also identified in lower concentrations (9.26%) than in this work. However, thymol and carvacrol were also identified [21]. 1,8 cineole was also identified in other species of its kind, such as *Lippia microphylla* Cham. [22], *Lippia alba* (Mill.) N. E. Brown [6, 23, 24] and *Lippia schomburgkiana* Schauer [24].

The results found in this work for *L. sidoides*, cultivated in an area of anthropic bushland, differ from that presented in the literature, where the thymol and carvacrol appear as major components [11–13]. What can be understood when considering that the environment of which the plant develops are factors such as temperature, relative humidity, exposure to the sun and wind, which exert a direct influence on the chemical composition of volatile oils [2, 23, 25]. Alteration in the yield of the essential oils, as well as the quantity of chemical constituents can still be observed due to the different phases of the plant's development [2, 6, 18] and at different times of the year [4, 23].

Qualitative and quantitative variations in the composition of the oils can also be observed in species that have chemotypes or chemical races, where botanically identical plants produce different chemical compounds, irrespective of their environment, as registered in *L. alba*, where three chemotypes from different regions and cultivated under the same conditions produce citral, carvone, and linalool, confirming that the variations occur in function of infraspecific variation [6]. Moreover, depending on the liability of the constituents of volatile oils, the method used to extract the essential oils could affect the identification [2, 26].

4. Conclusion

The chemical composition of essential oil described in this paper differs from that described in the literature for *L. sidoides* found in its native environment, highlighting the need for further studies to assess the variation in chemical composition of vegetal species in different environments, especially those who may have biological activity.

Acknowledgments

The authors are grateful to Brazilian agencies CNPq and CAPES for financial support and to Professor Dr. Marcos José da Silva, do Departamento de Biologia Geral do Instituto de Ciências Biológicas/UFG, for identification of the plant.

References

[1] A. T. Henriques, C. A. Simões-Pires, and M. A. Apel, "Óleos essenciais: importância e perspectivas terapêuticas," in *Química de Produtos Naturais, Novos Fármacos e a Moderna Farmacognosia*, R. A. Yunes and V. Cechinel-Filho, Eds., pp. 219–256, Itajaí, Univale, Brazil, 2009.

[2] C. M. O. Simões and V. Spitzer, "Óleos voláteis," in *Farmacognosia: Da Planta Ao Medicamento*, C. M. O. Simões, E. P. Schenkel, G. Gosmann, J. C. P. Mello, L. A. Mentz, and P. R. Petrovick, Eds., pp. 467–496, UFRGS, Porto Alegre, Brazil, 5th edition, 2004.

[3] E. M. Costa-Neto and M. V. M. Oliveira, "The use of medicinal plants in the county of Tanquinho, State of Bahia, Northeastern Brazil," *Revista Brasileira de Plantas Medicinais*, vol. 2, no. 2, pp. 1–8, 2000.

[4] M. R. A. Santos, R. Innecco, and A. A. Soares, "Caracterização anatômica das estruturas secretoras e produção de óleo essencial de *Lippia alba* (Mill.) N. E. Br. em função do horário de colheita nas estações seca e chuvosa," *Revista Ciência Agronômica*, vol. 35, no. 2, pp. 377–383, 2004.

[5] J. S. Aguiar and M. C. C. D. Costa, "*Lippia alba* (Mill.) N. E. Brown (Verbenaceae): survey of the publications in the chemical, agronomical and pharmacological area, published between 1974 to 2004," *Revista Brasileira de Plantas Medicinais*, vol. 8, no. 1, pp. 79–84, 2005.

[6] E. S. Tavares, E. S. Julião, H. D. Lopes, H. R. Bizzo, C. L. S. Lage, and S. G Leitão, "Análise do óleo essencial de folhas de três quimiotipos de *Lippia alba* (Mill.) N. E. Br. (Verbenaceae) cultivados em condições semelhantes," *Brazilian Journal of Pharmacognosy*, vol. 15, no. 1, pp. 1–5, 2005.

[7] J. E. B. P. Pinto, J. C. W. Cardoso, E. M. Castro, S. K. Bertolucci, L. A. Melo, and S. Dousseau, "Aspectos morfo-fisiológicos e conteúdo de óleo essencial de plantas de alfazema-do-Brasil em função de níveis de sombreamento," *Horticultura Brasileira*, vol. 25, no. 2, pp. 210–214, 2007.

[8] S. Froelich, M. P. Gupta, K. Siems, and K. Jenett-Siems, "Phenylethanoid glycosides from *Stachytarpheta cayennensis* (Rich.) Vahl, Verbenaceae, a traditional antimalarial medicinal plant," *Brazilian Journal of Pharmacognosy*, vol. 18, no. 4, pp. 517–520, 2008.

[9] S. L. Goulart and C. R. Marcati, "Anatomia comparada do lenho em raiz e caule de *Lippia salviifolia* Cham. (Verbenaceae)," *Revista Brasileira de Botanica*, vol. 31, no. 2, pp. 263–275, 2008.

[10] E. R. Martins, D. M. Castro, D. C. Castelanni, and J. E. Dias, *Plantas Medicinais*, Universidade Federal de Viçosa, Viçosa, Brazil, 2000.

[11] L. M. A. Macambira, C. H. S. Andrade, F. J. A. Matos, A. A. Craveiro, and R. Braz Filho, "Naphthoquinoids from *Lippia sidoides*," *Journal of Natural Products*, vol. 49, no. 2, pp. 310–312, 1986.

[12] S. M. O. Costa, T. L. G. Lemos, O. D. L. Pessoa, J. C. C. Assunção, and R. Braz-Filho, "Constituintes químicos de *Lippia sidoides* (Cham.) Verbenaceae," *Brazilian Journal of Pharmacognosy*, vol. 12, supplement 1, pp. 66–67, 2002.

[13] R. S. Nunes, H. S. Xavier, P. J. Rolim Neto, D. P. Santana, and U. P. Albuquerque, "Botanical standardization of *Lippia sidoides* Cham. (Verbenaceae)," *Acta Farmaceutica Bonaerense*, vol. 19, no. 2, pp. 115–118, 2000.

[14] M. A. Botelho, V. S. Rao, C. B. M. Carvalho et al., "*Lippia sidoides* and *Myracrodruon urundeuva* gel prevents alveolar bone resorption in experimental periodontitis in rats," *Journal of Ethnopharmacology*, vol. 113, no. 3, pp. 471–478, 2007.

[15] M. A. Botelho, N. A. P. Nogueira, G. M. Bastos et al., "Antimicrobial activity of the essential oil from *Lippia sidoides*, carvacrol and thymol against oral pathogens," *Brazilian Journal of Medical and Biological Research*, vol. 40, no. 3, pp. 349–356, 2007.

[16] E. Lacoste, J. P. Chaumont, D. Mandin, M. M. Plumel, and F. J. A. Matos, "Antiseptic properties of the essential oil of *Lippia sidoides* Cham: application to the cutaneous microflora," *Annales Pharmaceutiques Francaises*, vol. 54, no. 5, pp. 228–230, 1996.

[17] F. P. Oliveira, E. O. Lima, J. P. Siqueira-Júnior, E. L. Souza, B. H. C. Santos, and H. M. Barreto, "Effectiveness of *Lippia sidoides* Cham. (Verbenaceae) essential oil in inhibiting the growth of *Staphylococcus aureus* strains isolated from clinical material," *Brazilian Journal of Pharmacognosy*, vol. 16, no. 4, pp. 510–516, 2006.

[18] L. K. A. M. Leal, V. M. Oliveira, S. M. Araruna, M. C. C. Miranda, and F. M. A. Oliveira, "Análise de timol por CLAE na tintura de *Lippia sidoides* Cham. (alecrim-pimenta) produzida em diferentes estágios de desenvolvimento da planta," *Brazilian Journal of Pharmacognosy*, vol. 13, supplement 1, pp. 9–11, 2003.

[19] National Institute of Standards and Technology, *PC Version of the NIST/EPA/NIH Mass Spectral Database*, U.S. Department of Commerce, Gaithersburg, Md, USA, 1998.

[20] R. P. Adams, *Identification of Essential Oil Components by Gas Chromatography/Mass Spectroscopy*, Allured, Carol Stream, Ill, USA, 4th edition, 2007.

[21] R. K. Lima, M. G. Cardoso, J. C. Moraes, S. M. Carvalho, V. G. Rodrigues, and L. G. L. Guimarães, "composição química e efeito fumigante do óleo essencial de *Lippia sidoides* cham. e monoterpenos sobre *Tenebrio molitor* (L.) (coleoptera: tenebrionidae)," *Ciencia e Agrotecnologia*, vol. 35, no. 4, pp. 664–671, 2011.

[22] F. F. G. Rodrigues, H. D. M. Coutinho, A. R. Campos, S. G. de Lima, and J. G. M. da Costa, "Atividade antibacteriana e composição química do óleo essencial de *Lippia microphylla* cham," *Acta Scientiarum*, vol. 33, no. 2, pp. 141–144, 2011.

[23] F. M. C. Barros, E. O. Zambarda, B. M. Heinzmann, and C. A. Mailmann, "Variabilidade sazonal e biossíntese de terpenóides presentes no óleo essencial de *Lippia alba* (Mill.) n. e. brown (Verbenaceae)," *Quimica Nova*, vol. 32, no. 4, pp. 861–867, 2009.

[24] J. G. S. Maia and E. H. A. Andrade, "Database of the amazon aromatic plants and their essential oils," *Quimica Nova*, vol. 32, no. 3, pp. 595–622, 2009.

[25] T. S. Fiuza, S. M. T. Sabóia-Morais, J. R. Paula et al., "Composition and chemical variability in the essential oils of *Hyptidendron canum* (Pohl ex Benth.) Harley," *Journal of Essential Oil Research*, vol. 22, no. 2, pp. 159–163, 2010.

[26] X. Gu, Z. Zhang, X. Wan, J. Ning, C. Yao, and W. Shao, "Simultaneous distillation extraction of some volatile flavor components frompu-erh Tea samples—comparison with steam distillation-liquid/liquid extraction and soxhlet extraction," *International Journal of Analytical Chemistry*, vol. 2009, Article ID 276713, 6 pages, 2009.

12

Direct Electron Transfer of Hemoglobin on Manganese III Oxide-Ag Nanofibers Modified Glassy Carbon Electrode

Masoud Negahdary,[1] Gholamreza Mazaheri,[1] Somyyeh Rad,[1] Mohammadreza Hadi,[2] Roya Malekzadeh,[3] Mohammad Mahdi Saadatmand,[4] Saeed Rezaei-Zarchi,[3] Fariba Pishbin,[1] and Mojdeh Khosravian-hemami[2]

[1] *Department of Biology, Payam-e-Noor University, Tehran 7371719578, Iran*
[2] *Department of Biology, Shahrekord University, Shahrekord, Iran*
[3] *Department of Biology, Payam-e-Noor University, Taft, Iran*
[4] *Young Researchers Club, Islamic Azad University, Yazd Branch, Yazd, Iran*

Correspondence should be addressed to Masoud Negahdary, masoud_negahdary@yahoo.com

Academic Editor: Bahram Hemmateenejad

We investigated the electrochemical behavior of hemoglobin by glassy carbon electrode modified with Mn_2O_3-Ag nanofibers. The Mn_2O_3-Ag nanofibers were used as facilitator electron transfer between Hb and glassy-carbon-modified electrode. The Mn_2O_3-Ag nanofibers are studied by scanning electron microscopy (SEM) and transmission electron microscopy (TEM). The hemoglobin showed a quasireversible electrochemical redox behavior with a formal potential of −49 mV (versus Ag/AgCl) in 0.1 M potassium phosphate buffer solution at pH 7.0. The designed biosensor possesses good stability and reproducibility and achieves 95% of the steady-state current in less than five seconds.

1. Introduction

Hemoglobin is a remarkable molecular machine that uses motion and small structural changes to regulate its action. Oxygen binding at the four heme sites in hemoglobin does not happen simultaneously. Once the first heme binds oxygen, it introduces small changes in the structure of the corresponding protein chain. The main function of red blood cell is transfer of O_2 from lungs to tissue and transfer of CO_2 from tissue to lungs. To accomplish these functions, red cells have hemoglobin (Hb). The Hb has iron in center of its structure [1]. The iron has important role in Hb structure [2]. We too used Hb in this project. Glassy carbon derives its name from exhibiting fracture behavior similar to glass, from having a disordered structure over large dimensions (although it contains a graphitic microcrystalline structure), and because it is a hard shiny material, capable of high polish [3–5]. Glassy carbon is particularly useful in electrochemical applications because of its low electrical resistivity, impermeability to gases, high chemical resistance, and because it has the widest potential range observed for carbon electrode. Glassy carbon was first prepared by Yamada and Sato in 1962 by the hight emperature pyrolysis of phenolic resin, and later by Davison who used cellulose as the starting material [6, 7]. They concluded that glassy carbon consists of long microfibrils that twist, bend, and interlock to form interfibrillar bonds, and that these microfibrils are randomly oriented [8]. Because of the desire to impart selectivity to electrochemical reactions and to control electron transfer kinetics, several investigators have utilized adsorption or covalent bonding of catalysts to the glassy carbon surface. The intent of these modifications is to control the interaction of molecules and ions with the electrode surface [9]. Small fibers in the submicron range, in comparison with larger ones, are well known to provide better filter efficiency at the same pressure drop in the interception and inertial impaction regimes [10, 11]. Previous researches shown such effect by assuming nonslip flow at fiber surface [12]. While smaller fiber size leads to higher-pressure drop, interception and inertial impaction efficiencies will increase faster, more than compensating for the pressure drop increase. Thus, in the particle size of interest, that is, from submicron and up, better filter efficiency can be achieved at the same pressure drop, or conversely, the same filter efficiency at lower pressure drop can be achieved with smaller fiber sizes. Polymeric nanofibers can be made using the electrospinning process, which has been described in the literature [13] and in patents [14]. Electrospinning uses an electric field to draw a polymer solution from the tip of a capillary to a collector. A voltage is applied to the polymer solution, which causes a jet of the solution to be drawn toward a grounded collector. The fine jets dry to form polymeric fibers, which can be collected on a web. The electrospinning process has been documented using a variety of fiber-forming polymers [15, 16]. By choosing a suitable polymer and solvent system, nanofibers with diameters in the range of 40–2000 nm can be made. PVP is a kind of polymers which is able to adsorb through many points of the molecule on the surface of steel, copper, and zinc and reduce the corrosion rate of the material [17–19]. PVPs are considered as food additives and, hence, their addition will not raise health concerns for treating RO water. However, the effect of the molecular weight of polyvinylpyrrolidone, its concentration, and temperature on the corrosion properties of 316 L stainless steel in reverse osmosis water were not evaluated.

2. Material and Methods

2.1. Electrospinning Nanofibers. Electrospinning is a process of applying a high-voltage electric field (several to tens of kilovolts) to generate electrically charged jets from polymer solutions or melts and further to produce polymer (nano) fibers. This technique is quite similar with the commercial process for drawing microscale fibers; however, it is more suitable for generating nanofibers, because the elongation can be accomplished by a contactless scheme through the application of an external electric field [20]. Generally, generation of porous surface on a bulk electrospinning nanofiber can be realized through two different ways. The first one is based on the selective removal of a component from nanofibers made of a composite or blend material, while the other one involved the use of phase separation of different polymers during electrospinning under the application of proper sinning parameters [16]. Both the pore size and the density are controllable by changing the parameters. For instance, in PLA/PVP electrospinning nanofibers, more porosity can be generated when the two materials are loaded in equal amounts comparing to the 11 corresponding products by different proportion of PLA/PVP. It can be attributed to the rapid-phase separation and solidification in the spinning jet [14–17]. PLA is contraction name of polylactic acid, and it is a thermoplastic polymer made from lactic acid and has mainly been used for biodegradable products, such as plastic bags and planting cups [18]. The PLA/PVP/nanofibers have been prepared by electrospinning [19]. It was found that the average diameter of the PLA/PVP/nanofibers lead fibers became larger, and the morphology of the fibers became finer with the content of PLA increasing. The formation of pores is also affected by the solvent vapor pressure and the humidity in atmosphere. The cooling effect that comes from rapid evaporation of a highly volatile solvent might induce the polymers to separate into different phases in liquid jet. Because of evaporative cooling and condensation, water droplets could also be formed within the fibers to promote the formation of porous nanofibers [20–22]. The electrospun nanofibers exhibit several unique features, which enable the prevalent utilization of them. Because electrospinning is a continuous process without any contact force for elongation, the fibers can be as long as several kilometers and can be further assembled into a 3D mat with porous structure. At the same time, electrospun fibers can have a thinner diameter and surface-to-volume ratio, due to the presence of porous structure. In addition, due to the simple fabrication process and the diversity of suitable materials, the electrospinning technique and its resultant nanofiber product have attracted increasing attention. These properties potentiate the use of the electrospun nanofibers in various applications such as reinforced composites, nanofiber-based membranes, nanofiber-based support for enzyme, and catalyst [23–25].

2.2. Reagents. Hb, manganese(II) nitrate tetrahydrate, and poly(vinylpyrrolidone) (PVP, MW = 1,300,000) were purchased from Sigma. Silver nitrate and other materials were supplied by Merck. 0.1 M phosphate buffer solutions with various pH values were prepared by mixing stock standard solutions of Na_2HPO_4 and NaH_2PO_4 and adjusting the pH values with NaOH and H_3PO_4 solution. All solutions used in the experiments were prepared with deionized water generated by a Barnstead water system.

2.2.1. Preparation of Mn_2O_3-Ag Nanofibers. 44 wt% $Mn(NO_3)_2$, 11 wt% $AgNO_3$, and 44 wt% PVP were dissolved in DMF. The solution was kept under magnetic stirring for 2 h and then loaded into a plastic syringe equipped with a 23-gauge needle made of stainless steel. Electrospinning process was conducted at an applied voltage of 20 kV with a feeding rate of 0.3 mL/h and a collection distance of 15 cm. The nanofibers were collected on aluminum foil and then calcined under air atmosphere at 500°C for 3 h for the degradation of PVP and the decomposition of $Mn(NO_3)_2$ and $AgNO_3$:

$$\begin{gathered} 4Mn(NO_3) \xrightarrow{\Delta} 2Mn_2O_3 + 8NO_2 \uparrow + O_2 \uparrow \\ 2AgNO_3 \xrightarrow{\Delta} 2Ag + 2NO_2 \uparrow + O_2 \uparrow . \end{gathered} \tag{1}$$

Electrospinning is a process of applying a high-voltage electric field (several to tens of kilovolts) to generate electrically charged jets from polymer solutions or melts and further to produce polymer (nano) fibers. So, in this study, we used from this method to produce Mn_2O_3-Ag nanofibers.

2.2.2. Preparation of Mn_2O_3-Ag-Nanofibers-Modified Glassy Carbon Electrode. The most commonly used carbon-based electrode in the analytical laboratory is glassy carbon (GC). It is made by pyrolyzing a carbon polymer, under carefully controlled conditions, to a high temperature like 2000°C. An intertwining ribbon-like material results with retention of high conductivity, hardness, and inertness. Glassy carbon electrode (GCE, dia. 3 mm) was polished with 1 μm and

0.05 μm alumina slurries sequentially and then rinsed with DI water. After that, the electrode was sonicated in deionized water and finally dried under ambient conditions. To prepare the modified GCE, The Mn_2O_3-Ag nanofibers/glassy carbon electrode was placed into a fresh PBS including 5 mg mL^{-1} Hb (pH 7.0, 3 to 5°C) for 8 h. At the end, the modified electrode was washed in deionized water and placed in PBS (PH 7.0) at a refrigerator (3 to 5°C), before being employed in the electrochemical measurements as the working electrode.

2.2.3. Apparatus and Electrochemical Measurement. A JEOL 6335F field-emission scanning electron microscope (SEM) was used to examine the morphology and the size of the as-prepared nanofibers. More detailed morphology and selected area electron diffraction (SAED) patterns of Mn_2O_3-Ag nanofibers were obtained with a Tecnai T12 transmission electron microscope (TEM) operated at 120 kV. XRD pattern was obtained with Oxford diffraction XcaliburTM PX Ultra with ONYX detector to study the crystal structure of Mn_2O_3-Ag nanofibers measurements which were carried out with a potentiostat/galvanostat (Model 263A, EG&G, USA) using a single compartment voltammetric cell, equipped with a platinum rod auxiliary electrode. All experiments were conducted using a three-electrode electrochemical cell (10 mL volume with a working volume of 5 mL), with a working electrode, an Ag/AgCl reference electrode, and a platinum wire counter electrode. Electrochemical measurements were carried out with a potentiostat/galvanostat (Model 263A, EG&G, USA) using a single compartment voltammetric cell, equipped with a platinum rod auxiliary electrode.

3. Results and Discussion

3.1. Characterization of Mn_2O_3-Ag Nanofibers. SEM was first employed to investigate the morphology of the Mn_2O_3-Ag nanofibers. Figure 2(a) presents a typical SEM image of electrospun precursory PVP-$Mn(NO_3)_2$-$AgNO_3$ nanofibers. PVP was used for characterization of Mn_2O_3-Ag nanofibers, because PVP binds to polar molecules exceptionally well, owing to its polarity. This has led to its application in coatings for increase of photo quality in microscopic studies. The molecular structure of PVP can be observed in Figure 1. PVP is soluble in water and other polar solvents. When dry, it is a light flaky powder, which readily absorbs up to 40% of its weight in atmospheric water. In solution, it has excellent wetting properties and readily forms films. This makes it good as a coating or an additive to coatings. PVP is a branched polymer, that is, its structure is more complicated than linear polymer, but it too is in a two-dimensional plane. Moreover, because Mn_2O_3-Ag nanofibers have polar groups too, a good bind was created between this nanofiber and PVP.

After calcination, the as-prepared Mn_2O_3-Ag composite nanofibers (Figure 2(b)) exhibit a porous network structure, and their surfaces are no longer as smooth as the precursory nanofibers. Such feature endows the nanofibers with high surface-to-volume ratio, which could provide not only a large surface area for Hb loading but also a large interface for direct electron transfer of Hb. As a comparison, the nanofibers prepared by single metal salt ($Mn(NO_3)_2$) with PVP and its calcined product (Mn_2O nanofibers) are presented in Figures 2(c) and 2(d), respectively.

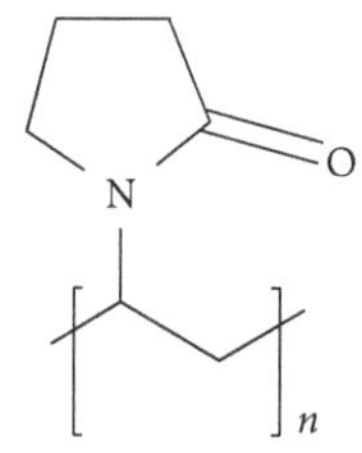

$(C_6H_9NO)_n$

FIGURE 1: Molecular structure of polyvinylpyrrolidone (PVP).

3.2. XRD Diffraction of Mn_2O_3-Ag Nanofibers. The composition and crystal structure were characterized by XRD (Figure 3). The XRD spectrum of Mn_2O_3-Ag nanofibers matches the combination of the standard spectrum of JCPDS 41-1442 (Mn_2O_3) and JCPDS 04-0783 (Ag). The formation of face-centered cubic crystalline Mn_2O_3 is revealed by the diffraction peaks at 2θ values of 32.951, 38.234, 45.178, 49.347, 55.189, and 65.806, corresponding to (111), (200), (220), (311), (222), and (400) crystal planes, respectively, while the diffraction peaks at 2θ values of 38.116, 44.277, and 64.426, which correspond to (111), (200), and (220) crystal planes, respectively, indicate the formation of cubic crystalline Ag.

3.2.1. Direct Electrochemistry of Hb/Mn_2O_3-Ag Nanofibers/GCE. Cyclic voltammetry (CV) was used to characterize the modification of electrodes. The cyclic voltammograms (CVs) of different modified electrodes were obtained in 0.1 M PBS with pH 7.0. No redox peak is observed for the CVs of bare GC electrode. Compared with bare GC electrode, the background current of Mn_2O_3-Ag-nanofibers-modified electrode is apparently larger, which indicates that the effective electrode surface area is significantly enhanced by use of Mn_2O_3-Ag nanofibers to modify the electrode. However, the CVs of Hb/Mn_2O_3-Ag nanofibers/GC electrode give a pair of well-defined redox peaks at −75 mV at scan rate of 100 mV/s, characteristic of heme Fe(III)/Fe(II) redox couples of Hb, suggesting that direct electron transfer has been achieved between Hb and modified electrode. The electron transfer of the proteins, at the bare electrodes is very slow so that the redox peak of proteins can usually be observed. Figure 4(a) shows a cyclic voltammogram (CV) of the bare glassy carbon electrode. Figure 4(b) shows a cyclic voltammogram of Hb/Mn_2O_3-Ag nanofibers/glassy carbon electrode in 0.1 M phosphate buffer at pH 7.0. The Hb showed quasireversible electrochemical behavior with a formal potential of −49 Mv (versus Ag/AgCl); cathodic and anodic peaks were not observed using the bare graphite electrode. This shows that Mn_2O_3-Ag nanofibers act as a facilitator of electron transfer from the redox species of Hb to the electrode surface and vice versa. These results are in line with the previous work that explains the behavior of nanoparticles as the facilitators of electron transfer.

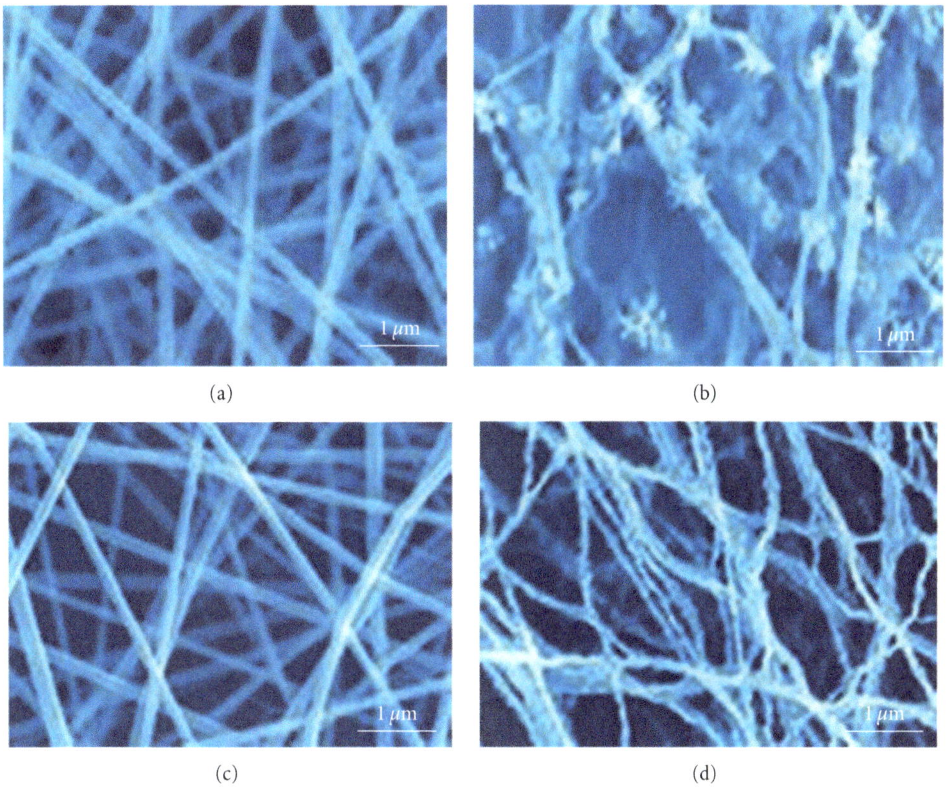

FIGURE 2: SEM images of (a) PVP-$Mn(NO_3)_2$-$AgNO_3$ nanofibers, (b) Mn_2O_3-Ag nanofibers, (c) PVP-$Mn(NO_3)_2$ nanofibers, and (d) Mn_2O_3 nanofibers, respectively.

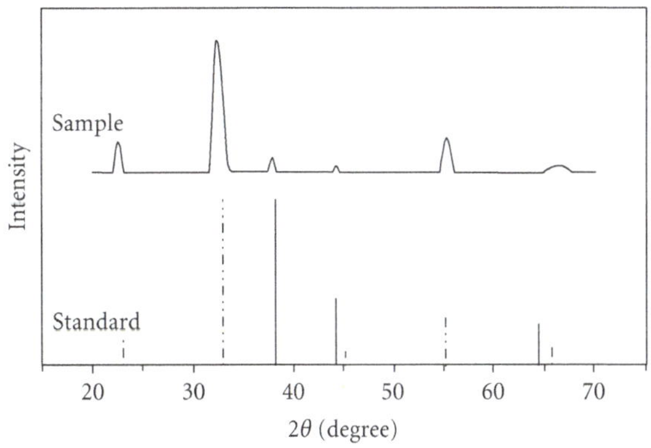

FIGURE 3: XRD patterns for the standard values of JCPDS 41-1442 (Mn_2O_3, dashed line), JCPDS 04-0783 (Ag, solid line), and the asprepared porous Mn_2O_3-Ag nanofibers.

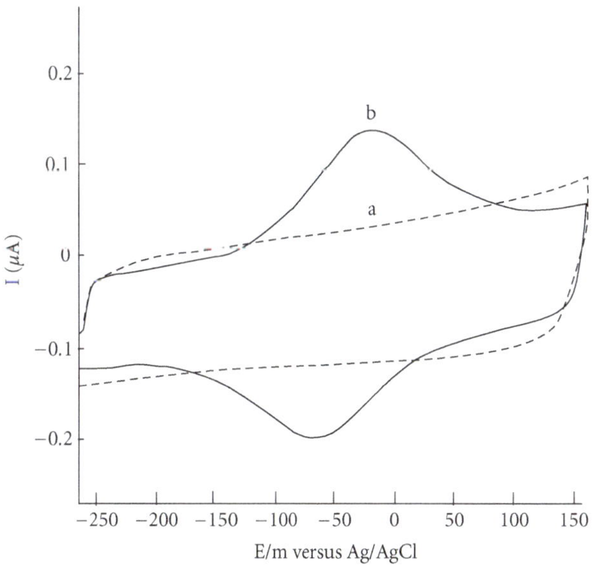

FIGURE 4: Cyclic voltammograms, using (a) bare glassy carbon electrode in 0.1 M phosphate buffer solution and (b) Hb/Mn_2O_3-Ag nanofibers/glassy carbon electrode in 0.1 M phosphate buffer solution (scanrate: 100 mV s^{-1}).

In Figure 4(a), one sees cyclic voltammogram for bare and modified glassy carbon electrode in 50, 100, 200, 300, 400, and 500 scan rates. These data showed that the Mn_2O_3-Ag nanofibers were successfully assembled on the glassy carbon electrode surface. The redox peak currents increase linearly with the square root of the scan rate. With an increasing scan rate, the anodic peak potential of adsorbed Hb shifted to a more positive value, while the cathodic peak current shifted in a negative direction. The redox peak currents were proportional to the scan rate (Figure 5); thus, the electrode reaction was typical of the surface-controlled quasireversible process. One can see in Figure 5(b) the dependence of the anodic and cathodic peak currents on the scan rates that the

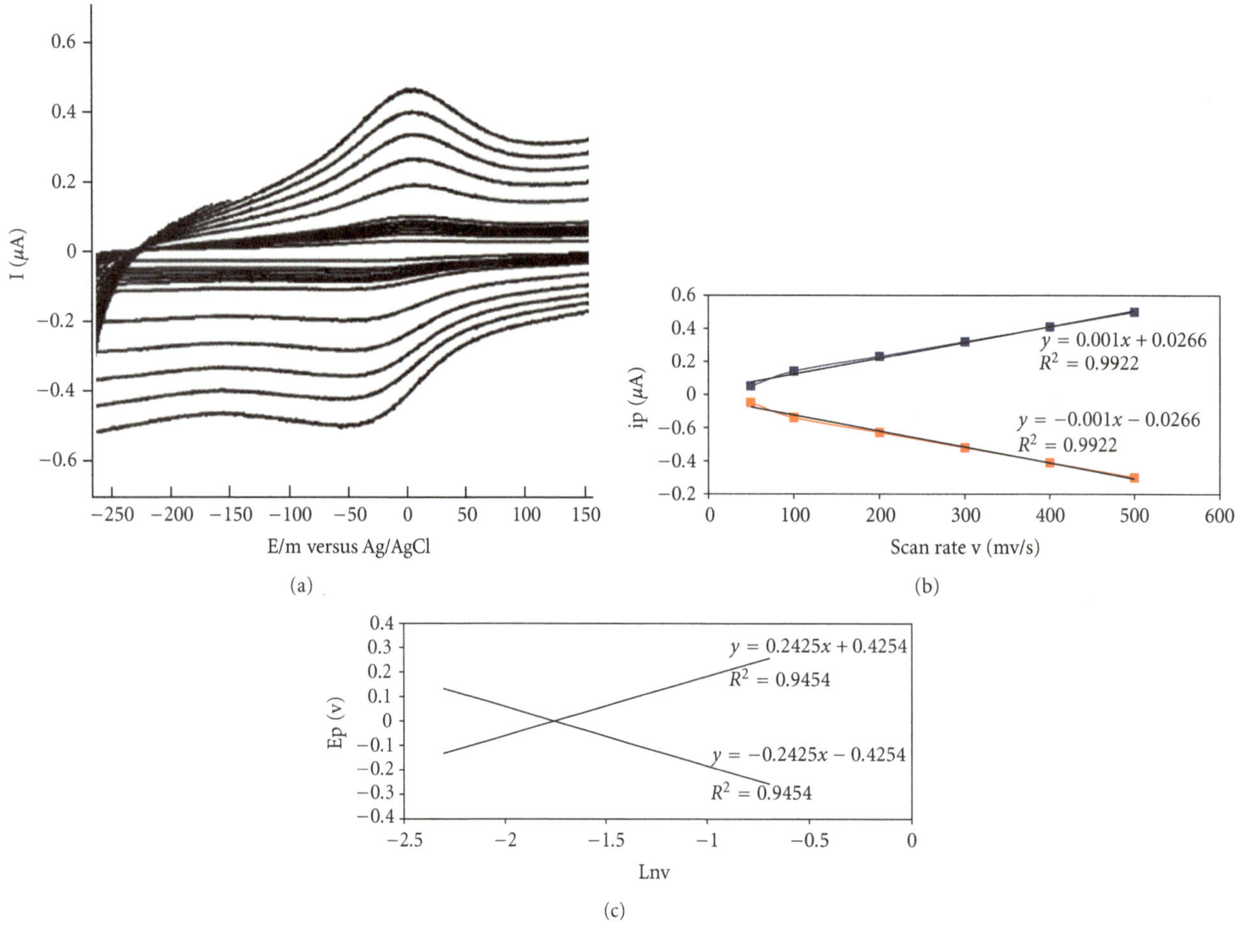

FIGURE 5: (a) Typical cyclic voltammograms of Hb/Mn_2O_3-Ag nanofibers/glassy carbon electrode at different scan rates. The voltammograms (from inner to outer) designate scan rates of 50, 100, 200, 300, 400, and 500 mV s^{-1}, respectively. (b) Dependence of the anodic and cathodic peak currents on the scan rates the red line is cathodic and the blue line is anodic. (c) Relationship between the peak potential (E_p) and the natural logarithm of scan rate ($\ln v$) for Hb/Mn_2O_3-Ag nanofibers/glassy carbon electrode. All the data were obtained at pH 7.0 and in 0.1 M phosphate buffer solution.

red line is cathodic and the blue line is anodic. The kinetics of the heterogeneous electron transfer was analyzed using the model of Laviron [9]. The plot of cathodic peak potentials versus the logarithm of scan rates gave a charge transfer coefficient α of 0.52. For scan rates from 50 to 500 mV/s, the electron transfer rate constant k_s was estimated to be $(1.9)\ s^{-1}$. These statements discussed below indicate that Hb is strongly adsorbed onto the surface of modified electrode. As could be seen in Figure 5(c), in the range from 200 to 500 mV s^{-1}, the catholic peak potential (E_{pc}) changed linearly versus $\ln v$ with a linear regression equation of $y = -0.2425x - 0.4254$, $R^2 = 0.945$, according to the following equation [26–28],

$$E_p = E^{\circ\prime} + \frac{RT}{\alpha nF} - \frac{RT}{\alpha nF} \ln v', \tag{2}$$

where α is the cathodic electron transfer coefficient, n is the number of electrons, R, T, and F are gas, temperature, and Faraday constant, respectively ($R = 8.314\ \text{J mol}^{-1}\text{K}^{-1}$, $F = 96493$ C/mol, $T = 298$ K), and αn is calculated to be 0.52. Given $0.3 < \alpha < 0.7$ in general [9], it could be concluded that $n = 1$ and $\alpha = 0.52$. From the width of the peak at midheight and low scan rate, we can also obtain $n = 1$ [9]. Therefore, the redox reaction between Hb and modified graphite electrode is a single electron transfer process. In order to calculate the value of apparent heterogeneous electron transfer rate constant (k_s), the following equation was used [26–28]:

$$\log k_s = \alpha \log(1-\alpha) + (1-\alpha)\log\alpha - \log\left(\frac{RT}{nFv}\right) - \alpha(1-\alpha)\frac{nF\Delta E_p}{2.3RT}. \tag{3}$$

The k_s was calculated to be $1.90\ s^{-1}$. Moreover, this k_s indicates that transfer of electron is good and fast in Hb-modified glassy carbon electrode.

3.2.2. Design a Hydrogen Peroxide Biosensor. The designed biosensor is introduced as a hydrogen biosensor because it has oxidation and reduction reactions in Hb; when Fe^{2+} changes to Fe^{3+}, the oxidation reaction occurred, and when Fe^{3+} changes to Fe^{2+}, the reduction reaction occurred. The

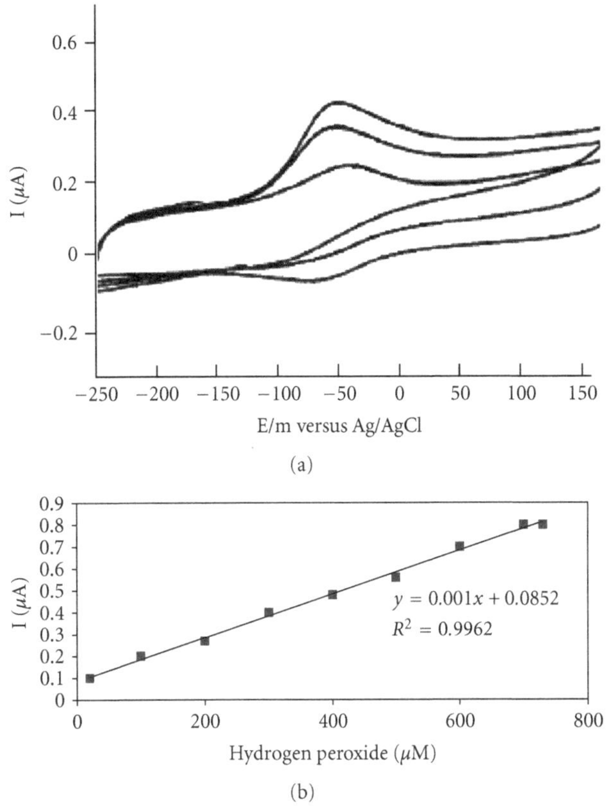

Figure 6: (a) Cyclic voltammograms obtained at an Hb/Mn_2O_3-Ag nanofibers/glassy carbon electrode in 50 μM phosphate buffer solution (pH 7.0) for different concentrations and (b) the relationship between cathodic peak current of Hb and different concentrations of H_2O_2 (scan rate: 100 mV s^{-1}).

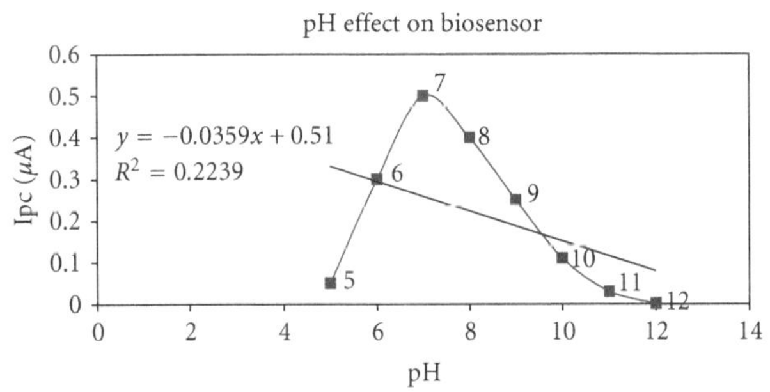

Figure 7: Dependence of the current response of Hb/Mn_2O_3-Ag nanofibers/GC electrode to 50.0 μM H_2O_2 on the pH of buffer solutions.

reduction state is an important matter for designing hydrogen peroxide biosensors [29].

The electrocatalytic reactivity of Hb/Mn_2O_3-Ag nanofibers/GC electrode toward H_2O_2 was investigated by CV. Figure 6 displays the cyclic voltammograms obtained for the hydrogen peroxide biosensor in PBS (pH 7.0) containing varied concentration of H_2O_2 in the absence of oxygen. The catalytic reduction of hydrogen peroxide at the biosensor can be seen clearly in Figure 5. With the addition of H_2O_2, the reduction peak current increases obviously, while the oxidation peak current decreases (Figures 6(a) and 6(b)), indicating a typical electrocatalytic reduction process of H_2O_2. However, no similar cathodic peak corresponding to the reduction of H_2O_2 can be observed at bare GC, Mn_2O_3-Ag nanofibers/GC electrode under the same condition, so it can be concluded that Hb immobilized on Mn_2O_3-Ag nanofibers/GC electrode shows good catalytic activity toward hydrogen peroxide [19]. Figure 6 shows the sensor response. The sensor was found to have sensitivity in the range of 20 to 700 μM based upon the mean of the slope found from the points on the response curve. As can be observed, the sensor response shows good linearity in this range. The correlation factor, R^2, was found to be 0.9962.

3.3. Influence of pH and Applied Potential on Biosensor Response. In order to obtain an efficient biosensor for H_2O_2, the influence of pH and applied potential on the response of Hb/Mn_2O_3-Ag nanofibers/GC electrode were investigated. The change of chronoamperometric current with the pH under constant hydrogen peroxide concentration (50.0 μM) is shown in Figure 7. As can be seen, the maximum response appears at pH 7.0. So the buffer solution of pH 7.0 was selected for experiments. This designed biosensor possesses good stability and reproducibility and achieves 95% of the steady-state current in less than 5 s.

4. Conclusion

A new biosensor for H_2O_2 was prepared based on Hb/Mn_2O_3-Ag nanofibers/glassy carbon electrode. Hb retained well in Mn_2O_3-Ag nanofibers/GCE, which combined the utilities of Mn_2O_3-Ag nanofibers facilitating the electron transfer. At Hb/Mn_2O_3-Ag nanofibers/GCE, the cyclic voltammogram exhibits a pair of redox peaks corresponding to a surface-controlled electrode process with a single-proton transfer. The designed biosensor displays a high affinity and sensitivity to H_2O_2. The sensor shows good reproducibility and stability.

References

[1] F. Davis and S. P. J. Higson, "Biofuel cells-recent advances and applications," *Biosensors and Bioelectronics*, vol. 22, no. 7, pp. 1224–1235, 2007.

[2] M. C. Potter, "Electrical effects accompanying the decomposition of organic compounds," *Proceedings of the Royal Society*, vol. 84, pp. 260–276, 1912.

[3] A. T. Yahiro, S. M. Lee, and D. O. Kimble, "Bioelectrochemistry. I. Enzyme utilizing bio-fuel cell studies," *Biochimica et Biophysica Acta*, vol. 88, no. 2, pp. 375–383, 1964.

[4] M. T. Sulak, O. Gökdoğan, A. Gülce, and H. Gülce, "Amperometric glucose biosensor based on gold-deposited polyvinylferrocene film on Pt electrode," *Biosensors and Bioelectronics*, vol. 21, no. 9, pp. 1719–1726, 2006.

[5] R. A. Bullen, T. C. Arnot, J. B. Lakeman, and F. C. Walsh, "Biofuel cells and their development," *Biosensors and Bioelectronics*, vol. 21, no. 11, pp. 2015–2045, 2006.

[6] K. G. Lim and G. T. R. Palmore, "Microfluidic biofuel cells: the influence of electrode diffusion layer on performance," *Biosensors and Bioelectronics*, vol. 22, no. 6, pp. 941–947, 2007.

[7] A. Heller, "Miniature biofuel cells," *Physical Chemistry Chemical Physics*, vol. 6, no. 2, pp. 209–216, 2004.

[8] Y. Kamitaka, S. Tsujimura, N. Setoyama, T. Kajino, and K. Kano, "Fructose/dioxygen biofuel cell based on direct electron transfer-type bioelectrocatalysis," *Physical Chemistry Chemical Physics*, vol. 9, no. 15, pp. 1793–1801, 2007.

[9] S. Rezaei-Zarchi, M. Negahdary, M. Doroudian et al., "Direct electron transfer of Myoglobin on nickel oxide Nanoparticles modified graphite electrode," *Advances in Environmental Biology*, vol. 5, no. 10, pp. 3241–3248, 2011.

[10] J. Wang and M. Musameh, "Carbon nanotube/Teflon composite electrochemical sensors and biosensors," *Analytical Chemistry*, vol. 75, no. 9, pp. 2075–2079, 2003.

[11] K. Wu, J. Fei, and S. Hu, "Simultaneous determination of dopamine and serotonin on a glassy carbon electrode coated with a film of carbon nanotubes," *Analytical Biochemistry*, vol. 318, no. 1, pp. 100–106, 2003.

[12] X.-H. Zhang and S.-F. Wang, "Determination of ethamsylate in the presence of catecholamines using 4-amino-2-mercaptopyrimidine self-assembled monolayer gold electrode," *Sensors and Actuators, B*, vol. 104, no. 1, pp. 29–34, 2005.

[13] J. X. Wang, X. W. Sun, A. Wei et al., "Zinc oxide nanocomb biosensor for glucose detection," *Applied Physics Letters*, vol. 88, no. 23, Article ID 233106, 2006.

[14] Y. Ding, Y. Wang, L. Su, M. Bellagamba, H. Zhang, and Y. Lei, "Electrospun Co_3O_4 nanofibers for sensitive and selective glucose detection," *Biosensors and Bioelectronics*, vol. 26, no. 2, pp. 542–548, 2010.

[15] J. Rubio Retama, E. López Cabarcos, D. Mecerreyes, and B. López-Ruiz, "Design of an amperometric biosensor using polypyrrole-microgel composites containing glucose oxidase," *Biosensors and Bioelectronics*, vol. 20, no. 6, pp. 1111–1117, 2004.

[16] X. Lu, J. Zhou, W. Lu, Q. Liu, and J. Li, "Carbon nanofiber-based composites for the construction of mediator-free biosensors," *Biosensors and Bioelectronics*, vol. 23, no. 8, pp. 1236–1243, 2008.

[17] A. J. Amalraj, M. Sundaravadivelm, A. P. P. Regis, and S. Rajendran, "Corrosion inhibition by polyvinylpyrrolidone," in *Proceedings of the 9th European Symposium on Corrosion Inhibitors*, p. 407, 2000.

[18] A. A. Gürten, M. Erbil, and K. Kayakirilmaz, "Effect of polyvinylpyrrolidone on the corrosion resistance of steel," *Cement and Concrete Composites*, vol. 27, no. 7-8, pp. 802–808, 2005.

[19] Y. Jianguo, W. Lin, V. Otieno-Alego, and D. P. Schweinsberg, "Polyvinylpyrrolidone and polyethylenimine as inhibitors for the corrosion of a low carbon steel in phosphoric acid," *Corrosion Science*, vol. 37, no. 6, pp. 975–985, 1995.

[20] K. Oksman, M. Skrifvars, and J. F. Selin, "Natural fibres as reinforcement in polylactic acid (PLA) composites," *Composites Science and Technology*, vol. 63, no. 9, pp. 1317–1324, 2003.

[21] J. Xu, J. Zhang, W. Gao, H. Liang, H. Wang, and J. Li, "Preparation of chitosan/PLA blend micro/nanofibers by electrospinning," *Materials Letters*, vol. 63, no. 8, pp. 658–660, 2009.

[22] T. W. Ebbesen and P. M. Ajayan, "Large-scale synthesis of carbon nanotubes," *Nature*, vol. 358, no. 6383, pp. 220–222, 1992.

[23] Y. B. Vassilyev, O. A. Khazova, and N. N. Nikolaeva, "Kinetics and mechanism of glucose electrooxidation on different electrode-catalysts. Part II. Effect of the nature of the electrode and the electrooxidation mechanism," *Journal of Electroanalytical Chemistry*, vol. 196, no. 1, pp. 127–144, 1985.

[24] A. Gorman, A. McGowan, and T. G. Cotter, "Role of peroxide and superoxide anion during tumour cell apoptosis," *FEBS Letters*, vol. 404, no. 1, pp. 27–33, 1997.

[25] Z. Zhuang, X. Su, H. Yuan, Q. Sun, D. Xiao, and M. M. F. Choi, "An improved sensitivity non-enzymatic glucose sensor based on a CuO nanowire modified Cu electrode," *Analyst*, vol. 133, no. 1, pp. 126–132, 2008.

[26] A. J. Bard and L. R. Faulkner, "Electrochemical methods," in *Fundamentals and Applications*, p. 241, John Wiley & Sons, New York, NY, USA, 2nd edition, 2001.

[27] E. Laviron, "General expression of the linear potential sweep voltammogram in the case of diffusionless electrochemical systems," *Journal of Electroanalytical Chemistry*, vol. 101, no. 1, pp. 19–28, 1979.

[28] E. Laviron, "The use of linear potential sweep voltammetry and of a. c. voltammetry for the study of the surface electrochemical reaction of strongly adsorbed systems and of redox modified electrodes," *Journal of Electroanalytical Chemistry*, vol. 100, no. 1-2, pp. 263–270, 1979.

[29] R. Banerjee and R. Cassoly, "Preparation and properties of the isolated α and β chains of human hemoglobin in the ferri state. Investigation of oxidation-reduction equilibria," *Journal of Molecular Biology*, vol. 42, no. 2, pp. 337–349, 1969.

13

Tin Content Determination in Canned Fruits and Vegetables by Hydride Generation Inductively Coupled Plasma Optical Emission Spectrometry

Sanda Rončević,[1] Anica Benutić,[2] Ivan Nemet,[1] and Buga Gabelica[2]

[1] *Laboratory of Analytical Chemistry, Department of Chemistry, Faculty of Science, University of Zagreb, Horvatovac 102 a, 10000 Zagreb, Croatia*
[2] *Croatian National Institute of Public Health, Enviromental Health Service Rockefellerova 7, 10000 Zagreb, Croatia*

Correspondence should be addressed to Sanda Rončević, roncevic@chem.pmf.hr

Academic Editor: Hian Kee Lee

Tin content in samples of canned fruits and vegetables was determined by hydride generation inductively coupled plasma atomic emission spectrometry (HG-ICP-OES), and it was compared with results obtained by standard method of flame atomic absorption spectrometry (AAS). Selected tin emission lines intensity was measured in prepared samples after addition of tartaric acid and followed by hydride generation with sodium borohydride solution. The most favorable line at 189.991 nm showed the best detection limit (1.9 μg L^{-1}) and limit of quantification (6.4 μg kg^{-1}). Good linearity and sensitivity were established from time resolved analysis and calibration tests. Analytical accuracy of 98–102% was obtained by recovery study of spiked samples. Method of standard addition was applied for tin determination in samples from fully protected tinplate. Tin presence at low-concentration range was successfully determined. It was shown that tenth times less concentrations of Sn were present in protected cans than in nonprotected or partially protected tinplate.

1. Introduction

A significant quantity of food and beverages arrive on the market in a robust form of tinplate packaging. Hermetically sealed can allow minimization of headspace oxygen and also keep a long and safe shelf life with minimal use of preservatives. However, the use of tinplate for food and beverage packing will result in some tin dissolving into food content. Tinplate corrosion depends on many factors including can material, nature of the can linings and coatings, and nature and acidity of the contacting foodstuff. Only limited reports are available on the toxicological effects of inorganic tin as present in canned foods, resultant from dissolution on the tin coating [1]. The main potential hazard from acute ingestion seems to be gastric irritation in some individuals exposed to high levels [2]. The maximum limit of 250 mg kg^{-1} for tin in canned foods and 150 mg kg^{-1} for tin in canned beverages was recommended by World Health Organization [3]. Within the European Union large discrepancies exist between national regulations and maximum permitted levels range from 50 mg kg^{-1} to 250 mg kg^{-1}.

Determination of tin in canned food became very important because it gives information about the contamination process and provides help to increase canned food quality and safety. In order to evaluate tin concentration in canned food and beverages, several analytical methods are described in the recent literature: fluorimetric with use of surfactant reagents to increase the sensitivity [4, 5], flame (FAAS) and electrothermal atomic absorption spectrometry (ETAAS) [6–8], and inductively coupled plasma optical emission spectrometry (ICP-OES) [9–11]. There are also several studies concerning the quality and resistance of tinplate after interaction with foodstuff which performed the use of X-ray microanalysis (EDS) and scanning electron microscopy (SEM) [12, 13].

Atomic spectrometry methods in analysis of tin level in canned foods presume decomposition of food samples which is usually performed by acid digestion or dry ashing methods following by measurements in flame or graphite furnace mode of AAS or ICP-OES. Although the preparation step could influence the complete recovery of Sn, the sensitivity of applied measurement technique is also essential in correct evaluation of tin content in foods. For example, flame atomic absorption is recommended by European Committee for Standardization (CEN) as a standard analytical method for the determination of tin in fruit and vegetables preserved in cans [14]. In most of routine applications where Sn content fall below 3–5 $\mu g\,g^{-1}$, the flame AAS shows limited detection power. In combination with hydride generation system (HG), flame AAS technique becomes more sensitive, but it is greatly complicated by the dependency of the hydride species on acidity of sample solution [6, 8, 15]. Analytical performances of ICP-OES technique allow measurements of numerous elements and/or more emission lines of the particular element. Consequently, tin content could be determined in combination with other trace elements present in food samples [11, 16]. Comparable results of different sample preparation procedures followed by ICP-OES measurement of tin content were reported for matrices containing more than 30 $mg\,kg^{-1}$ of tin [9]. However, interference effects in direct measurements of prepared solutions were not avoided completely, especially those caused by complex matrix. The technique of hydride generation with inductively coupled plasma optical emission spectrometry (HG-ICP-OES) is widely used to enhance the sensitivity of semimetal determination and to eliminate the majority of matrix interferences in plasma. The literature survey still suffers from the lack of data concerning the tin determination in canned food by this technique.

In the present work a method of the tin determination in canned food samples by HG-ICP-OES method is described. The aim of study was to select the most appropriate measurement conditions for tin determination, especially at low-concentration range where FAAS shows insufficient sensitivity. Commercially available samples of fruits and vegetables analyzed by both methods were preserved in variously manufactured tinplate, that is, interior surfaces of cans were fully or partially lacquered. Among the chemical and physical factors which influence tin transfer into canned food, a quality of tinplate protection is also essential. Therefore, HG-ICP-OES as sensitive method for low-content tin determination in canned fruit and vegetable samples should be also helpful in estimating the efficiency and resistance of tinplate package.

2. Materials and Methods

2.1. Instruments. A *Prodigy High Dispersion* inductively coupled plasma optical emission spectrometer (Teledyne Leeman Labs, Hudson, NH, USA) was used for the tin content determination in all samples. The operating parameters of instrument and hydride generation device are given in Table 1. Emission lines used in this work were chosen as most prominent lines given in instrument line library file.

A *SpectrAA 220 FS* atomic absorption spectrometer (Varian, Australia) equipped with deuterium background correction and UltrAA Sn lamp was used for flame AAS determinations. Tin analysis was performed in acetylene/nitrous oxide flame; measurements of absorption were made at 235.5 nm with 0.5 nm slit aperture.

For the homogenization of food samples a *Mixer Büchi B400* (Büchi Labortechnik AG, Switzerland) was used; a *Hot Block CS 54* (Environmental Express, USA) was used for the digestion of previously homogenized samples.

2.2. Reagents and Solutions. High-purity-deionised water (Milli-Q Element system, Millipore, USA) was used for the preparation of standard solutions and dilution of samples. In the sample digestion procedures, a hydrochloric acid of suprapure grade (30% m/v) was used. Single-element standard solutions of Sn 1000 $mg\,L^{-1}$ (Plasma Pure, Leeman Labs, Hudson, NH, USA) was used for the preparation of calibration standard solutions and control of plasma line positioning.

All calibration standards for AAS measurement were prepared during instrument run by automated dilution of reference tin solution of 50 $mg\,L^{-1}$ in 10% v/v HCl.

For the determination of tin by HG-ICP-OES, a fresh solution of $NaBH_4$ 0.8% m/v in NaOH 0.5% m/v was prepared. Tartaric acid for the reduction of tin was prepared by dissolution of 10 g of solid compound in 1 L of ultrapure water. Calibration solutions of tin were prepared in the range of 0.1–50.0 $mg\,L^{-1}$ by dilution with 1% m/v tartaric acid solution to appropriate volume. Calibration blank contained only aqueous solution of tartaric acid. Method of standard addition (MSA) included aliquots of prepared samples in which a standard solution of tin was added. MSA solutions were diluted with tartaric acid solution. The final concentration range in MSA sample solutions was 0.1–50.0 $mg\,L^{-1}$ of tin.

2.3. Samples. The twenty five samples of canned fruits and vegetables stored in original package at room temperature were opened, and the whole content was transferred into a *Mixer Büchi B400* where homogenization step was performed. Homogenized samples were stored in polypropylene containers until digestion step. An amount of 5 g of homogenized sample was transferred into a hot block vessel, and 10 mL of 50% v/v HCl was added. Samples were incubated in a Hot Block CS 54 at 80°C for two hours with occasional shaking of vessel content. Solution was then transferred by filtrating through filter paper (84 g/m^2) into 50 mL volumetric flask and filled up to a volume with ultrapure water. Reagent blanks were included in each series of digestions. In order to prevent influence of oxidation and losses of analyte, samples stored in glass vessels should be submitted to measurement within 6 hours from

TABLE 1: ICP-AES operating conditions.

Instrument	Prodigy high-dispersive ICP
Spectrometer	High resolution echelle polychromator Large-format programmable array detector (L-PAD)
RF generator	40 MHz "free running"
Argon flow	Coolant: 18 L min^{-1} Auxiliary: 0.8 L min^{-1} Nebulizer: 36 psi
Nebulizer	Pneumatic (glass concentric)
Spray chamber	Glass cyclonic
Hydride generator	Leeman Labs. Inc. Part No. 130-1070 Three channel peristaltic pump (0.9 mL min^{-1}) T connector Reaction coil Output power (1.3 kW) Plasma viewing (Radial)
Replicates for each analysis run	3
Sample uptake delay	50 s
Integration time	40 s
Emission lines	Sn (II) 189.991 nm Sn (I) 224.605 nm Sn (I) 235.484 nm Sn (I) 283.999 nm

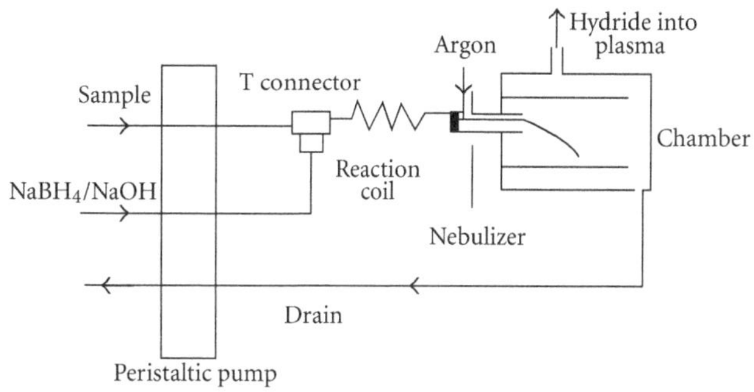

FIGURE 1: Hydride generator (HG) schematic.

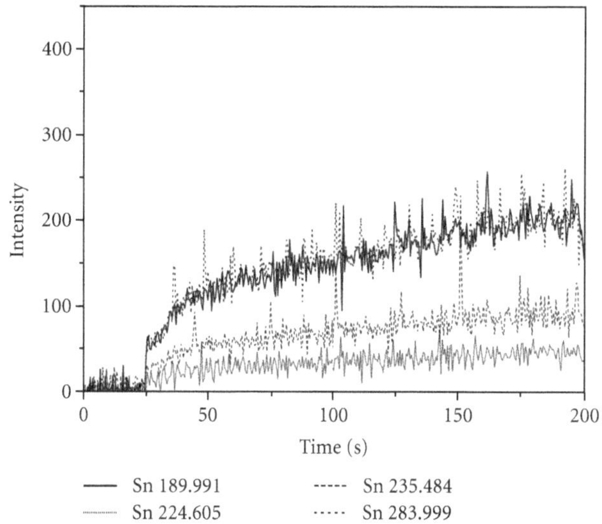

FIGURE 2: Time resolved analysis (TRA) of Sn solution (1 mg L^{-1} in tartaric acid) by HG-ICP-OES.

preparation. Otherwise, they should be stored hermetically into polypropylene containers.

2.4. Measurement. Atomic absorption measurements were performed in accordance with recommended standard procedure for the tin determination described in CEN/TS 15506:2007 [14]. Calibration of AAS instrument was performed with five solutions which were automatically diluted from reference standard. Control of calibration by external standard was performed after eight measuring steps, and recalibration procedure was repeated after twenty fourth measuring step. Characteristic concentration (c_c) of tin at 1% of signal absorption was 1.24 mg L^{-1}. Blank sample was measured prior to every set of samples. Each of two replicates of samples was measured three times subsequently.

For the purpose of HG-ICP-OES measurements, the calibration standards and 10-fold diluted samples were mixed with a solution of $NaBH_4$ by three-channel peristaltic pump at a rate of 0.9 mL min^{-1}. Hydride generator assembly scheme is shown in Figure 1. Solutions were introduced into a reaction coil where stannane gas was generated and introduced into plasma through nebulizer and nebulization chamber. The RF power and nebulizer gas flow rate were optimized to give the maximum intensity of Sn lines on L-PAD detector, that are, 1300 W and 0.9 L min^{-1}, respectively. Sample uptake delay before starting of intensity measurement was set on 50 seconds. This value was selected after collecting a signal by option of time-resolved analysis (TRA) of instrument software. Figure 2 shows signals of chosen Sn lines for solution of 1 mg L^{-1} in TRA mode. The signals started to rise after 25 seconds and plateau of signals was reached after 50 seconds from aspiration start.

Intensity measurements were performed in triplicate. The precision of signal measurements expressed as relative standard deviation was established in the range 0.1%–4.0%. Calibration curves for each Sn line were shown in Figure 3. A good linearity of all observed Sn lines was obtained (R^2 = 0.9995–0.9999). Intensity measurement in MSA mode was performed with those samples which were initially analyzed by AAS and showed less than 5 mg L^{-1} of Sn content. An example of obtained MSA curves for a selected sample was shown in Figure 4. The intensity values in MSA operation mode were one order of magnitude greater comparing to those obtained in standard calibration. Linear coefficients

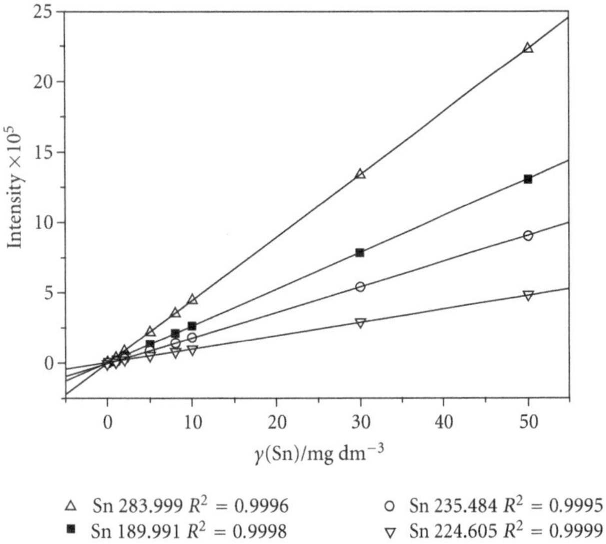

Figure 3: Calibration curves (HG-ICP-OES) in conventional mode.

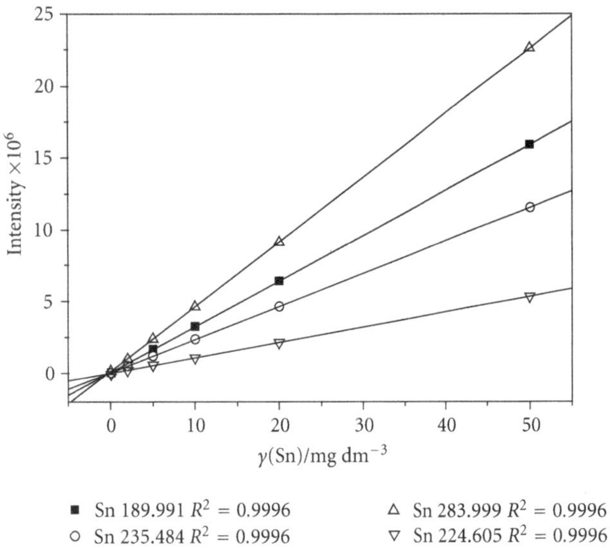

Figure 4: Calibration curves (HG-ICP-OES) in method of standard addition mode (MSA).

comprised the same values (R^2 = 0.9996) at each observed line. The method detection limits (c_L), which were calculated using 3σ criterion, that is, the concentration equivalent to three times standard deviation (3σ) of the signal (n = 10) of the reagent blank solution, are given in Table 2. The limits of quantification (LOQ) were calculated from calibration curves and expressed in μg kg^{-1} of sample mass.

3. Results and Discussion

HG-ICP-OES measurements of tin content in samples were performed at all selected emission lines. The precision of measurement based on RSD calculations from three replicates showed the smallest disturbance of signal at 235.484 nm

Table 2: Method detection limits (c_L) and limits of quantification (LOQ) for HG-ICP-OES.

Wavelength/nm	c_L/μg L^{-1}	LOQ/μg kg^{-1}
Sn (II) 189.991	1.9	6.4
Sn (I) 224.605	1.4	4.5
Sn (I) 235.484	6.3	20.0
Sn (I) 283.999	6.5	20.7

(RSD 0.1%) and 189.991 nm (RSD 0.4%), compared to 283.999 nm (RSD 2.6%) and 224.605 nm (RSD 4.0%). The sensitivity of measurements should be examined from the obtained calibration curves and TRA signals. By aspirating the solution of low tin concentration, that is, 1 mg L^{-1} the better sensitivity of signal was noted at 283.999 nm and 189.991 nm lines (Figure 2). Comparison of detection capability of tin at those lines showed lower detection limit at ionic line of 189.991 nm than at atomic line of 283.999 nm. Lower detection limits, good linearity in the observed range of concentrations, and the better precision of measurements favoured the determination of tin content at 189.991 nm. Comparing with literature data, the same conclusions from ICP-OES determination of tin content without use of hydride generation were already derived [9], but an order of magnitude better sensitivity and detection capability is obtained here by use of HG-ICP-OES system. Hydride generation from samples where tartaric acid as interference-retarding agent is added also provides a selective formation of stannane gas which enters into plasma [17]. It also reduces the need to perform extensive study of possible interfering effects.

Tin content measured by flame AAS and HG-ICP-OES in different samples of canned fruits and vegetables is shown in Table 3. All results are expressed as a milligram portion of Sn per kg of sample mass, along with standard deviation of results (σ). Description of tinplate protection of particular can, that is, fully or partially lacquered interior is also inserted in Table 3. Accuracy of results was tested by spiking of selected samples with Sn standard solution after homogenization step. The selected samples were foodstuffs from complete-or partially lacquered can (nos. 1 and 13). The recoveries of Sn measured by AAS were 98% for both samples. HG-ICP-OES measurement of spiked samples gave recoveries of 99% for Sample 1 and 102% for Sample 13. Statistical comparison of results obtained by reference AAS method and tested HG-ICP-OES method were performed by two-paired t-test at significance level of P = 0.05. Calculated t-value of 1.09 is lower than critical value of 2.06, which means that two methods were not significantly different.

Generally, the results of flame AAS and HG-ICP-OES measurements showed that tin content in all samples did not exceed maximum permissive level of 200 mg kg^{-1} in foodstuffs. Only one sample (no. 11) of pineapples compote reached this level. Comparing the results obtained from two applied techniques, it should be noticed that for the most of samples measured by AAS and where tin content was above 5 mg kg^{-1}, the similar results by HG-ICP-OES

TABLE 3: Tin content in canned fruit determined by AAS and HG-ICP-OES (mean of two replicate samples, $n = 3$).

No.	Sample	Tinplate protection	AAS $w(Sn) \pm \sigma/mg\,kg^{-1}$	HG-ICP-OES $w(Sn) \pm \sigma/mg\,kg^{-1}$
(1)	Peach compote, less sweat	Lacquered	68.9 ± 1.2	69.6 ± 1.1
(2)	Tomato puree, double concentrated	Lacquered	<5	2.5 ± 0.1
(3)	White beans, sterilized	Lacquered	<5	5.8 ± 0.1
(4)	Red beans, sterilized	Lacquered	<5	3.5 ± 0.1
(5)	Champignons	Lacquered	<5	0.44 ± 0.02
(6)	Champignons, sterilized	Lacquered	77.2 ± 0.8	77.2 ± 0.7
(7)	Peeled plum tomatoes, sterilized	White lacquer	<5	10.9 ± 0.2
(8)	Fruit salad	Yellow lacquer	<5	1.21 ± 0.04
(9)	Pineapples compote	Tin side and bottom, lacquered lid and seam	25.6 ± 0.1	28.8 ± 0.1
(10)	Pineapples compote	Tin side and bottom, lacquered lid and seam	37.6 ± 0.2	40.5 ± 0.2
(11)	Pineapples compote	Tin side and bottom, lacquered lid and seam	199.2 ± 3.2	199.2 ± 3.2
(12)	Pineapples compote	Tin side and bottom, lacquered lid and seam	37.7 ± 3.7	41.2 ± 3.7
(13)	Pineapples compote	Tin side and bottom, lacquered lid and seam	74.5 ± 1.1	75.5 ± 1.3
(14)	Pineapples compote	Tin side and bottom, lacquered lid and seam	23.3 ± 1.1	25.7 ± 1.1
(15)	Pineapples compote	Tin side and bottom, lacquered lid and seam	47.8 ± 0.5	47.8 ± 0.7
(16)	Pineapples compote	Tin side and bottom, lacquered lid and seam	40.3 ± 0.7	40.4 ± 0.7
(17)	Peach halves peeled, compote	Tin side and bottom, lacquered lid and seam	28.6 ± 0.5	29.6 ± 0.4
(18)	Peach halves in syrup, less sweat compote	Tin side and bottom, lacquered lid and seam	82.8 ± 1.4	104.9 ± 1.8
(19)	Compote of mixed fruits	Tin side and bottom, lacquered lid and seam	53.5 ± 0.9	54.82 ± 0.8
(20)	Pineapples compote	Tin side and bottom, lacquered lid and seam	37.7 ± 0.7	41.2 ± 0.7
(21)	Apricot compote, less sweat	Tin side and bottom, lacquered lid and seam	115.9 ± 1.4	115.6 ± 1.3
(22)	Apricot compote	Tin side and bottom, lacquered lid and seam	81.0 ± 1.1	76.7 ± 1.4
(23)	Fruit cocktail in light syrup	Tin side and bottom, lacquered lid and seam	117.3 ± 1.3	115.8 ± 1.4
(24)	Fruit cocktail in light syrup	Tin side and bottom, lacquered lid and seam	108.6 ± 1.2	107.3 ± 1.3
(25)	Mandarin oranges whole segments	Tin side and bottom, lacquered lid and seam	105.5 ± 1.3	104.8 ± 1.2

were obtained. The exceptions were samples of pineapples compote (nos. 9, 10, 12, 14, and 20) and one sample of peeled plum tomatoes (No 7) with slightly higher tin concentration obtained by HG-ICP-OES. By knowing that samples for HG-ICP-OES determination include a tartaric acid which also prevents a hydrolysis of possibly present Sn (IV) species, observed difference is quite understandable. A sample of peach halves compote (no. 18) showed slightly higher Sn concentration when measured by HG-ICP-OES compared to concentration obtained by AAS. Since that longer delay from sample preparation to start of AAS measurement could lead to hydrolysis of analyte and forming of insoluble Sn (IV) compounds, a significant loss of Sn absorption signal might occur. HG-ICP-OES measurements include the use of highly reactive reducing agents which convert the majority of tin species in Sn (II) and favour the stannane gas formation. Therefore, the much more reliable value in such case is denoted to HG-ICP-OES measurement.

Tin content in samples from complete lacquered cans (nos. 2–5, 7, 8) measured by AAS shows limitation of method capability at low-concentration range (Table 3). Considering the results where AAS measurements are limited at 5 mg kg^{-1} of tin in canned food, the method of standard addition in HG-ICP-OES technique showed exclusive advantage. Tin content measured in diluted samples was an order of magnitude higher than the LOQ of 6.4 $\mu g\,kg^{-1}$. It must be mentioned that tin determination in standard calibration mode gave approximately similar results, but with very poor precision (>20% RSD). This is mainly caused by matrix effect of solution with organic matter remains. Changes of surface tension, relative volatility, and viscosity of such solution affect the droplet distribution in HG introduction system and consequently, the amount of tin that reaches the plasma. The complexity of organic matter impact is accomplished by entering solution into plasma where changes in excitation temperature and electronic density of plasma take place. Generally, organic matter present in solution can affect the emission which results with enhancement or suppression of signal [18, 19]. Therefore, method of standard addition allows compensation of matrix effects which is also evident from the improved precision of measurements. Applied technique allows the determination of tin presence at low-concentration range where FAAS, XRF, or ICP-OES without hydride system suffer from poor sensitivity [6, 7, 9, 10, 20–22].

The results obtained for the food samples from fully protected inner walls of cans are very useful in predicting a quality of protection. It is shown that tenth times less portions of Sn are present in protected cans than in non-protected or partially protected tinplate. The exceptions, measured as approximately 70 mg kg^{-1} of tin, are established in Samples 1 and 6. The studies concerning the quality of tinplate by SEM and EDS suggest that defects of lacquer result in tin exposure [12, 13] but in much less extent than that was measured in those two samples (nos. 12, 13). Therefore, such concentrations might be attributed to the addition of tin chloride which also acts as preserving agent in manufacturing of canned foods [1, 2].

In a summary, analysis of tin content in canned fruits and vegetables was performed by use of flame AAS which is recommended as a standard analytical method and also by HG-ICP-OES. Analytical efficiency of HG-ICP-OES method was tested on several Sn emission lines. Low-detection limit, (1.9 $\mu g\,L^{-1}$) and LOQ of 6.4 $\mu g\,kg^{-1}$, good linearity in the observed range of concentrations, and the better precision of measurements favoured the determination of tin content at 189.991 nm. Comparable results of tin content for samples from partially protected tinplate cans were obtained by use of AAS and HG-ICP-OES. Method of standard addition in HG-ICP-OES technique was applied on samples from complete lacquered cans where AAS method is limited at 5 mg kg^{-1}. The reliable results for tin content at low-concentration range were established. This analytical advantage could utilize HG-ICP-OES technique as a helpful tool in certain examination of tinplate protection efficiency.

Acknowledgment

The investigation was performed as a part of project 119-1191342-1083 financed by Republic Croatia Ministry of Science, Education, and Sport.

References

[1] S. Blunden and T. Wallace, "Tin in canned food: a review and understanding of occurrence and effect," *Food and Chemical Toxicology*, vol. 41, no. 12, pp. 1651–1662, 2003.

[2] P. J. Boogaard, M. Boisset, S. Blunden, S. Davies, T. J. Ong, and J. P. Taverne, "Comparative assessment of gastrointestinal irritant potency in man of tin(II) chloride and tin migrated from packaging," *Food and Chemical Toxicology*, vol. 41, no. 12, pp. 1663–1670, 2003.

[3] FAO/WHO Expert Committee on Food Additives, "Evaluation of certain food additives and contaminants," WHO Technical Report Series 776, Thirty-Third Report of the Joint FAO/WHO Expert Committee on Food Additives, Geneva, Switzerland, 1989.

[4] J. L. Manzoori, M. Amjadi, and D. Abolhasani, "Spectrofluorimetric determination of tin in canned foods," *Journal of Hazardous Materials*, vol. 137, no. 3, pp. 1631–1635, 2006.

[5] E. S. B. Morte, M. G. A. Korn, M. L. M. F. S. Saraiva, J. L. F. C. Lima, and P. C. A. G. Pinto, "Sequential injection fluorimetric determination of Sn in juices of canned fruits," *Talanta*, vol. 79, no. 4, pp. 1100–1103, 2009.

[6] F. E. Khansari, M. Ghazi-Khansari, and M. Abdollahi, "Heavy metals content of canned tuna fish," *Food Chemistry*, vol. 93, no. 2, pp. 293–296, 2005.

[7] A. G. Cameron and J. E. Brittain, "The measurement of tin in canned beans by atomic absorption spectrometry," *International Journal of Food Science & Technology*, vol. 6, no. 2, pp. 187–192, 2007.

[8] H. O. Haug and L. Yiping, "Automated determination of tin by hydride generation using in situ trapping on stable coatings in graphite furnace atomic absorption spectrometry," *Spectrochimica Acta B*, vol. 50, no. 11, pp. 1311–1324, 1995.

[9] L. Perring and M. Basic-Dvorzak, "Determination of total tin in canned food using inductively coupled plasma atomic emission spectroscopy," *Analytical and Bioanalytical Chemistry*, vol. 374, no. 2, pp. 235–243, 2002.

[10] R. E. S. Froes, W. Borges Neto, R. L. P. Naveira, N. C. Silva, C. C. Nascentes, and J. B. B. da Silva, "Exploratory analysis and inductively coupled plasma optical emission spectrometry (ICP OES) applied in the determination of metals in soft drinks," *Microchemical Journal*, vol. 92, no. 1, pp. 68–72, 2009.

[11] A. Ikem and N. O. Egiebor, "Assessment of trace elements in canned fishes (mackerel, tuna, salmon, sardines and herrings) marketed in Georgia and Alabama (United States of America)," *Journal of Food Composition and Analysis*, vol. 18, no. 8, pp. 771–787, 2005.

[12] A. V. Pournaras, M. I. Prodromidis, A. P. Katsoulidis, A. V. Badeka, D. Georgantelis, and M. G. Kontominas, "Evaluation of lacquered tinplated cans containing octopus in brine by employing X-ray microanalysis and electrochemical impedance spectroscopy," *Journal of Food Engineering*, vol. 86, no. 3, pp. 460–464, 2008.

[13] M. G. Kontominas, M. I. Prodromidis, E. K. Paleologos, A. V. Badeka, and D. Georgantelis, "Investigation of fish product-metal container interaction using scanning electron microscopy-X-ray microanalysis," *Food Chemistry*, vol. 98, no. 2, pp. 225–230, 2006.

[14] P CEN/TS 15506, Foodstuffs-"Determination of trace elements-Determination of tin in fruit and vegetables preserved in cans by flame atomic absorption spectrometry (AAS)", 2007.

[15] G. H. Alvarez and S. G. Capar, "Determination of tin in foods by hydride generation-atomic absorption spectrometry," *Analytical Chemistry*, vol. 59, no. 3, pp. 530–533, 1987.
[16] A. S. Ribeiro, A. L. Moretto, M. A. Z. Arruda, and S. Cadore, "Analysis of powdered coffee and milk by ICP OES after sample treatment with tetramethylammonium hydroxide," *Mikrochimica Acta*, vol. 141, no. 3-4, pp. 149–155, 2003.
[17] H. Wiltsche, I. B. Brenner, G. Knapp, and K. Prattes, "Simultaneous determination of As, Bi, Se, Sn and Te in high alloy steels - Re-evaluation of hydride generation inductively coupled plasma atomic emission spectrometry," *Journal of Analytical Atomic Spectrometry*, vol. 22, no. 9, pp. 1083–1088, 2007.
[18] R. I. Botto, "Matrix interferences in the analysis of organic solutions by inductively coupled plasma-atomic emission spectrometry," *Spectrochimica Acta B*, vol. 42, no. 1-2, pp. 181–199, 1987.
[19] T. D. Hettipathirana, A. P. Wade, and M. W. Blades, "Effects of organic acids in low power inductively coupled argon plasma-optical emission spectroscopy," *Spectrochimica Acta B*, vol. 45, no. 3, pp. 271–280, 1990.
[20] M. Muntenau, E. Chirila, G. Stanciu, and N. Marin, "Tin determination in canned fruits," *Ovidius University Annals of Chemistry*, vol. 21, pp. 79–82, 2010.
[21] J. Knapek, V. Herman, R. Buchtová, and D. Vosmerova, "Determination of tin in canned foods by atomic absorption spectrometry," *Czech Journal of Food Sciences*, vol. 27, pp. S407–S409, 2009.
[22] Y. Mino, "Determination of tin in canned foods by X-ray fluorescence spectrometry," *Journal of Health Science*, vol. 52, no. 1, pp. 67–72, 2006.

14

Ultra High-Mass Resolution Paper Spray by Fourier Transform Ion Cyclotron Resonance Mass Spectrometry

Kevin D. Quinn, Charmion I. Cruickshank, and Troy D. Wood

Department of Chemistry, State University of New York at Buffalo, Buffalo, NY 14260, USA

Correspondence should be addressed to Troy D. Wood, twood@buffalo.edu

Academic Editor: Charles L. Wilkins

Paper Spray Ionization is an atmospheric pressure ionization technique that utilizes an offline electro-osmotic flow to generate ions off a paper medium. This technique can be performed on a Bruker SolariX Fourier transform ion cyclotron resonance mass spectrometer by modifying the existing nanospray source. High-resolution paper spray spectra were obtained for both organic and biological samples to demonstrate the benefit of linking the technique with a high-resolution mass analyzer. Error values in the range 0.23 to 2.14 ppm were obtained for calf lung surfactant extract with broadband mass resolving power ($m/\Delta m_{50\%}$) above 60,000 utilizing an external calibration standard.

1. Introduction

Since the first practical implementation of electrospray ionization (ESI) with mass spectrometry reported by Yamashita and Fenn in 1984, a number of different atmospheric pressure ionization (API) techniques have been developed [3]. Many API techniques require extensive sample preparation as well as costly consumables to obtain mass spectra. One of the most cost effective and widely available media for chemical separation and analysis is paper. It has been shown that an electro-osmotic flow similar to that of static electrospray can be utilized to interface a triangle of wetted paper to a mass spectrometer (MS); this form of analysis is called paper spray (PS) [4–9].

The use of a chromatographic medium such as paper allows for the analysis of complex mixtures since separations will occur as the analyte is being driven along the paper medium [7]. Separating analytes on a paper medium provides the benefits of reduced probability of ion suppression and rapid analysis. The paper medium allows for analysis of complex samples such as biological tissue and dried blood spots which are not suitable for traditional infusion techniques due to their complex matrices [4, 7].

One limitation to PS, like many MS techniques that utilize incomplete chromatographic separations is the detector being utilized to perform the analysis. When dealing with complex samples or unknown samples, there is a need for high-quality data in order to determine what is being analyzed. Fourier transform ion cyclotron resonance (FT-ICR) mass spectrometers are powerful tools for analysis of complex mixtures due to their unsurpassed capability to obtain broadband high-resolution mass spectra with high mass accuracy. Utilizing PS with a Fourier transform ion cyclotron resonance (FT-ICR) mass spectrometer as a detector allows for rapid analysis of both unknowns and/or complex mixtures.

In this present work, we show the suitability of a 12 Tesla FT-ICR MS to obtain high-resolution broadband mass spectra utilizing a coffee filter paper spray medium. High-resolution mass spectra were first obtained for polypropylene glycol (PPG) and polyethylene glycol (PEG) standards to test the source. Second, urine and calf lung surfactant extract (CLSE) were analyzed to test the PS source for analysis of complex mixtures not suitable for direct infusion.

2. Materials, Methods, and Instrumentation

2.1. Chemicals. All solvents used were HPLC grade obtained from Sigma Aldrich (St. Louis, MO, USA). PPG was obtained in the form of a tuning standard from Applied Biosystems (Bedford, MA, USA). PEG standard was made by dissolving

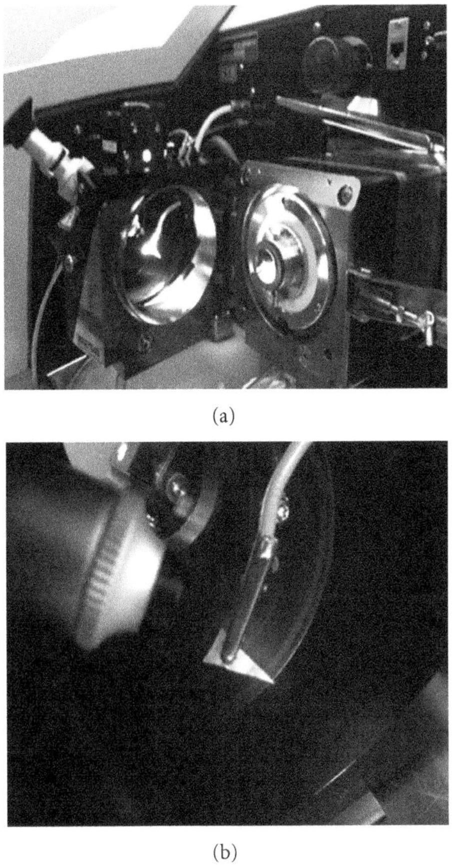

FIGURE 1: PS source utilizing a modified Bruker nano-ESI source, a copper clip, and cable to ground the paper triangle to the chasse of the instrument.

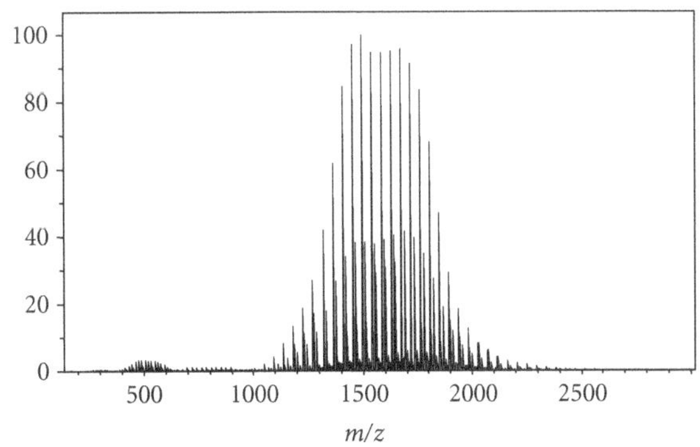

FIGURE 2: Polyethylene glycol paper spray mass spectrum; identification of oligomers as given in the literature [1].

1 mg of PEG 2000 obtained from Sigma Aldrich (St. Louis, MO, USA) in 1 mL of HPLC grade water. Coffee filter, Wegmans brand (Rochester, NY, USA), was cut by hand into an isosceles triangle with dimensions of approximately 5 mm (base) by 10 mm (height).

2.2. PPG and PEG. Samples for PPG and PEG PS analysis were prepared by wetting the coffee filter triangle with 10 μL of the standard solutions. The wet filter paper samples were allowed to dry under ambient conditions before PS analysis was performed. Mass spectra were then obtained by wetting the dried paper triangle with 10 μL of 50 : 50 methanol : water and using the method described in Section 2.5.

2.3. CLSE. CLSE was obtained from ONY Inc. (Amherst, NY, USA). Prior to analysis, the sample was stored in chloroform in the refrigerator. Samples for CLSE analysis were produced by wetting a coffee paper triangle with 2 μL of CLSE and allowing the sample to dry under ambient conditions before PS analysis was performed. Mass spectra were then obtained by wetting the paper with 10 μL of 50 : 50 methanol : water using the instrument method described in Section 2.5.

2.4. Urine. Urine samples were collected from two healthy adult males into sterile polypropylene centrifuge tubes (VWR Scientific) and immediately placed in the freezer. The samples were collected in accordance with the Health Sciences Institutional Review Board at the University at Buffalo (HSIRB no. CHE0050411E). Prior to analysis, the samples were thawed to room temperature and 1 mL was pipetted into 2 mL sample vials (Agilent, Santa Clara, CA, USA) for use. Ten μL was injected directly onto a triangular coffee filter and allowed to dry under ambient conditions. Once dry, the filter paper was attached to the mass spectrometer via a copper clip and rewetted using 50 : 50 methanol : water for PS analysis using the instrument method described in Section 2.5.

2.5. Paper Spray FT-ICR Method. High-resolution spectra were obtained on a 12T Bruker SolariX FT-ICR MS (Bruker Daltonics Inc., Billerica, MA, USA) utilizing a modified Bruker nanoelectrospray source. The positive potential was obtained by grounding the paper triangle using a copper clip and applying a negative voltage at the inlet to the mass spectrometer (Figure 1 shows an image of the source). Mass spectra were obtained at various mass to charge ranges with a pre-FT-ICR-trap hexapole accumulation time of 0.400 seconds and an FT-ICR trap accumulation time of 0.025 seconds. All mass spectra were obtained in the positive mode at a spray voltage of 2 kV. Mass resolving power ($m/\Delta m_{50\%}$) was defined in accordance with the literature [10].

3. Results and Discussion

3.1. Polymer Standards. Polymeric standards were utilized to determine the suitability of PS with high-resolution FT-ICR MS. In FT-ICR MS the resolution of the mass spectra is relative to *m/z* as well as the number of data points that can be obtained during the detection process. High-resolution spectra require a time domain signal stable for longer than several hundred milliseconds. Polyethylene glycol and polypropylene glycol produce a broad *m/z* range of ions ideal for testing the suitability of PS with high-resolution mass spectrometry. High-resolution mass spectra of PEG and PPG are displayed in Figures 2 and 3, respectively. Identifies of the oligomer ion detected and their associated errors is given in Tables 1 and 2, respectively.

Table 1: PEG standards identified using paper spray mass spectrometry. The PEG mass spectrum was obtained with both an excitation and detection frequency range of 1.5 MHz.

PEG	Experimental m/z	Theoretical m/z	Mass measurement error (Δppm)	Mass resolving power ($M/\Delta M_{50\%}$)	$[H(OCH_2CH_2)_nOH+Na]^+$
	1009.577639	1009.57651	−0.12	46309	$n = 22$
	1053.60274	1053.60273	0.01	42870	$n = 23$
	1097.62886	1097.62894	−0.07	40306	$n = 24$
	1141.65527	1141.65515	0.02	38791	$n = 25$
	1185.68128	1185.68137	−0.08	37199	$n = 26$
	1229.70779	1229.70758	0.17	35953	$n = 27$
	1273.73391	1273.7338	0.09	34776	$n = 28$
	1317.76035	1317.76001	0.26	33645	$n = 29$
	1361.78640	1361.78623	0.12	32579	$n = 30$
	1405.81232	1405.81244	−0.09	31288	$n = 31$
	1449.83868	1449.83866	0.01	30101	$n = 32$
	1493.86539	1493.86487	0.35	29219	$n = 33$
	1537.89172	1537.89109	0.41	28219	$n = 34$
	1581.91742	1581.9173	0.08	27254	$n = 35$
	1625.94358	1625.94352	0.04	26391	$n = 36$
	1669.96968	1669.96973	−0.03	25567	$n = 37$
	1713.99520	1713.99595	−0.44	24691	$n = 38$

Table 2: PPG standards identified using paper spray mass spectrometry. The PPG mass spectrum was obtained with excitation and detection frequency range of 1.5 MHz.

PPG	Experimental m/z	Theoretical m/z	Mass measurement error (Δppm)	Mass resolving power ($M/\Delta M_{50\%}$)	$[C_3O_2H_7(C_3H_6O)_nC_6H_{13}O_2+Na]^+$
	1666.17197	1666.17200	−0.02	22686	$n = 25$
	1724.21224	1724.21386	−0.94	23629	$n = 26$
	1782.25781	1782.25573	1.17	23817	$n = 27$
	1840.29829	1840.29760	0.37	23808	$n = 28$
	1898.34113	1898.33946	0.88	23139	$n = 29$
	1956.37992	1956.38133	−0.72	22968	$n = 30$
	2014.41937	2014.42319	−1.90	22638	$n = 31$
	2072.46849	2072.46505	1.66	21991	$n = 32$
	2130.50088	2130.50692	−2.84	21503	$n = 33$
	2188.55370	2188.54878	2.25	20679	$n = 34$
	2246.59219	2246.59065	0.69	19685	$n = 35$
	2304.63182	2304.63254	−0.31	17868	$n = 36$
	2362.67370	2362.67438	−0.29	16365	$n = 37$

Signal-to-noise obtained for PEG and PPG after tuning spectra was in excess of 100/1 with absolute intensities that varied shot-to-shot by less than 20%. This allowed ample time for adjustment of the filter paper towards the inlet of the mass spectrometer so as to optimize instrument setup and analysis. Utilizing polymers of known M_n values (provided by the manufacturers; polydispersity not given) such as PEG 2000 and PPG 2000 provides a detection window several hundred m/z wide for broad mass accuracy calibrations for FT-ICR MS.

3.2. Calf Lung Surfactant Extract. Calf lung surfactant extract (CLSE), which is commercially available in drug form as Infasurf, is used to treat respiratory distress syndrome [11]. CLSE is obtained by doing a total lipid extraction on the lavage fluid from a slaughtered calf lung [11, 12]. Phosphatidylcholine (PC) makes 79% of the molar distribution of lipids and consists of a choline head group attached to the phospholipid with varying degrees of unsaturation in the fatty acid chains [2]. Based on the optimized setup determined from previously analyzing the standards, CLSE was then

Table 3: CLSE compounds identified using paper spray mass spectrometry. The phospholipids detected are denoted using the lipid numbers nomenclature to indicate degrees of unsaturation, and PC refers to phosphatidylcholine. The CLSE mass spectrum was obtained with excitation and detection frequency range of 1.5 MHz.

Experimental *m/z*	Theoretical *m/z*	Mass measurement error (Δppm)	Mass resolving power ($M/\Delta M_{50\%}$)	Compound
728.52151	728.52008	1.96	61637	16 : 0/14 : 0 PC $[M+Na]^+$
732.55505	732.55378	1.73	60967	16/16 PC $[M+H]^+$ 1 degree of unsaturation
734.57100	734.56943	2.14	62820	16 : 0/16 : 0 PC $[M+H]^+$
754.53660	754.53572	1.17	57074	16/16 PC $[M+Na]^+$ 1 degree of unsaturation
756.55257	756.55137	1.59	57597	16 : 0/16 : 0 PC $[M+Na]^+$
760.58591	760.58508	1.09	55950	16/18 PC $[M+H]^+$ 1 degree of unsaturation
782.56728	782.56702	0.33	52802	16/18 PC $[M+Na]^+$ 1 degree of unsaturation
798.54114	798.54096	0.23	48188	16/18 PC $[M+K]^+$ 1 degree of unsaturation
808.58322	808.58267	0.68	48764	18/18 PC $[M+Na]^+$ 2 degrees of unsaturation

Table 4: Compounds identified using paper spray mass spectrometry on urine sample from 27-year-old volunteer. The urine mass spectrum was obtained with both an excitation and detection frequency range of 1.5 MHz.

Compound	Species	Formula	Experimental *m/z*	Theoretical *m/z*	Mass measurement error (Δppm)
3-Hexaprenyl-4,5-Dihydroxybenzoic acid	[M + Na]	$C_{37}H_{54}O_4Na$	585.39288	585.39143	2.5
3′-hydroxy-e,e-caroten-3-one	$[M + CH_3OH + H]$	$C_{41}H_{59}O_3$	599.44790	599.44587	3.4
Vitamin D2 3-glucuronide	$[M + CH_3OH + H]$	$C_{35}H_{57}O_8$	605.40328	605.40480	2.5
Coenzyme Q10	[M + K]	$C_{59}H_{90}O_4K$	901.64549	901.64707	1.8

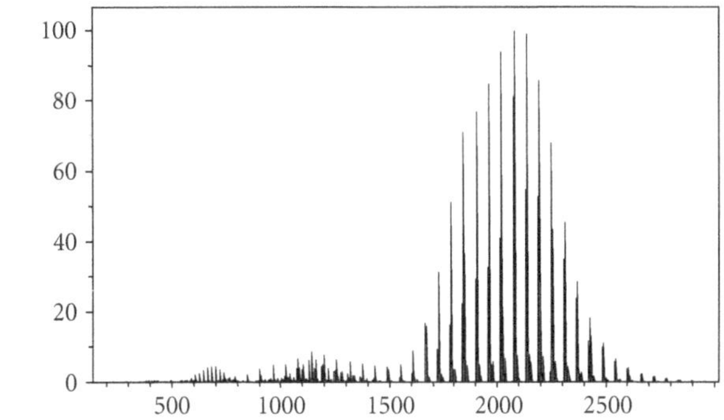

Figure 3: Polypropylene glycol paper spray mass spectrum; identifications of oligomers as given in the literature [2].

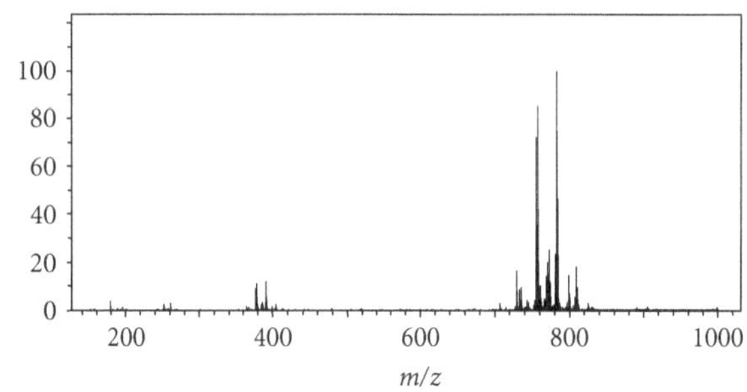

Figure 4: Broadband PS mass spectrum of 2 μL CLSE, the spectrum was obtained by signal averaging 48 scans (arbitrary).

analyzed. PS mass spectra of CLSE are given in Figures 4 and 5, and the identities with associated error provided in Table 3.

3.3. Urine. Due to the complex nature of urine, many small molecules and metabolites were detected at varying intensities using PS with FT-ICR. PS mass spectra, showing numerous lipids and other metabolites obtained from the analyzed urine specimens of two healthy adult male subjects, are shown in Figure 6; an inset of one of these PS mass spectra is shown in Figure 7 while the other is shown in Figure 8. A few of the metabolites identified in the 27-year-old volunteer using the exact mass capability of FT-ICR are provided in Table 4. Metabolites identified in the 44-year-old volunteer using the exact mass capability of FT-ICR are

Table 5: Compounds identified using paper spray mass spectrometry on urine sample from 44-year-old volunteer. The urine mass spectrum was obtained with excitation and detection frequency range of 1.5 MHz.

Compound	Species	Formula	Experimental *m/z*	Theoretical *m/z*	Mass measurement error (Δppm)
Vitamin K1	[M + K]	$C_{31}H_{46}O_2K$	489.31176	489.31294	2.4
16-alpha, 17-beta-estriol 17-beta-D-glucuronide	[M + CH_3OH + H]	$C_{25}H_{37}O_{10}$	497.23617	497.23812	3.9
Tetrahydroaldosterone-3-glucuronide	[M + H]	$C_{27}H_{41}O_{11}$	541.26279	541.26434	2.9
25-Hydroxyvitamin D2-25-glucuronide	[M + CH_3OH + H]	$C_{35}H_{57}O_9$	621.39946	621.39971	0.4

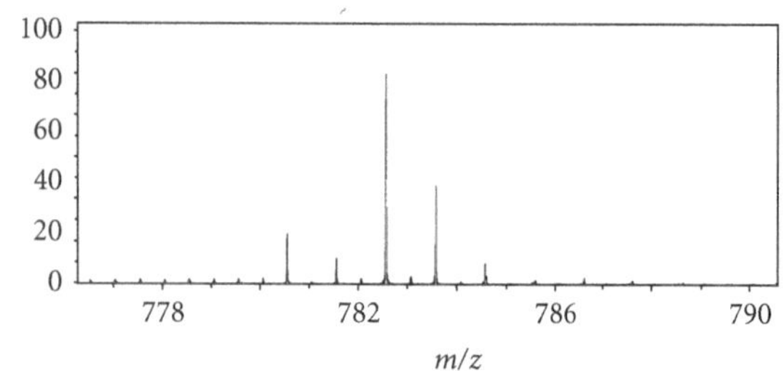

Figure 5: Inset of broadband PS mass spectrum of CLSE over *m/z* 776–791.

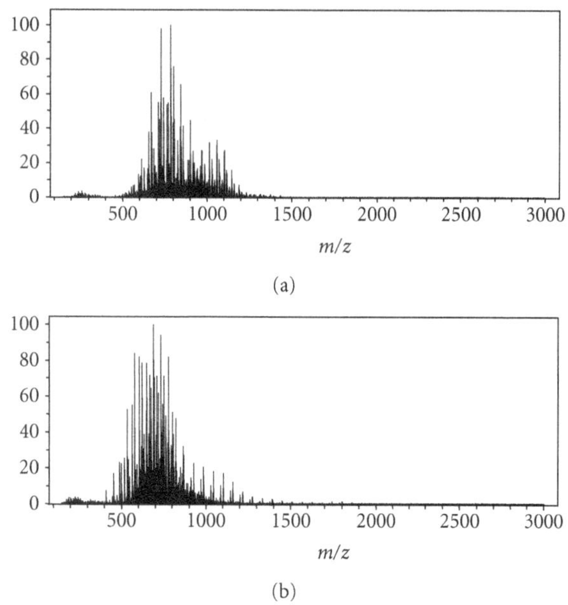

Figure 6: PS mass spectra of urine specimens from two healthy male individuals: 27 year-old (a) and 44 year-old (b).

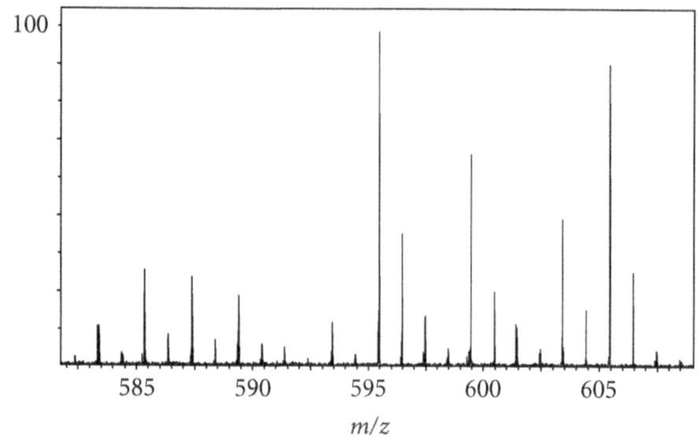

Figure 7: Inset of PS mass spectrum for urine specimen from 27-year-old volunteer over *m/z* 582–608.

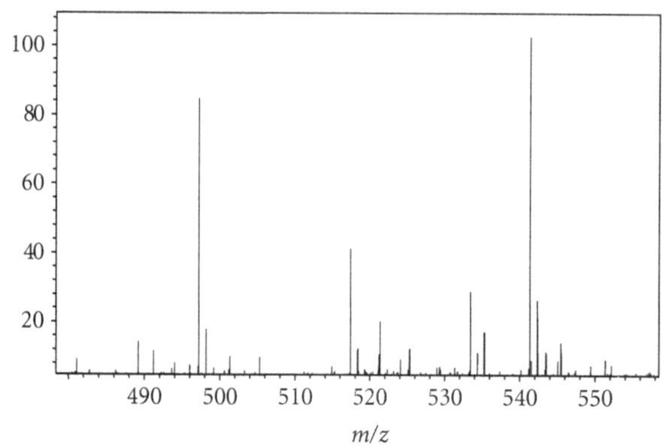

Figure 8: Inset of PS mass spectrum for urine specimen from 44-year-old volunteer being treated with prescription vitamin D supplement over *m/z* 480–560.

given in Table 5. Of note, the 44-year-old volunteer, who was taking prescribed vitamin D supplements, showed indications of the vitamin D metabolite 25-Hydroxyvitamin D2-25-glucuronide in the urine. Interestingly, both adult male subjects had indications of one or more glucuronides in urine. This may signal a particularly strong propensity for glucuronides to ionize via the PS ionization mechanism; this capability of PS should be further explored.

4. Conclusions

We developed a PS source for Bruker FT-ICR instrumentation that can be utilized for a broad range of chemical species. The PS source was tested on a range of samples and determined the source was suitable for both polymer and biological samples. The PS source provided an API medium for the analysis of complex mixtures while mitigating the possibility of cross contamination due to the disposable nature of coffee filter paper.

Nomenclature

ESI: Electrospray ionization
API: Atmospheric pressure ionization
MS: Mass spectrometer
PS: Paper spray
FT-ICR: Fourier transform ion cyclotron resonance
PPG: Polypropylene glycol
PEG: Polyethylene glycol
CLSE: Calf lung surfactant extract
PC: Phosphatidylcholine
HSIRB: Health Sciences Institutional Review Board.

Acknowledgments

Donation of the CLSE sample from ONY, Inc. and funding from the National Institutes of Health for the 12-tesla FT-ICR (S10RR029517 to T.D.W.) is gratefully acknowledged. This paper is dedicated in remembrance of Dr. Bruce A. Holm.

References

[1] G. Montaudo, M. S. Montaudo, C. Puglisi, and F. Samperi, "Characterization of polymers by matrix-assisted laser desorption ionization—time of flight mass spectrometry. End group determination and molecular weight estimates in poly(ethylene glycols)," *Macromolecules*, vol. 28, no. 13, pp. 4562–4569, 1995.
[2] A. T. Jackson, S. E. Slade, K. Thalassinos, and J. H. Scrivens, "End-group characterisation of poly(propylene glycol)s by means of electrospray ionisation-tandem mass spectrometry (ESI-MS/MS)," *Analytical and Bioanalytical Chemistry*, vol. 392, no. 4, pp. 643–650, 2008.
[3] M. Yamashita and J. B. Fenn, "Electrospray ion source. Another variation on the free-jet theme," *Journal of Physical Chemistry*, vol. 88, no. 20, pp. 4451–4459, 1984.
[4] H. Wang, N. E. Manicke, Q. Yang et al., "Direct analysis of biological tissue by paper spray mass spectrometry," *Analytical Chemistry*, vol. 83, no. 4, pp. 1197–1201, 2011.
[5] L. Anyin, H. Wang, Z. Ouyang, and R. G. Cooks, "Paper spray ionization of polar analytes using non-polar solvents," *Chemical Communications*, vol. 47, no. 10, pp. 2811–2813, 2011.
[6] J. Liu, H. Wang, N. E. Manicke, J. M. Lin, R. G. Cooks, and Z. Ouyang, "Development, characterization, and application of paper spray ionization," *Analytical Chemistry*, vol. 82, no. 6, pp. 2463–2471, 2010.
[7] H. Wang, J. Liu, R. Graham Cooks, and Z. Ouyang, "Paper spray for direct analysis of complex mixtures using mass spectrometry," *Angewandte Chemie*, vol. 49, no. 5, pp. 877–880, 2010.
[8] R. G. Cooks, N. E. Manicke, A. L. Dill et al., "New ionization methods and miniature mass spectrometers for biomedicine: DESI imaging for cancer diagnostics and paper spray ionization for therapeutic drug monitoring," *Faraday Discussions*, vol. 149, pp. 247–267, 2011.
[9] R. Haddad, H. M. S. Milagre, R. R. Catharino, and M. N. Eberlin, "Easy ambient sonic-spray ionization mass spectrometry combined with thin-layer chromatography," *Analytical Chemistry*, vol. 80, no. 8, pp. 2744–2750, 2008.
[10] A. G. Marshall, C. L. Hendrickson, and G. S. Jackson, "Fourier transform ion cyclotron resonance mass spectrometry: a primer," *Mass Spectrometry Reviews*, vol. 17, no. 1, pp. 1–35, 1998.
[11] R. H. Notter, E. A. Egan, and M. S. Kwong, "Lung surfactant replacement in premature lambs with extracted lipids from bovine lung lavage: effects of dose, dispersion technique, and gestational age," *Pediatric Research*, vol. 19, no. 6, pp. 569–577, 1985.
[12] E. G. Blugh and W. J. Dyer, "A rapid method of total lipid extraction and purification," *Canadian Journal of Biochemistry and Physiology*, vol. 37, no. 8, pp. 911–917, 1959.

15

Live Cells as Dynamic Laboratories: Time Lapse Raman Spectral Microscopy of Nanoparticles with Both IgE Targeting and pH-Sensing Functions

Kristy L. Nowak-Lovato[1] and Kirk D. Rector[2]

[1] *Decision Applications Division, Los Alamos National Laboratory, Los Alamos, NM 87545, USA*
[2] *Chemistry Division, Los Alamos National Laboratory, Los Alamos, NM 87545, USA*

Correspondence should be addressed to Kirk D. Rector, kdr@lanl.gov

Academic Editor: Bahram Hemmateenejad

This review captures the use of live cells as dynamic microlaboratories through implementation of labeled nanoparticles (nanosensors) that have both sensing and targeting functions. The addition of 2,4-ε-dinitrophenol-L-lysine (DNP) as a FcεRI targeting ligand and 4-mercaptopyridine (4-MPy) as a pH-sensing ligand enables spatial and temporal monitoring of FcεRI receptors and their pH environment within the endocytic pathway. To ensure reliability, the sensor is calibrated *in vivo* using the ionophore nigericin and standard buffer solutions to equilibrate the external $[H^+]$ concentration with that of the cell compartments. This review highlights the nanosensors, ability to traffic and respond to pH of receptor-bound nanosensors (1) at physiological temperature (37°C) versus room temperature (25°C), (2) after pharmacological treatment with bafilomycin, an H^+ ATPase pump inhibitor, or amiloride, an inhibitor of Na^+/H^+ exchange, and (3) in response to both temperature and pharmacological treatment. Whole-cell, time lapse images are demonstrated to show the ability to transform live cells into dynamic laboratories to monitor temporal and spatial endosomal pH. The versatility of these probes shows promise for future applications relevant to intracellular trafficking and intelligent drug design.

1. Introduction

The chemical concentrations, within the lumens of cellular organelles, such as endosomes, lysosomes, mitochondria, the Golgi apparatus, and the endoplasmic reticulum, are frequently different from the cytoplasm. The locations and chemical differences in these cellular compartments are now being individually interrogated with an ever-expanding suite of techniques. The vast majority of these reagents should be considered fluorescence labels as their spectral responses are integrated to provide contrast in the confocal image [1, 2]. Frequently, however, it is desirable to have a sensor, rather than a label, in order to measure metabolite concentrations, including Ca^{2+}, NO, Na^+, and pH [3]. Generally, the sensor must be employed in the spectral domain as features in the spectrum change in response to the changing analyte concentration, chemical or oxidative state.

It is also possible to perform experiments using Raman scattering as the contrast mechanism in cellular imaging. Raman spectroscopy or imaging provides chemically specific information based upon unique vibrational signatures of the species present. Therefore, recording of the spectra of a cell can give detailed information of the chemical species present. Unfortunately, the Raman effect is weak, with molecular cross-sections in the 10^{-29} to 10^{-32} cm^2 range, significantly lower than those of good fluorescence labels which are in the 10^{-16} cm^2 range [4]. An inherent benefit of Raman microscopy over fluorescence is that it records the natural abundance of chemicals present and, therefore, is a label-free technique. However, if one was willing to forego this advantage, it becomes possible to use Surface-enhanced Raman scattering (SERS)-based nanoparticles as labels or sensing agents [5–13]. SERS provides an increase in the

Raman scattering signal by many orders of magnitude [9, 14–16]. This enhancement makes it possible to access the chemical specificity inherent in Raman spectroscopy with low detection limits and/or fast acquisition speeds. In addition, it enables nearly limitless flexibility to design experiments based upon the ligands chosen for the nanoparticle surface. In general, multiple ligands are possible, including ligands that serve to cause the cell to deliver nanoparticles to specific locations or systems and ligands which record the concentrations of specific metabolites when present. The creation of a targeted nanosensor, one which has both functions, creates a novel tool to turn cells into dynamic laboratories. Further, the most significant advantage may ultimately be the ability to record time-lapse spectral microscopy of whole live cells as they respond to stimuli for substantially longer periods of time over fluorescence microscopy that tend to suffer from photobleaching.

Nanosensors are created based upon choices of a colloidal particle platform, a targeting ligand, and a reporter molecule. The nanoparticle platforms generally include noble metals, particularly of Ag [17], and Au [18, 19], as well as a variety of geometries including hollow shell [20] and core/shell [21, 22] particles. For targeting ligands, generally researchers have used localization peptides [23] or antibodies (IgE) to deliver the nanoparticles to specific locations or subcellular systems [24–26]. For the reporter ligands, the most demonstrated analyte concentration studied with SERS nanosensors is pH. Roughly speaking, the sensitivity of the nanosensor will be within a couple orders of magnitude of the dissociation constant, k_D, of the reporter molecule used. The ligands chosen are based upon the ability to coat the metal surface and have unique Raman vibrational signatures as well as sensitivity and specificity to the analyte of interest. For pH, various researchers have demonstrated basic [27], neutral [17, 20], and acidic-range [10–12, 28, 29] sensors based on different ligands. In particular, for measurements of acidic concentrations ($3 < pH < 7$), it has been shown that use of 4-mercaptopyridine (4-MPy) is an effective strategy [7, 21]. The focus of this work is on the cell's dynamic regulation of acidic pH within the endosome-lysosome system, which is necessary to support degradative enzymes as well as the proper sorting of receptors and ligands following endocytosis [30–32]. Endosomal sequestration of weakly basic compounds such as daunorubicin, possibly through a pH-partitioning mechanism, may be a factor in drug resistance [33]. Low pH within the endosomal-lysosomal system can also play a role in development of disease states [34–37].

To be able to employ live cells as dynamic micro-laboratories, it is important that entire hyperspectral maps be recorded. However, point spectral Raman mapping is generally too slow to be relevant for most biological systems. One approach to this goal is the use of a so-called "push-broom" Raman microscope [38], where the excitation light is cylindrically, rather than spherically, focused to a line image across the sample, and the resulting Raman line image is then dispersed onto a CCD camera. The cylindrically focused line image allows for the collection of hundreds of spectra, along a single spatial line, to be recorded simultaneously. The sample is then stepped across the line to build up the other spatial axis, resulting in an entire hyperspectral 3D map in a couple of minutes, depending on signal strength. The spectral maps consist of two spatial dimensions (x, y), defining the area of interest, and the spectral dimension (λ) used to identify the chemical composition of the material at each pixel of the image. The map is then reacquired sequentially to build up dynamic spectral movies (x, y, λ, t). These time lapse spectral movies allow for the identification of cells in *statis* as well as the recording of them responding to external stimuli. The ability to monitor changes in intraluminal pH, in the context of cargo moving through the endosomal-lysosomal system or in response to causal on/off external stimuli, could lead to novel strategies toward drug development and understanding of pathogen survival and other disease states [39–43].

Recently, it has been shown that a targeted SERS nanosensor with both IgE-targeting and pH-sensing functions is capable of measuring the pH of intracellular compartments harboring internalized IgE receptors, using the RBL-2H3 mast cell line as a model system. This sensor is based upon a Au@Ag core shell particle with MPy as a reporter molecule and 2,4 ε-dinitrophenol-L-lysine (DNP) as a targeting molecule for IgE bound to the high affinity IgE receptor, FcεRI [10, 12]. Subsequently, the application of this nanosensor to measure the pH of endosomal compartments as sensor-bound receptors traffics through the endosome-lysosome system while the cell exposed to dynamic external stimuli has been reported [11]. It was demonstrated that the nanosensors are resilient to changes in temperature, by lowering the temperature of cultured cells from the physiologically relevant temperature of 37°C to 25°C; this procedure also slows the transit of endocytosed receptors. More importantly, it was also demonstrated that the targeted nanosensors report changes in luminal pH after treatment of cells with the H^+ flux inhibiting drugs, amiloride and bafilomycin. Importantly, the FcεRI-targeted pH SERS nanosensors enable the direct localization of vesicles bearing FcεRI receptors and also report the dynamic changes in endosomal vesicle pH in response to these drugs. The compartments recover their normal pH after the removal of these drugs from the media on the timescale of 10s of minutes. This technology has the potential to create new techniques useful in the fields of medicine, pharmacology, and biotechnology.

2. Experimental Section

2.1. Chemicals and Materials. Au colloids at 60 nm diameter were purchased from Ted Pella, *Inc.* (p/n 15708-6, Redding, CA) at 2.6×10^{10} particles per mL. The Au colloids were enhanced with a Ag coating by the addition of LI silver enhancing and initiator agents from Nanoprobes, Inc. (p/n 2013, Yaphank, NY, USA). Silver enhancement was performed at a 1 : 1 ratio of both initiator and enhancer. The enhancement solution was added to the colloids for two minutes, centrifuged at 13,000 rpm for five minutes, and then washed with deionized H_2O. MPy (p/n 148202, Sigma Aldrich Chemicals, St. Louis, MO) and DNP (MP Biomedical, Solon, OH, USA) used without further purification and aqueous solutions were prepared at 1 mM. BSA-DNP and IgE antibody was acquired via a generous gift from the Oliver-Wilson laboratory at the University of New Mexico. Nigericin, KCl, $MgCl_2$, $CaCl_2$, and glucose were purchased from Sigma Aldrich chemicals (St. Louis, MO, USA) and used without further purification. Bafilomycin (Sigma) was dissolved in EtOH and used at a concentration 200 nM [44]. Amiloride (MP Biomedical, *Inc.*) was dissolved in dH_2O and used at a concentration of 1 mM [45].

2.2. Cell Culture. For cellular experiments, RBL-2H3 cells were cultured in MEM media supplemented with 10% fetal clone III (p/n SH3010902, Thermo Scientific Hyclone, Logan, UT), 90 I.U./mL of penicillin (ATCC, Manassas, VA, USA) and 90 μg/mL streptomycin (ATCC, Manassas, VA, USA), with an additional 2 μM of L-glutamine (Sigma Aldrich, St. Louis, MO, USA). RBL-2H3 cells are an adherent cell line that was grown in a 5% supplied CO_2 incubator at 37°C to 80% confluency and then passaged. Cells were primed with anti-DNP-specific IgE (1 μg/mL) 18–24 hrs and kept at 37°C and 5% CO_2 prior to use. Imaging experiments with the targeted nanosensors employ sonication of colloids for five minutes and filtering through a 0.2 μm filter and added to the cells at 10–100x particles per cell number. After five minutes, the cells are then washed 2x with phosphate-buffered solution, to eliminate most of the residual nanoparticles left outside of the cell, and then freshly warmed Hanks buffer is applied. The buffer consists of Hanks, HEPES, BSA, glucose, $MgSO_4$, and $NaHCO_3$. The cover slips with adherent cells are then added to the environmental chamber vessel (Carl Zeiss, Inc., Thornwood, NY, USA) equilibrated at 25°C or 37°C and 5% CO_2 for imaging.

2.3. Degranulation Assay. To demonstrate biological activity of the IgE receptor-mediated endocytosis nanosensor, a degranulation assay was performed. In this assay, β-hexosaminidase was used to hydrolytically cleave substrate p-nitrophenyl-N-acetyl-β-D-glucosaminide (N9376, Sigma) and cause a color change that can be detected. The assay was performed in 24 well culture plates. Positive control for the assay is done by lysing whole cells with 1% Triton.

2.4. Microscope Layout and Imaging. The microscope-imaging system is based upon a Carl Zeiss Axiovert 135TV inverted microscope with an Epiplan 10x (N.A. 0.20), a LD Plan-achroplan 40x (N.A. 0.60 and a C-apochromat 63x (N.A. 1.2) water immersion objective. It was this last objective primarily used for the SERS experiments. A xenon arc lamp (XBO 75, Carl Zeiss, Thornwood, NY, USA) was used to illuminate the sample for bright field visualization in transmission mode, by video camera, using transfer optics in the trinocular head. (Optem 70XL, Labtek, Campbell, CA) The total zoom factor of the transfer optics was 0.5x in addition to the objective used. An Infinity 2-2 monochrome (Lumenera Corp., Ottawa, ON, USA) camera was used to acquire the visible image using an acquisition rate of approximately 200 ms/frame and no detector gain. The visible image resolution was 1616 × 1216 pixels. The camera was connected to the personal computer through a USB 2.0 connection, and the image was captured using the Infinity Capture software and saved as a .tiff file format. Raman experiments were performed using 514.5 nm light from a Spectra Physics 177-G01 air-cooled argon ion laser. The excitation light was separated from other Ar^+ laser lines using a 300 ln/mm dispersive grating and a ~1 mm spatial filter several feet away. The laser light is expanded and collimated using a 1 : 2 Keplerian telescope. It is coupled into the back of the microscope objective with an antireflection-coated plano-convex BK-7 150 mm focal length cylindrical lens (CKX150AR.14, Newport Corp. Irvine, CA, USA), a Raman edge dichroic (z514rdc no. 105366, Chroma Technology, Rockingham, VT, USA) and a 50.8 mm diameter, 150 mm focal length biconvex spherical anti-reflection coated doublet tube lens (PAC086AR.14, Newport Corporation, Irvine, CA). The laser light is focused to a line which is approximately 0.5 μm wide and 50 μm tall and a total power at the sample of 10 mW for the experiments detailed in this paper. The Raman signal is captured with the same objective and tube lens and is passed through the edge dichroic. The signal is transferred using two additional 150 mm spherical lenses and reimaged onto the 100 μm-wide slit of a 0.25 m f/2.2 imaging spectrograph (Holospec 2.2, Kaiser Optical Systems, Inc, Ann Arbor, MI, USA). The signal is dispersed by a volume holographic grating (HSG-514.4-LF, Kaiser) and imaged with a LN_2-cooled CCD array detector (LN/CCD-1024E, Princeton Instruments, Trenton, NJ, USA). Each image of the CCD records spectral and "*Y*" spatial information of the sample. The spectral calibration of the CCD was performed using a Ne standard lamp (Neon 6032, Newport, Irvine, CA, USA) and confirmed with an Hg standard lamp (Newport). The spectral resolution of the spectrograph is approximately 2 cm^{-1}. The spatial axis was calibrated using a resolution test target, USAF-1951 using lines in the 6th group. The spatial axis was corrected for spherical aberrations using the Hg standard lamp and custom-written Labview code (National Instruments, Austin, TX, USA). In addition, the creation of a flat field correction was performed to remove the roughly Gaussian-shaped power dependence along the laser line focus using a NIST Relative Intensity Correction Standard (NIST 2243). Individual spectra were acquired by integration along the vertical axis, and background spectra were acquired under the same conditions using a microscope coverslip only. The data were acquired using custom Labview code and stored in ASCII format on a personal computer. In order to acquire spectral image cubes, the frequency and "*Y*" spatial axis were acquired and saved, and the sample was moved across the excitation source by stepping the computer controlled microscope stage (MS-2000, Applied Scientific Instrumentation, Eugene, OR, USA), to sequentially build up the "*X*" spatial access of the cube. The synchronization of the CCD camera and the stage were performed using the same Labview code. All cellular experiments were performed with an independently controlled environmental chamber (Carl Zeiss, Inc., Thornwood, NY, USA) mounted to the stage.

2.5. Electron Microscopy. Scanning electron micrographs were acquired using a FEI Quanta 200FEG operating at an accelerating voltage of 30 kV. All imaging was performed

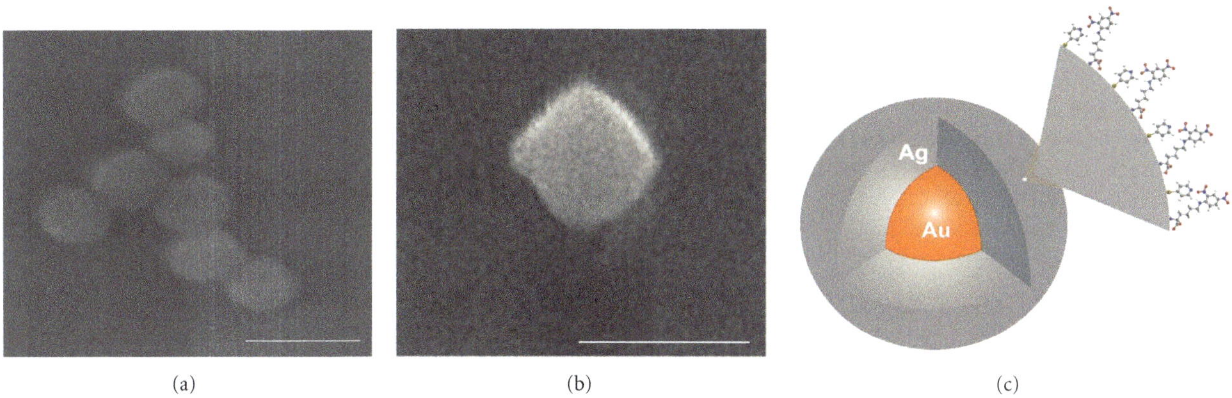

FIGURE 1: (a) Scanning electron microscope images of a few of the commercial Au colloidal particles. Scale bar is 60 μm. (b) Composite SEM image of "developed" gold-silver core-shell colloidal particles. Ag layer varies from a few nm to a few 10s of nm in thickness, resulting in sizes varying from 70 to 150 nm. Scale bar is 100 μm. (c) Schematic of the targeted nanosensor. Au@Ag nanoparticle with DNP and 4-MPy ligands, not drawn to scale. (a and b) Reprinted with permission from Nowak-Lovato et al. [11] (copyright 2009 Society for Applied Spectroscopy).

with a chamber pressure of ~100 Pa of water vapor to avoid charging and to reduce sample dehydration. Images were taken to examine the size and morphology of the nanoparticles during development. Samples were mounted on glass slides. For transmission electron microscopy (TEM), cells were fixed with 2% glutaraldehyde in cacodylate buffer (pH 7.4), postfixed with osmium tetroxide, dehydrated in ascending alcohols, and embedded in Epon. Due to the size of the Au@Ag nanosensors, samples were sectioned with a 100–300 nm thickness on a Leica Ultracut UCT. Samples were stained with uranyl acetate and lead citrate and observed on a Hitachi 7500 electron microscope.

2.6. Surface-Enhanced Raman Scattering pH Calibration. In vivo calibration experiments employ the addition of 10 μm nigericin, 140 mM KCl, 1 mM $MgCl_2$, 2 mM $CaCl_2$, 5 mM glucose, and 20 mM phosphate-citrate buffers of varying pH (4–8). Once the calibration solution was added, cells were returned to the incubator for ten minutes. After pH equilibration, targeted nanosensors were added to the cells, and whole images were taken of all particles found to gain entry inside the cells after ten minutes of addition. Cellular images representing approximately 100 0.5 μm × 0.5 μm-sized pixels with nanosensors were used. These images were processed to obtain individual spectra as indicated in the data analysis section. Once individual data points were collected for each pH value, the data was fit to a sigmoidal curve.

2.7. Data Analysis Procedures. The data analysis procedures were performed using custom written Labview code. All pixels of the hyperspectral image cube were subject to the same correction procedures including wavelength calibration, flat field correction, background subtraction, CCD bias voltage subtraction, and spherical aberration removal. The images were thresholded by eliminating any spectral pixel information with a signal to noise ratio of less than 2.5. The signal-to-noise ratio was calculated by taking the highest intensity count from the integrated peak at 1572 cm^{-1} to 1599 cm^{-1} subtracted from the baseline value at that point, divided by the highest intensity count from the noise at 1800 cm^{-1} to 1827 cm^{-1} subtracted from the baseline value at that point. Any pixel with a signal-to-noise ratio below 2.5 was considered to not have any useful information and was removed from the image. The two spectral peaks of interest were then integrated from 1572 cm^{-1} to 1599 cm^{-1} for the nonprotonated peak and 1599 cm^{-1} to 1626 cm^{-1} for the protonated peak. Because the Raman spectra often occur on a weak fluorescence background, the integrated limits were evaluated against a linearly sloping background defined as the spectral intensity at 1000 cm^{-1} and 2000 cm^{-1}. Because the peaks of interest are so close spectrally, variably sloping baselines that may arise from the cellular autofluorescence can be effectively removed without changing the intrinsic ratio of the peaks.

3. Results and Discussion

Scanning electron micrographs of the nanoparticles are shown in Figure 1 before (a) and after (b) development with silver. These core-shell particles are denoted Au@Ag based on similar geometries reported elsewhere [21]. After development, the particles ranged in size from approximately 70 nm to 150 nm. Any possible larger aggregates were removed via filtration using a 200 nm syringe filter. The silver development strategy proved to be a rapid, cost-effective method for generating large quantities of silver-based particles in the 100 nm range. Figure 1(c) is a schematic illustration of the nanosensor design. The drawing is not to scale, but demonstrates the Au core with Ag surface coating as well as the addition of multiple 4-MPy ligands and multiple DNP ligands on the Ag surface. It may be of interest to put an outside boundary on the numbers of ligands on the surface of the nanoparticle. The widths of 4-MPy (between the meta hydrogens) is 0.42 nm, while the DNP width (between the nitro moiety and the H para to that) is 0.60 nm. Assuming a uniform spherical nanoparticle with a diameter of 100 nm yields an effective surface area of 31,415 nm^2. Given equal numbers of both ligands on the surface', and that they are rigid, space filling, the maximum numbers of ligands is approximately 30,800 ligand molecules per nanoparticle. As there are many assumptions here in this idealized case, it is most likely that this calculation represents the maximum case, and the actual number is probably less than that. For the creation of targeted nanosensors with both a targeted and sensing capability, it is important to determine the degree with which either of the ligands interferes with the other's functionality.

3.1. Testing the Nanoparticle Targeting Ligand. To first test the effect of the presence of 4-MPy on the ability of a DNP particle to be targeted to the endosomal pathway of cell lines, a fluorescence-based degranulation assay was performed. Cellular degranulation upon stimulation of the IgE FcεRI-mediated pathway will result in the cell's secretion of constituents including the granular enzyme β-hexosaminidase [46]. Release of β-hexosaminidase is used in this assay to hydrolytically cleave an externally added substrate molecule, p-nitrophenyl-N-acetyl-(3-D-glucosamine), and cause a detectable color change. The assay was used to compare Au@Ag-DNP particles with previously measured bovine serum albumin (BSA)-DNP particles.

Assay data has been reported previously [10]. Spontaneous release resulting from IgE-sensitized cells, without the addition of BSA-DNP, shows a response at 5%, which is used as the negative control. A 5% release comparable to spontaneous release was demonstrated with the use of Au@Ag-4MPy nanoparticles, lacking the targeting DNP component. Decreasing concentrations of BSA-DNP show a corresponding drop in percentage release which serves as a positive control and generally agree to literature values for this cell line [47, 48]. Particles of Au@Ag-DNP show a 55% granule release from whole cells, which is comparable to that seen with 10 μg of BSA-based DNP particles. When DNP: 4-MPy mixtures of 1 : 1, 3 : 1, and 1 : 3 were used to coat the Au@Ag nanoparticle surface, degranulation percentages were seen at 30%, 37%, and 27%, respectively, which is comparable to 10 ng of BSA-DNP. This drop in efficiency can be attributed to the decrease in concentration of the DNP on the particle surface because of the presence of 4-MPy. The 1 : 1 ratio (DNP/4-MPy) nanoparticles were chosen for these experiments, as a reasonable balance between the ability of DNP to target the nanosensor to the endosomal pathway, with high enough SERS signal levels from 4-MPy, to enable reasonably fast image acquisition. Figure 2 demonstrates internalization of nanosensors using TEM images. Figure 2(a) shows a high-resolution image of a RBL-2H3 cell after 5 min exposure to the nanosensors. The arrow points to the location of an individual nanosensor, seen as a black (electron-dense) dot bound to the surface of the cell, due to specific binding of DNP portion of the nanosensor to at least one IgE-bound receptor. Figure 2(b) illustrates a similar image of a cell exposed to a five-minute pulse of nanosensors, followed

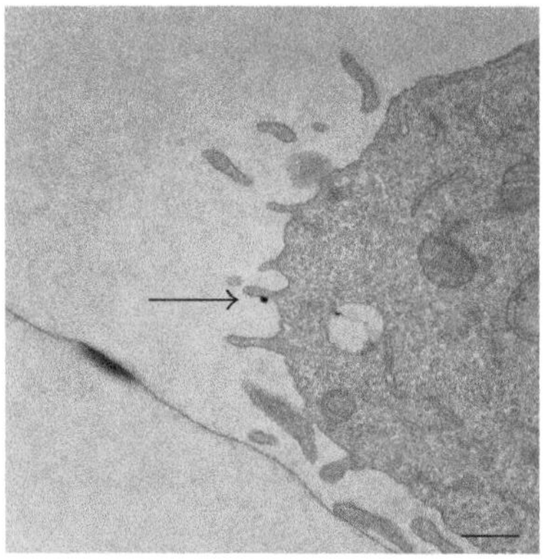

(a)

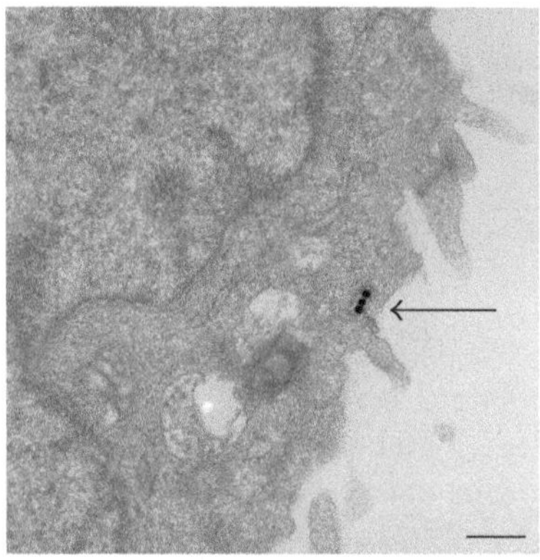

(b)

Figure 2: Transmission electron microscopy images of RBL-2H3 cells (a) 5 min after incubation of nanosensors and (b) 60 min after incubation with nanosensors. *Arrows* are added to pinpoint locations of nanosensors within the cells. *Bars* = 500 nm. Reprinted with permission from Nowak-Lovato et al. [11] (copyright 2010 Springer).

by a sixty-minute chase to allow for internalization of sensor-bound receptors prior to fixation and processing for TEM. In this photo, the arrow points to three internalized nanosensors within an endocytic compartment just under the membrane surface.

3.2. Testing the Nanoparticle Sensing Reporter Molecule. To complement the previous studies regarding 4-MPy ligand interference with targeting, the next studies demonstrated minimal interference of DNP with the nanosensors ability to sense pH. Figure 3 shows representative *in vitro* SERS spectra of targeted nanosensors at various stages of development. Figure 3(A) shows a spectrum of unmodified Au@Ag nanoparticles which has no significant Raman signatures. Figure 3(B) shows the spectrum of Au@Ag-4MPy which shows a variety of peaks including the peaks near 1600 cm^{-1} of primary interest and generally agrees with those reported elsewhere [7, 21]. Figure 3(C) represents the SERS spectrum of Au@Ag-DNP which demonstrates weak Raman peaks; however, these peaks are of a decreased intensity most likely due to the fact that DNP is conjugated to lysine, and further in distance from the nanoparticle

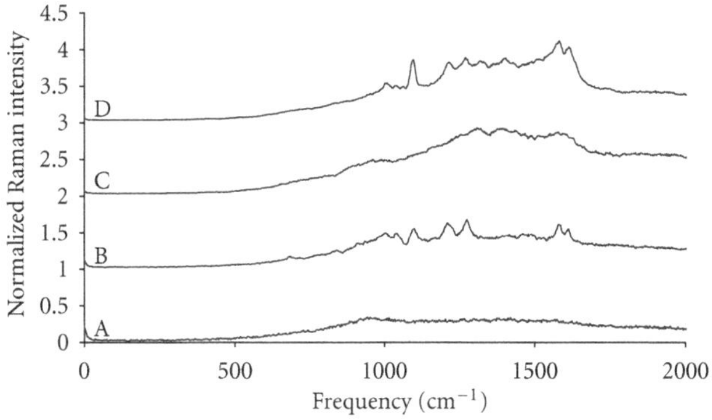

FIGURE 3: Normalized and offset Raman spectra during *in vitro* sensor development. (A) Raman spectrum of Au@Ag colloids. (B) Raman spectrum of 4-MPy adherent to Au@Ag colloids. (C) Raman spectrum of DNP adherent to Au@Ag colloids. (D) Raman spectrum of targeted nanosensor with DNP and 4-MPy ligands. *Y* axis is arbitrary intensity units. Reprinted with permission from Nowak-Lovato et al. [11] (copyright 2009 Society for Applied Spectroscopy).

surface. The distance of DNP from the particle could be contributing to the weakened DNP Raman signals [49]. Figure 3(D) represents a SERS spectrum of Au@Ag-4MPy-DNP which shows general agreement with Figure 3(B) with a mild contribution from DNP.

Many bands associated with 4-MPy seem to display a certain degree of pH sensitivity [7]. In particular, vibrational modes at frequencies of 1580 cm^{-1} and 1612 cm^{-1} display proportional ratiometric effects dependent upon pH. The vibrational modes of the pH-sensitive 4-MPy species are based on protonation and deprotonation of the ring N atom in the 4-MPy molecule [7]. Protonation of the molecule has been assigned to vibrational mode 1612 cm^{-1} while the unprotonated form has been assigned to 1580 cm^{-1}. The peaks do not appear to spectrally shift with changing pH in this range. Calculating the ratio between the two frequency modes demonstrates a value that can be correlated to the pH of the molecule's environment. Since the two peaks to be rationed are spectrally close and completely dependent upon the same element (nitrogen) and whether it is bound or unbound to hydrogen, this sensor can accurately report pH without being tainted by the quantity of molecules present or by other molecular bond orientations. As is demonstrated in Figure 4(a), the intensity of the 1612 cm^{-1} peak is greatest at low pH and decreases while the pH is increased. The peak at 1580 cm^{-1} is greatest at high pH and decreases with decreasing pH. The spectra are representative of *in vitro* Au@Ag nanoparticles coated with 4-MPy on the surface.

3.3. In Vitro and In Vivo Calibration. The *in vitro* calibration curve is shown in Figure 4(b) as closed circles. There is a potential that the pKa values of 4-MPy on the nanosensors will be different with and without the DNP added and also within cells. It is noted that previous work of measuring the point at which there is equal contributions of protonated and unprotonated species for *in vitro* nanosensors was 3.9 [50] and 4.0 [7]. It is known that the *in vitro* generation of calibration curves (signal versus concentrations) for biological measurements is of only limited use because of possible probe interactions with natural species inside live cells [40]. Measurements of dissociation constants, k_D, show that, for example, Ca^{2+} binding of dyes *in vivo* can be up to 5x than *in vitro* [51]. Fortunately, the generation of *in vivo* calibration curves is routine in fluorescence microscopy, and procedures for this application were similar. Briefly, multiple measurements are performed with varying levels of external concentrations of the metabolite of interest in the bulk media. The ionophore nigericin serves to equilibrate the external $[H^+]$ concentration with that of the cell [52]. To obtain *in vivo* calibration data, many measurements similar to those seen in Figure 4(a) are used to generate a calibration curve as shown in Figure 4(b) as the open circles. The error bars for the frequency ratios represent the standard deviation of the repeated measurements and generally are more precise at higher pH values. Note the dramatic shift in the apparent isosbestic point of the two curves. In addition, the ratio of the bands does not appear to flatten at zero with high acidity. This is likely arising from the background signal from DNP shown in Figure 3(D). However, all calibrations performed were with the addition of DNP to account for this effect. Vesicles formed along the endocytic pathway have been shown to display pH concentrations from 4.5 to 7.5. This range agrees quite well with our calibration curve showing a relevant pH range from 4.0 to 8.0 pH units. These data were fit to a sigmoidal curve with the values shown in (1). This equation was used to convert SERS spectra to pH maps in the subsequent experiments as follows:

$$y = 0.80 - \frac{0.24}{1 + \exp((x - 6.1)/0.69)}. \qquad (1)$$

Time-lapse, hyperspectral SERS imagery showing the evolution of targeted nanosensor delivery into the cells at 25°C is shown in Figure 5. One weakness of the microscope is that it only has the ability to acquire brightfield images at the beginning of a time-lapse image cycle. In addition, the brightfield image is a transmission, whole cell view at lower resolution than the pseudoconfocal, epi-acquired SERS image. Thus, the brightfield image only serves as a rough guide for cell structure and location prior to the initial Raman image. Figure 5(a) shows the presence of targeted nanosensors twenty-eight minutes after the particles were added to the cells. The color-coded pixels represent the calculated pH values of the spectrum using the calibration curve shown in Figure 4(b). The colors correspond to the scale bar, with red and orange signifying neutral and near-neutral pH while blue and purple represent the most acidic locations. At the twenty-eight minute time point, most of the particles are near the edge of the cell membrane. There is a slight registration error between the brightfield image and the SERS image because the live cells, though adherent, are not immobile. Brownian motions on the order of 1-2 μm can occur over the time course of a given experiment. However, it is noted that these nanoparticles have been or are about to be internalized because free-floating particles move too quickly to provide significant signal during the

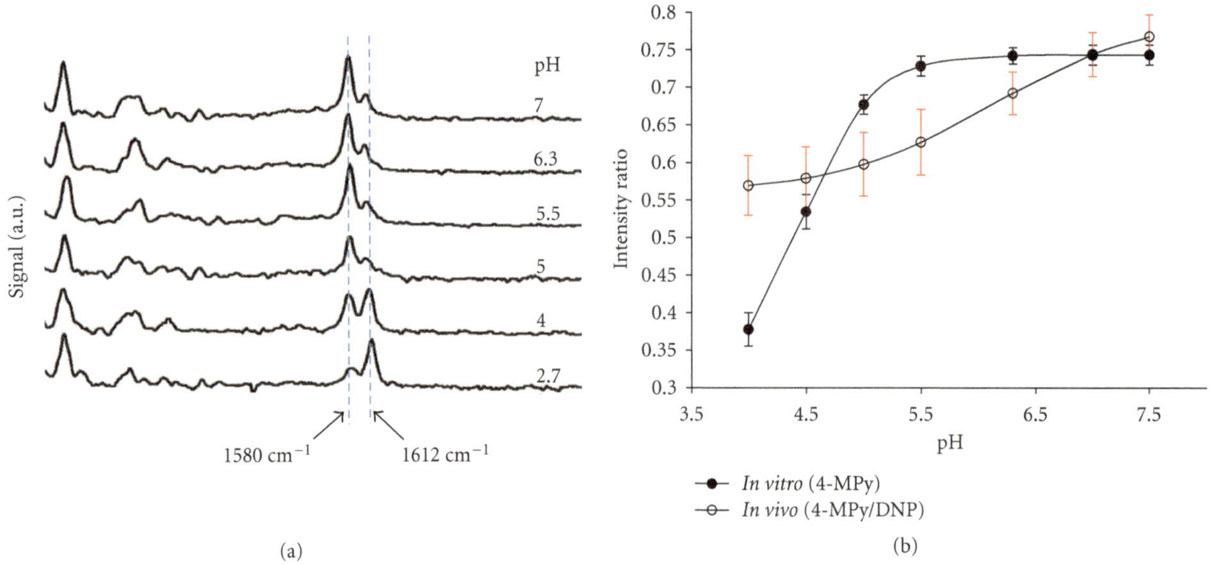

FIGURE 4: (a) Normalized and offset Raman spectra of Au@Ag-4MPy nanoparticles in varying pH-buffered solutions. (b) *In vitro* (closed circle) calibration curve of Au@Ag-4MPy nanosensors and *in vivo* (open circle) calibration curves of Au@Ag-4MPy-DNP nanosensors inside RBL-2H3 cells in pH-buffered media containing nigericin. *Y*-axis represents ratio of integration of peak centered at 1580 cm^{-1} versus the additive integration of both the peak centered around 1580 cm^{-1} and 1612 cm^{-1}. Vertical error bars are dependent on approximately 100 pixels per image, repeated five times.

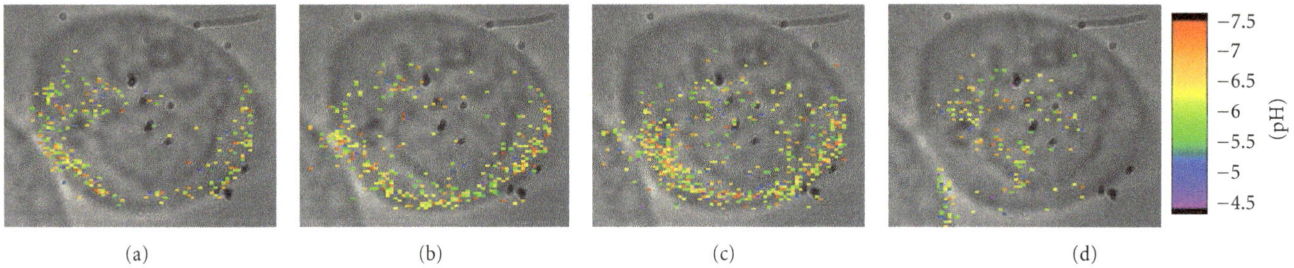

FIGURE 5: Time-lapse pH-calibrated images acquired every 3.5 min upon nanosensor addition up to 90 min. Calibrated pixels demonstrating pH are overlaid with bright field images. (a) 28 minutes, (b) 45.5 minutes, (c) 59.5 minutes and (d) 80.5 minutes. Reprinted with permission from Nowak-Lovato et al. [11] (copyright 2009 Society for Applied Spectroscopy).

integrated acquisition time of the experiment. Figure 5(b) shows the same scene at forty-five and a half minutes after addition. At this time point, there are significantly more particles internalized, although still the preponderance of the targeted nanosensors near the edge of the cell. In Figure 5(c), taken at fifty-nine and a half minutes, yet more particles have been delivered to the cells, and there is an increase of particles near the center of the cell. It is noted that the large nuclei of these cells could represent a "dead space" area where these targeted nanosensors, being inside endosomes, would not be found. The nucleus of the cell is just below center in the greyscale image but may be above the area of interrogation so that a more uniform distribution of particles is seen towards the surface of the cell. The SERS image at eighty and a half minutes, Figure 5(d), shows the majority of particles are away from the surface of the cell. These types of distributions of particles inside cells are also seen in the optical scattering experiments.

The number of pixels having significant SERS signal is plotted in Figure 6(a), representative on the secondary *Y* axis. There is most likely not a one-to-one correspondence between pixels measured and particles internalized. The measured size of each pixel is 0.5 μm × 0.5 μm which is also roughly the size of endosomes seen in this cell type [53]. The targeted nanosensors are below 0.2 μm in size. Therefore, the targeted nanosensors encompass single pixels on average. However, it is uncertain if multiple nanosensors are present within an individual pixel. At fifteen minutes, approximately 400 pixels have SERS signal above the threshold. Over the course of an hour, the number of pixels with SERS signal swells to its peaks at approximately 1200. Over the subsequent thirty minutes, the number of pixels measured decreases to just fewer than 400. The rise in pixels over the first hour after nanoparticle delivery is explained by the slower cell processes expected to occur at 25°C compared to normal cellular growth temperature of 37°C.

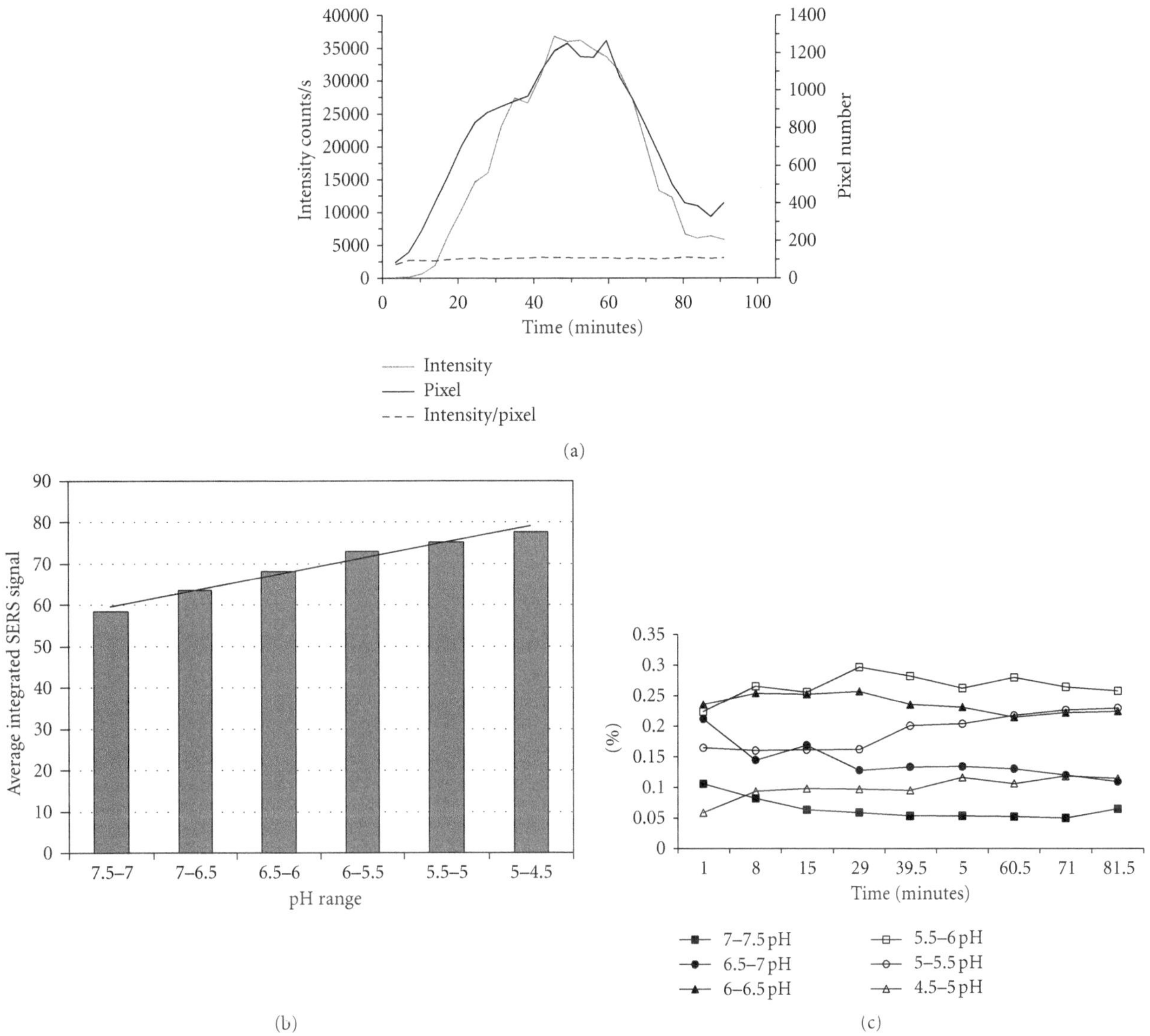

Figure 6: (a) Whole-cell image data ranging from 3.5 to 90 minutes. Total nanosensor intensity per image is depicted with gray line, corresponding to primary y axis. Total pixel count per image is shown with black line corresponding to secondary y axis. Average intensity per pixel is displayed with dashed line and also corresponds to secondary y axis. (b) Integrated SERS signal in the $1612\,cm^{-1}$ range as a function of pH normalized to total pyridine signal and averaged overtime. Line is linear fit to the data, to guide the eye. (c) Percentage of pixels representing each pH grouping. (a and c) Reprinted with permission from Nowak-Lovato et al. [11] (copyright 2009 Society for Applied Spectroscopy), (b) Reprinted with permission from Nowak-Lovato and Rector [12] (copyright 2009 Thomson Corporation).

The nanosensors delivered suggest that the washing step did not completely remove free nanosensors. The drop in total pixels after one hour could possibly result from the degranulation process, where the nanosensors are excreted from the cells over time. In addition, it is possible that some of the nanosensors are being destroyed by the interrogation of the laser or being chemically modified by the harsh environment of the endosome. The possibility also exists that the nanosensors are aggregating, causing a decrease in pixel number without a loss of nanosensors. The total intensity of SERS signal per pixel and how this measurement corresponds to the numbers of nanosensors present was performed. The total signal intensity of the two pH sensitive 4-MPy peaks from $1500\,cm^{-1}$ to $1626\,cm^{-1}$ with a baseline subtraction from $900\,cm^{-1}$ to $2200\,cm^{-1}$ at each pixel was generated. All pixels with relevant SERS spectrum were used in the analysis. The primary Y axis is representative of the total SERS intensity counts for each image at the respective time point. The average intensity per pixel is also plotted in Figure 6(a) and aligns with the secondary Y axis just as the pixel number does. The trends for SERS intensity and pixel number appear to be quite similar. The average intensity per pixel plot depicts a nearly linear 1 : 1 ratio of SERS signal to pixel number. This data lends to elimination of nanosensors from the cells rather than aggregation. Since the ratio is not exactly 1 : 1, there might be a minor contribution of nanosensor aggregation or a slight difference in nanosensor size leading to minor signal enhancements.

Comparison of integrated intensities of nanosensors over time and space indicated that most of the decrease in pixels seen is from elimination. Further, it would be ideal to be able to more specifically quantify the number of nanosenors within each endosome compartment so that endosome-specific information can be attributed, or that single endosome compartments can be followed over space and time. In Figure 6(b), the pH dependence of the integrated SERS $1612\,cm^{-1}$ signal has been plotted and normalized to total pyridine signal and averaged over time. The pH dependence shows a near-linear trend that the integrated SERS signal increases with increasing acidity by approximately 23% when changed from 7.0–7.5 range to 4.5–5.0 range. The percentages of particles within pH categories which correspond to hypothetical states in the endosome maturation pathway are recorded over time [54]. These data are illustrated in Figure 6(c). The pH of the media used is approximately 7.4. It is assumed that, immediately after internalization, the endosome would have the same pH as the surrounding media. In the cell under interrogation, immediately there are about 10% of the pixels in the pH = 7.0–7.5 category and 23% of the pixels in the pH = 6.5–7.0 category. In both categories, the percentages of particles in these categories drop by 80 minutes to 6% and 11%, respectively. This trend can be attributed to internalized endosome maturing to lower pH values and recycled receptors having less particles available for new internalization processes. The particles representing the category of pH = 6.0–6.5, for the early endosome, begins the experiment at 23% of the population and remains constant throughout as particles entering and leaving this pH range are nearly balanced. The category of pH = 5.5–6.0, representing possibly the sorting endosome/early endosome, is the most populous category that begins in the twenties and increases to 30% over thirty minutes and remains increased throughout the experiment. The categories of pH = 5.0–5.5 and pH = 4.5–5.0, representing the late endosome/lysosomal stage, begins the experiment at 17% and 6%, respectively. Over time, both categories increase in relative abundance ending at 23% and 10%, respectively. This increase in abundance suggests that the bulk of the particles have been processed to the late endosome/lysosomal stage.

3.4. Temperature. SERS-targeted nanosensors were added to cells held at either 37°C or 25°C. Figure 7(a) depicts brightfield cellular images overlaid with color-calibrated SERS images and cells incubated at the two temperatures. The brightfield images are all recorded at initial time points before the first corresponding SERS image; minor alignment issues in the overlays correct for modest movements of the adherent cells. At the physiological temperature of 37°C, FcεRI crosslinking leads to marked endocytosis of the receptors over a 2–12-minute time period.

As expected, there was a considerable delay in receptor internalization in cells incubated at 25°C (Figure 7(a)). Accordingly, the imaging time was extended to over sixty minutes. The images in Figure 7(a) give a qualitative picture of the nanosensors and the pH of their localized environment. Figure 7(b) averages data from three cells, with the goal of reporting the total nanosensor numbers in each pH grouping as a percentage of the whole. Figure 7(b) also illustrates the expeditious movement of nanosensors through the endocytic pathway when incubated at 37°C. The nanosensor groupings devised from the images show that, at twelve minutes, 80% of the nanosensors are in pH groupings indicative of late endosomal/lysosomal vesicles. At later times, the overall number of sensors is reduced, consistent with trafficking of nanosensors to the cell surface via both the recycling compartments and by exocytic release of secretory lysosomes, followed by dissociation into the medium. A pool of nanosensors is detected with pH 6.0–7.0 compartments at these later times, suggestive of their retention within early sorting and recycling endosomes.

Experiments performed at 25°C demonstrate the temperature dependence of both initial receptor-mediated internalization of sensors and their slower progress through the endocytic pathway. When compared to cells incubated at 37°C, it is notable that nanosensors tend to accumulate in the pre-endosomal/recycling endosomes at the lower temperature. There is also a notable retention of the total number of nanosensors within the lowest pH compartments, as expected if the low temperature slows the arrival of receptor-nanosensor complexes into secretory lysosomes and their subsequent return to the plasma membrane by exocytosis.

3.5. Pharmacologics. The targeted nanosensors were validated through a pharmacologic approach. First, the ability of the nanosensor to detect *increases* in pH within endosomes of cells treated with bafilomycin, an inhibitor of the H^+ ATPase pump that pumps H^+ into the endosomal vesicle [55–57], was tested. These results were compared with those in cells treated with amiloride, a known inhibitor of the Na^+/H^+ exchanger. This results in impairment of H^+ exit from the endosomal vesicle and a *decrease* in endosomal pH [58–60]. In both cases, cells were incubated at 37°C with or without drugs for two hours prior to ten-minute exposure of cells to nanosensors and imaging in the continued presence of drugs. Results are reported in Figure 8.

Figure 8(a) shows the total uptake of nanosensors under each condition, reported as the average number of Raman active pixels, at each time point, for at least three different cells. As stated previously, based on nanosensor size ($<0.2\,\mu$M) and pixel size ($0.5\,\mu M \times 0.5\,\mu M$), it is assumed that there is at least one nanosensor per active pixel with a possibility for multiple nanosensors to be present per pixel (Figure 6(a)). Untreated cells accumulated approximately 100 nanosensors per cell during a 10-min exposure estimated based on assuming a 1 : 1 correlation of active pixel to the presence of at least one nanosensor. This value slowly decreased in number after washout, likely due to a combination of (1) recycling back to the surface, (2) directed trafficking to secretory lysosomes followed by egress of receptor-bound nanosensors through exocytosis, and (3) lysosomal degradation. Although the relative contributions

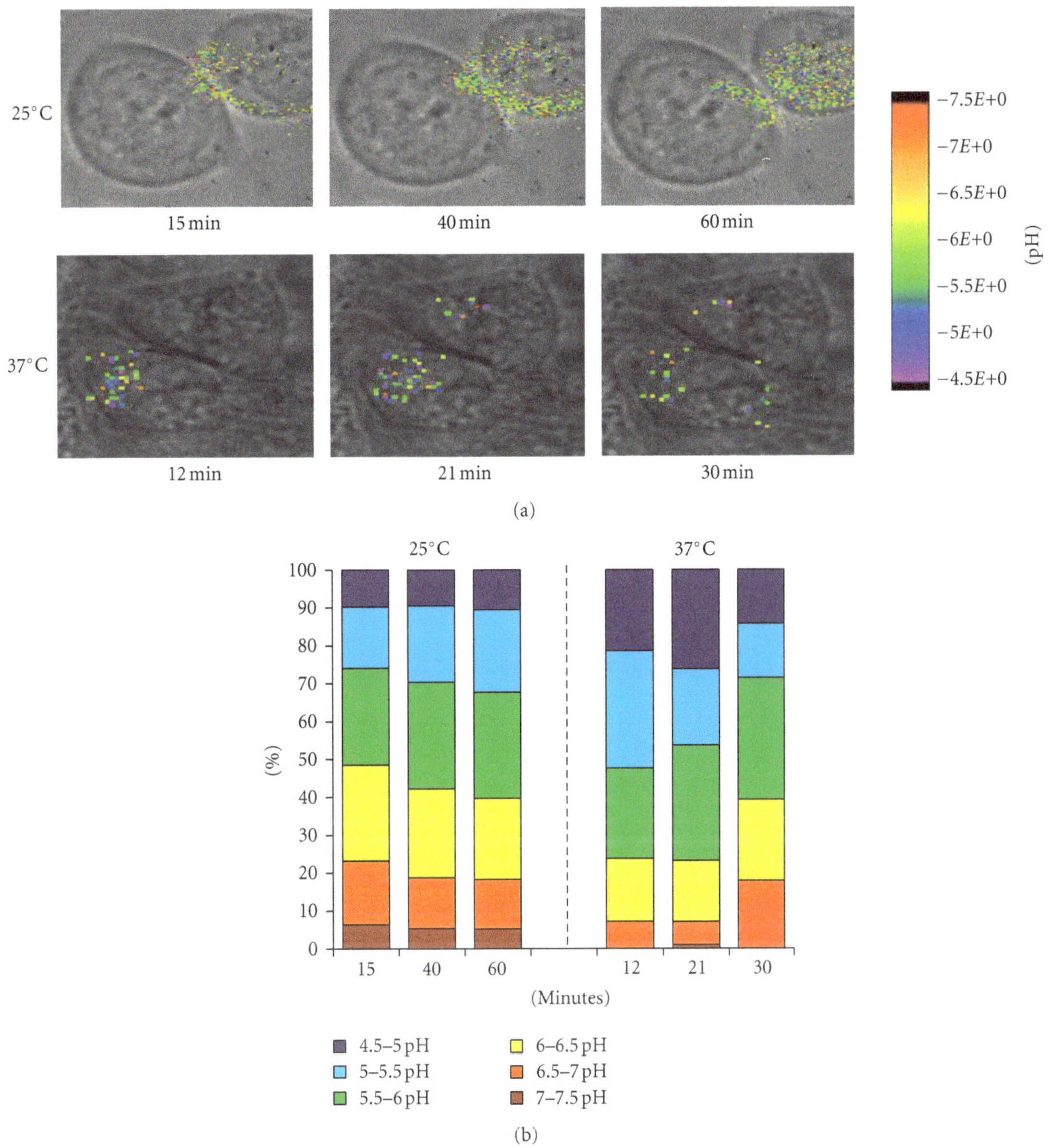

FIGURE 7: Temperature effects on nanosensors at 37°C versus 25°C. (a) Whole-cell pH image maps of cells incubated at 25°C with 10-minutes incubation of nanosensors, followed by imaging at 15′, 40′, and 60′ after the initial nanosensor addition. Bottom series are cells incubated at 37°C with 10-minute incubation with nanosensors followed by imaging at 12′, 21′, and 30′ after initial nanosensor addition. (b) Bar graphs demonstrating percentage of nanosensors from whole, in each pH grouping, at specified time points. Cells were incubated for ten minutes with nanosensors and then imaged at specified time points. Reprinted with permission from Nowak-Lovato et al. [11] (copyright 2010 Springer).

of these two pathways are not distinguishable under these experimental conditions, it is possible to infer the locations of nanosensors from their distribution within compartments with specific pH at intervals during the chase (Figure 8(b)). At 12 minutes of chase, 70% of nanosensors were in compartments with pH of 4.5–5.0 (40%) or 5.0–5.5 (30%). Again, selective loss of this pool is consistent with arrival in the secretory lysosome, where they may be degraded or routed to the surface by exocytosis.

Remarkably, amiloride-treated cells accumulated 4-fold more nanosensors (400/cell) within the same time of exposure (Figure 8(a)). It can be speculated that this large increase in number may be attributed to increased acidification of early and recycling endosomes, shifting the balance of receptors from the recycling pathway to the lysosome-directed pathway. Disruption of the Na^+/H^+ exchanger may also partially block the internalization process, leaving sensors bound to receptors at the cell surface where the medium is buffered at pH 7.2. This is illustrated in Figure 8(b), where 10–18% of sensors register a pH of 7.0–7.5 for up to 20 min of chase. Other notable observations in amiloride-treated cells include a detectable increase in nanosensors within very low pH compartments (pH 4.5–5) [58–60] and marked overall decline in number of nanosensors over the 40 min chase

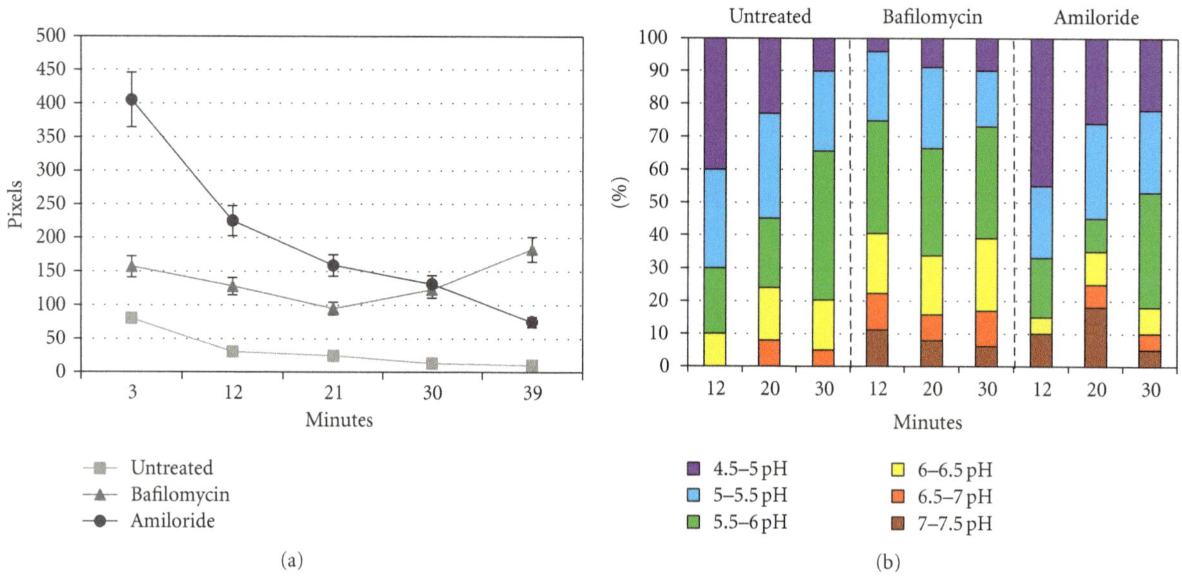

FIGURE 8: Pharmacologic effects on nanosensors at 37°C. Cells incubated at 37°C with the addition of bafilomycin or amiloride for two hours, followed by the addition of nanosensors. (a) Demonstration of total pixel/nanosensor number averaged from three individual cells. Cells are incubated at 37°C untreated or treated for two hours with drug, followed by the addition of nanosensors for ten minutes, then imaged. (b) Demonstration of the percentage of nanosensors averaged from three cells, from the whole, within the assigned pH groupings. Reprinted with permission from Nowak-Lovato et al. [11] (copyright 2010 Springer).

period. This is consistent with degradation with low pH compartments and/or exocytosis.

Bafilomycin-treated cells took up an average of 150 nanosensors during the 10 min exposure, with no obvious decline over the 40-min chase period (Figure 8(a)). As shown in Figure 8(b), these nanosensors are trapped in compartments with significantly higher pH, consistent with blockade of the proton pump. Approximately 40% of the sensors reside in pH environments of 6.0–7.5 (red-orange-yellow bars in Figure 8(b) for the duration of the experiment). This may reflect occupancy in endosomal and recycling compartments with aberrant pH or retention of some receptors at the cell surface [61, 62]. Others have suggested that bafilomycin affects endosomal acidification primarily from the late endosomal to lysosomal stage [63–65]. Accordingly, less than 10% of sensors sense typical lysosomal pH in the range of pH 4.5–5.0. It seems likely that the sustained levels of nanosensors are due to either poor delivery to lysosomes and/or protection from degradation because of the pH dependency for lysosomal enzyme activity.

The protocol was varied in two ways: by (1) uptake of nanosensors by untreated cells at 37°C, followed by the addition of drugs and imaging or (2) treatment with drugs and nanosensors, followed by the removal of the drugs and imaging to follow recovery of compartments. Figure 9(a) shows the distribution of nanosensors immediately after bafilomycin was added to the cells exposed to ten minute incubation with nanosensors. Graphs in this figure show that the effects of bafilomycin are apparent within minutes. This is most dramatic when comparisons are made at the 12-minute time point where only 25% of sensors are in lowest pH (4.5–5.5) environments compared to 70% in normal cells (see Figure 8(b)). Importantly, acute exposure to bafilomycin after uptake prevents loss of nanosensors. This is demonstrated in Figure 9(b), where the total number of nanosensors is stabilized within 12 minutes of bafilomycin treatment. This effect is reversible, as shown in Figure 9(b), where cells were treated for 30 min with bafilomycin followed by exchange with fresh media without drug.

Acute effects of amiloride treatment on cells were also examined in Figure 9(c). Cells were again incubated with nanosensors for ten minutes at 37°C, followed by wash and addition of amiloride to the media (dashed line). Although 30% of nanosensors reach a low pH (4.5–5.0) environment by 3 min, there is a significant shift to higher pH environments within 12 minutes. This change appears to be complete within 20 minutes since the range of pH compartments is similar for both 20 min and 30 min of amiloride treatment.

3.6. Pharmacologics and Temperature. The final set of experiments compared the uptake and transiting of nanosensors in cells pretreated with amiloride or bafilomycin but held at 25°C for all treatments. Under these conditions, untreated cells and amiloride-treated cells show a slow uptake of nanosensors, reaching similar plateau levels at approximately 45 min (Figure 10(a)). In contrast, bafilomycin-treated cells accumulate markedly fewer nanosensors per cell (Figure 10(a)) when held at room temperature.

Analysis of the pH environment for internalized receptors under each condition suggests complex effects of low temperature on receptor trafficking through the endosome-lysosome systems. A spatial view of nanosensor localization within the untreated and treated cells at 15, 40, and 60

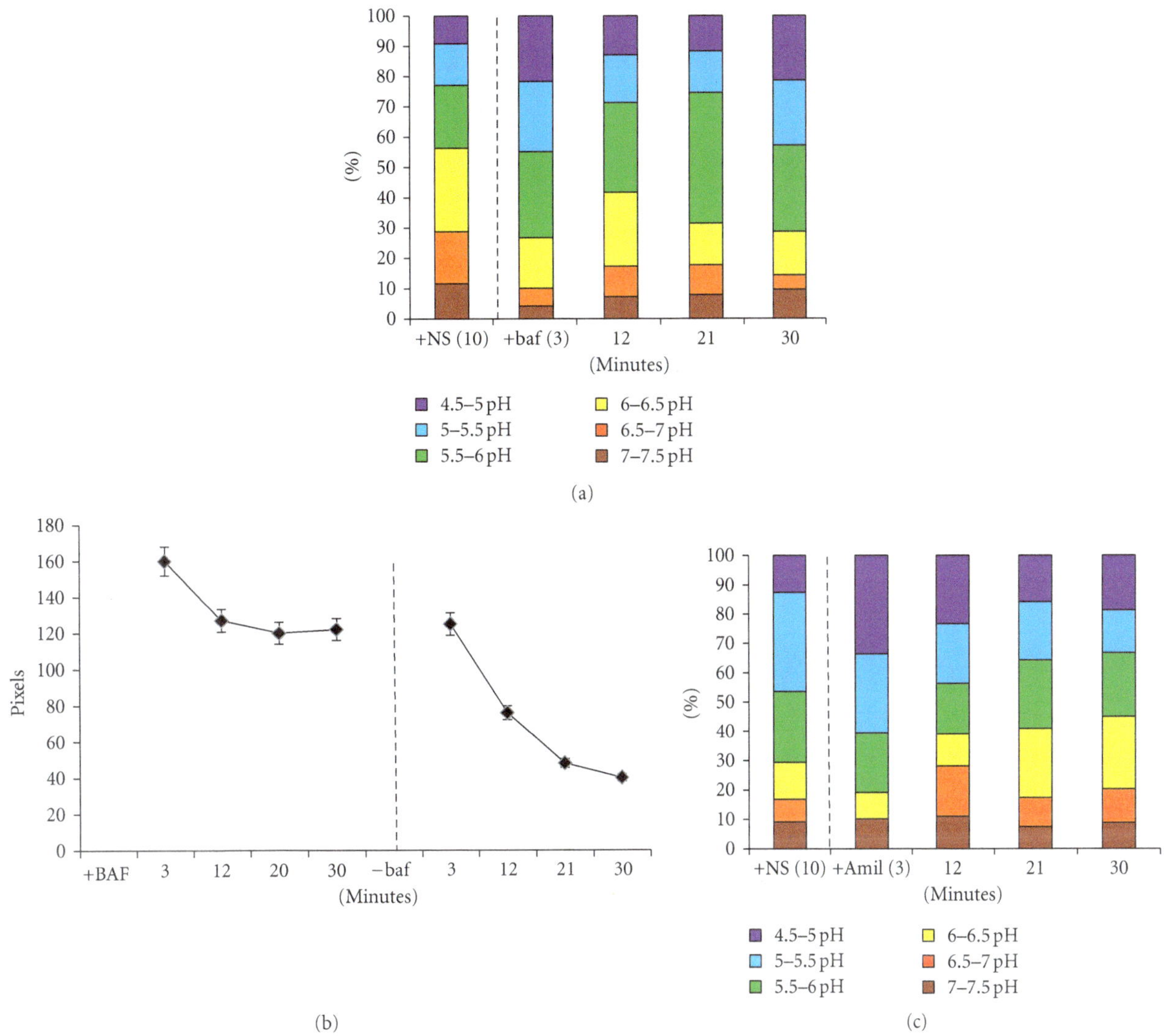

Figure 9: Nanosensor recovery from pharmacological treatment. (a) Representation of cells incubated at 37°C with nanosensor addition and incubation for ten minutes. The first bar of the graph is representative of nanosensors inside cells without chemical addition. After the dashed line, the bar graph is representative to nanosensor ph after the addition of bafilomycin to the cells. (b) Total pixel/nanosensor count within cells. Time points after dashed line are representative to the removal of bafilomycin and addition of fresh media. (c) Representation of nanosensor pH in cells incubated at 37°C with nanosensor incubation for ten minutes followed by the addition of amiloride. The first bar of the graph is cells incubated with nanosensors for ten minutes. The data after the dashed line is the addition of amiloride to the media followed by imaging at respective time points. Reprinted with permission from Nowak-Lovato et al. [11] (copyright 2010 Springer).

minutes is demonstrated (Figure 10(b)). In untreated cells, the slow uptake is accompanied by marked lack of delivery to low pH (4.5–5.5) compartments. Even after 60 minutes, about 10% of receptors would appear to have reached late endosomes or lysosomes (Figure 10(c)). Interpretation of receptor progression in bafilomycin-treated cells is complicated by the poor overall uptake. However, it is notable that over 35% of nanosensors taken up by amiloride-treated cells reside in a pH 4.5–5.5 environment within 60 min of uptake at 25°C. It cannot be distinguished whether these nanosensors have reached lysosomal compartments at this stage or whether they reside in endosomal compartments that have dramatically acidified after amiloride blockage of the Na^+/H^+ exchanger on their membranes.

4. Conclusion

The development of targeted SERS pH nanosensors and the first applications of this technology have been demonstrated in live cells using time lapse Raman spectral microscopy. These experiments demonstrate that targeted nanosensors can be taken up by receptor-mediated internalization, providing a mechanism to simultaneously measure endocytic vesicle pH and follow the temporal and spatial progression of receptors as they traffic through the endosome-lysosome system. The ligands used for targeting and sensing have been shown to not significantly interfere with each other's function. As is demonstrated, confocal hyperspectral SERS whole-cell imagery can be rapidly acquired overtime. The

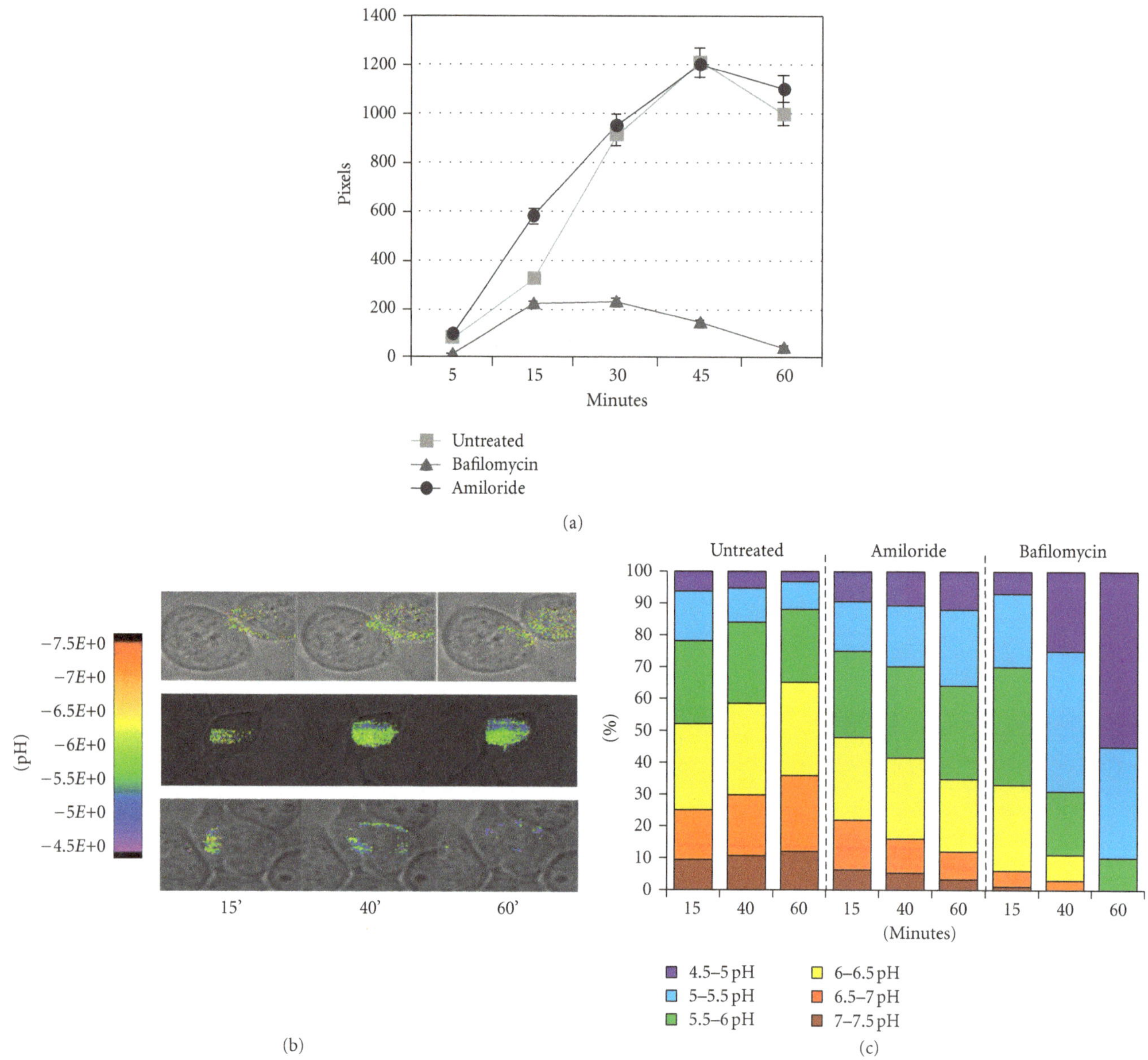

Figure 10: Pharmacologic effects on nanosensors at 25°C. Cells incubated at 25°C with the addition of bafilomycin or amiloride. Cells are incubated at 25°C with simultaneous addition of nanosensors and respective drug. (a) Is a demonstration of total nanosensor number displayed as pixels, averaged from three individual cells. (b) Whole-cell pH image maps of cells incubated at 25°C with the addition of nanosensors alone (untreated), with bafilomycin, or amiloride, followed by imaging at the corresponding time points below. (c) Bar graph demonstration of the percentage of nanosensors averaged from three cells, from the whole, that fall into each pH grouping in cells that are either untreated and incubated at 25°C, or treated with drug and incubated at 25°C. Reprinted with permission from Nowak-Lovato et al. [11] (copyright 2010 Springer).

imagery, with *in vivo* calibration data, can be used to create whole-cell, time-dependant localized pH maps. The targeted nanosensors show the markedly slow progression of receptors through the endocytic pathway when cells are held at 25°C. The targeted nanosensors reported the effects of H^+ flux inhibiting drugs amiloride and bafilomycin on the endocytic compartments, providing reliable measurements of changes in lumenal pH. This technology represents a novel approach to use live cells as microlaboratories by developing nanosensors based on a nanoparticle platform and an abundance of ligand possibilities for sensing and targeting. The versatility of these probes shows promise for future applications relevant to intracellular trafficking and intelligent drug design.

Acknowledgments

This manuscript has been authored by Los Alamos National Security under Contract no. DE-RP52-05NA25396 with the US Department of Energy. The United States Government retains and the publisher, by accepting the article for publication, acknowledges that the United States Government retains a nonexclusive, paid-up, irrevocable, and worldwide license to publish or reproduce the published form of this manuscript, or allow others to do so, for United States Government purposes. They acknowledge Los Alamos National Laboratory Directed Research and Development grant no. 20080001DR (to K. D. Rector) for support of this project. They acknowledge use of the UNM-HSC Electron Microscopy Facility and the UNM Cancer Center Fluorescent Microscopy Facility. They also thank Professors Bridget Wilson, Janet Oliver, Diane Lidke and Keith Lidke (University of New Mexico) for advice and input. The authors would also like to thank Dr. Stephen Joyce (Los Alamos National Laboratory) for his SEM imaging experiments, and Dr. David Kimball and Maida Trujillo (Los Alamos National Laboratory) for their help in the artwork of Figure 1(c). Finally, we thank Brian Dyer of Emory University for useful discussions.

References

[1] H. Komatsu, T. Miki, Y. Shindo, Y. Kitamura, K. Oka, and K. Suzuki, "Single molecular multi analyte fluorescent probes for bioimaging," *Abstracts of Papers of the American Chemical Society*, vol. 230, pp. U409–U410, 2005.

[2] H. Tadakuma, J. Yamaguchi, Y. Ishihama, and T. Funatsu, "Imaging of single fluorescent molecules using video-rate confocal microscopy," *Biochemical and Biophysical Research Communications*, vol. 287, no. 2, pp. 323–327, 2001.

[3] R. Yotter and D. Wilson, "Sensor technologies for monitoring metabolic activity in single cells—Part II: nonoptical methods and applications," *IEEE Sensors Journal*, vol. 4, no. 4, pp. 412–429, 2004.

[4] M. Kall, X. Hongxing, and P. Johansson, "Field enhancement and molecular response in surface-enhanced Raman scattering and fluorescence spectroscopy," *Journal of Raman Spectroscopy*, vol. 36, no. 6-7, pp. 510–514, 2005.

[5] A. Campion and P. Kambhampati, "Surface-enhanced Raman scattering," *Chemical Society Reviews*, vol. 27, no. 4, pp. 241–250, 1998.

[6] M. Fleischmann, P. Hendra, and A. McQuillan, "Raman spectra of pyridine adsorbed at a silver electrode," *Chemical Physics Letters*, vol. 26, no. 2, pp. 163–166, 1974.

[7] J. Hu, B. Zhao, W. Xu, B. Li, and Y. Fan, "Surface-enhanced Raman spectroscopy study on the structure changes of 4-mercaptopyridine adsorbed on silver substrates and silver colloids," *Spectrochimica Acta A*, vol. 58, no. 13, pp. 2827–2834, 2002.

[8] J. Kneipp, H. Kneipp, and K. Kneipp, "SERS-a single-molecule and nanoscale tool for bioanalytics," *Chemical Society Reviews*, vol. 37, no. 5, pp. 1052–1060, 2008.

[9] M. Moskovits, "Surface-enhanced spectroscopy," *Reviews of Modern Physics*, vol. 57, no. 3, pp. 783–826, 1985.

[10] K. L. Nowak-Lovato and K. D. Rector, "Targeted surface-enhanced Raman scattering nanosensors for whole-cell pH imagery," *Applied Spectroscopy*, vol. 63, no. 4, pp. 387–395, 2009.

[11] K. L. Nowak-Lovato, B. S. Wilson, and K. D. Rector, "SERS nanosensors that report pH of endocytic compartments during FcεRI transit," *Analytical and Bioanalytical Chemistry*, vol. 398, no. 5, pp. 2019–2029, 2010.

[12] K. L. Nowak-Lovato and K. D. Rector, "Use of targeted SERS nanosensors in whole cell pH analysis," *Microscopy and Microanalysis*, vol. 15, supplement 2, no. 2, pp. 52–53, 2009.

[13] Z. Tian, "Surface-enhanced Raman spectroscopy: advancements and applications," *Journal of Raman Spectroscopy*, vol. 36, no. 6-7, pp. 466–470, 2005.

[14] K. Kneipp, Y. Wang, H. Kneipp et al., "Single molecule detection using surface-enhanced Raman scattering (SERS)," *Physical Review Letters*, vol. 78, no. 9, pp. 1667–1670, 1997.

[15] S. Nie and S. Emory, "Probing single molecules and single nanoparticles by surface-enhanced Raman scattering," *Science*, vol. 275, no. 5303, pp. 1102–1106, 1997.

[16] A. Otto, I. Mrozek, H. Grabhorn, and W. Akemann, "Surface-enhanced Raman scattering," *Journal of Physics*, vol. 4, no. 5, pp. 1143–1212, 1992.

[17] C. Talley, L. Jusinski, C. W. Hollars, S. M. Lane, and T. Huser, "Intracellular pH sensors based on Surface-enhanced Raman Scattering," *Analytical Chemistry*, vol. 76, no. 23, pp. 7064–7068, 2004.

[18] W. Jin, J. M. Costa-Fernández, R. Pereiro, and A. Sanz-Medel, "Surface-modified CdSe quantum dots as luminescent probes for cyanide determination," *Analytica Chimica Acta*, vol. 522, no. 1, pp. 1–8, 2004.

[19] S. W. Joo, "Adsorption of bipyridine compounds on gold nanoparticle surfaces investigated by UV-Vis absorbance spectroscopy and surface enhanced Raman scattering," *Spectroscopy Letters*, vol. 39, no. 1, pp. 85–96, 2006.

[20] A. M. Schwartzberg, T. Y. Oshiro, J. Z. Zhang, T. Huser, and C. E. Talley, "Improving nanoprobes using surface-enhanced Raman scattering from 30-nm hollow gold particles," *Analytical Chemistry*, vol. 78, no. 13, pp. 4732–4736, 2006.

[21] R. Jensen, J. Sherin, and S. Emory, "Single nanoparticle based optical pH probe," *Applied Spectroscopy*, vol. 61, no. 8, pp. 832–838, 2007.

[22] Q. Wu, C. Zhang, and F. Li, "Preparation of spindle-shape silver core-shell particles," *Materials Letters*, vol. 59, no. 28, pp. 3672–3677, 2005.

[23] A. G. Tkachenko, H. Xie, D. Coleman et al., "Multifunctional gold nanoparticle-peptide complexes for nuclear targeting," *Journal of the American Chemical Society*, vol. 125, no. 16, pp. 4700–4701, 2003.

[24] X. Qian, X. H. Peng, D. O. Ansari et al., "*In vivo* tumor targeting and spectroscopic detection with surface-enhanced Raman nanoparticle tags," *Nature Biotechnology*, vol. 26, no. 1, pp. 83–90, 2008.

[25] W. Doering and S. Nie, "Spectroscopic tags using dye-embedded nanoparticles and surface-enhanced Raman scattering," *Analytical Chemistry*, vol. 75, no. 22, pp. 6171–6176, 2003.

[26] G. R. Souza, C. S. Levin, A. Hajitou, R. Pasqualini, W. Arap, and J. H. Miller, "*In vivo* detection of gold-imidazole self-assembly complexes: NIR-SERS signal reporters," *Analytical Chemistry*, vol. 78, no. 17, pp. 6232–6237, 2006.

[27] J. K. Lim and S. W. Joo, "Gold nanoparticle-based pH sensor in highly alkaline region at pH>11: surface-enhanced Raman scattering study," *Applied Spectroscopy*, vol. 60, no. 8, pp. 847–852, 2006.

[28] J. Kneipp, H. Kneipp, B. Wittig, and K. Kneipp, "One-and two-photon excited optical pH probing for cells using surface-enhanced Raman and Hyper-Raman nanosensors," *Nano Letters*, vol. 7, no. 9, pp. 2819–2823, 2007.

[29] Z. Wang, A. Bonoiu, M. Samoc, Y. Cui, and P. N. Prasad, "Biological pH sensing based on surface enhanced Raman scattering through a 2-aminothiophenol-silver probe," *Biosensors and Bioelectronics*, vol. 23, no. 6, pp. 886–891, 2008.

[30] M. Bradley, L. Alexander, K. Duncan, M. Chennaoui, A. C. Jones, and R. M. Sánchez-Martín, "pH sensing in living cells using fluorescent microspheres," *Bioorganic & Medicinal Chemistry Letters*, vol. 18, no. 1, pp. 313–317, 2008.

[31] J. Liu, W. Lu, D. Reigada, J. Nguyen, A. M. Laties, and C. H. Mitchell, "Restoration of lysosomal pH in RPE cells from cultured human and ABCA4(−/−) mice: pharmacologic approaches and functional recovery," *Investigative Ophthalmology & Visual Science*, vol. 49, no. 2, pp. 772–780, 2008.

[32] B. Varghese, H. Barriere, C. J. Carbone et al., "Polyubiquitination of prolactin receptor stimulates its internalization, post-internalization sorting, and degradation via the lysosomal pathway," *Molecular and Cellular Biology*, vol. 28, no. 17, pp. 5275–5287, 2008.

[33] Y. Gong, M. Duvvuri, and J. Krise, "Separate roles for the Golgi apparatus and Lysosomes in the sequestration of drugs in the multidrug-resistant human leukemic cell line HL-60," *The Journal of Biological Chemistry*, vol. 278, no. 50, pp. 50234–50239, 2003.

[34] R. Haugland, *The Handbook: A Guide to Fluorescence Probes and Labeling Technologies*, Invitrogen Corporation, 10th edition, 2005.

[35] N. Kokkonen, A. Rivinoja, A. Kauppila, M. Suokas, I. Kellokumpu, and S. Kellokumpu, "Defective acidification of intracellular organelles results in aberrant secretion of cathepsin D in cancer cells," *The Journal of Biological Chemistry*, vol. 279, no. 38, pp. 39982–39988, 2004.

[36] Y. Su and P. Chang, "Acidic pH promotes the formation of toxic fibrils from β-amyloid peptide," *Brain Research*, vol. 893, no. 1-2, pp. 287–291, 2001.

[37] S. J. Wood, B. Maleeff, T. Hart, and R. Wetzel, "Physical, morphological and functional differences between pH 5.8 and 7.4 aggregates of the Alzheimer's amyloid peptide A β," *Journal of Molecular Biology*, vol. 256, no. 5, pp. 870–877, 1996.

[38] C. J. D. Grauw, C. Otto, and J. Greve, "Line-scan Raman microspectrometry for biological applications," *Applied Spectroscopy*, vol. 51, no. 11, pp. 1607–1612, 1997.

[39] R. J. Lee, S. Wang, and P. S. Low, "Measurement of endosome pH following folate receptor-mediated endocytosis," *Biochimica et Biophysica Acta*, vol. 1312, no. 3, pp. 237–242, 1996.

[40] H. Lin, P. Herman, and J. Lakowicz, "Fluorescence lifetime-resolved pH imaging of living cells," *Cytometry A*, vol. 52, no. 2, pp. 77–89, 2003.

[41] S. Ohkuma and B. Poole, "Fluorescence probe measurement of the intralysosomal pH in living cells and the perturbation of pH by various agents," *Proceedings of the National Academy of Sciences of the United States of America*, vol. 75, no. 7, pp. 3327–3331, 1978.

[42] M. Serresi, R. Bizzarri, F. Cardarelli, and F. Beltram, "Real-time measurement of endosomal acidification by a novel genetically encoded biosensor," *Analytical and Bioanalytical Chemistry*, vol. 393, no. 4, pp. 1123–1133, 2009.

[43] J. A. Thomas, R. N. Buchsbaum, A. Zimniak, and E. Racker, "Intracellular pH measurements in Ehrlich ascites tumor cells utilizing spectroscopic probes generated in situ," *Biochemistry*, vol. 18, no. 11, pp. 2210–2218, 1979.

[44] D. W. MacGlashan Jr., "Endocytosis, recycling, and degradation of unoccupied FcεRI in human basophils," *Journal of Leukocyte Biology*, vol. 82, no. 4, pp. 1003–1010, 2007.

[45] R. F. Stump, J. M. Oliver, E. J. Cragoe, and G. G. Deanin, "The control of mediator release from RBL-2H3 cells: roles for Ca^{2+}, Na^{+}, and protein kinase C," *The Journal of Immunology*, vol. 139, no. 3, pp. 881–886, 1987.

[46] L. B. Schwartz, K. F. Austen, and S. I. Wasserman, "Immunologic release of β-hexosaminidase and β-glucuronidase from purified rat serosal mast cells," *The Journal of Immunology*, vol. 123, no. 4, pp. 1445–1450, 1979.

[47] G. Labrecque, D. Holowka, and B. Baird, "Antigen-triggered membrane potential changes in IgE-sensitized rat basophilic leukemia cells: evidence for a repolarizing response that is important in the stimulation of cellular degranulation," *The Journal of Immunology*, vol. 142, no. 1, pp. 236–243, 1989.

[48] S. Narita, R. M. Goldblum, C. S. Watson et al., "Environmental estrogens induce mast cell degranulation and enhance IgE-mediated release of allergic mediators," *Environmental Health Perspectives*, vol. 115, no. 1, pp. 48–52, 2007.

[49] M. Tsen and L. Sun, "Surface-enhanced Raman scattering from functionalized self-assembled monolayers. Part 1. Distance dependence of enhanced Raman scattering from a terminal phenyl group," *Analytica Chimica Acta*, vol. 307, no. 2-3, pp. 333–340, 1995.

[50] H. Z. Yu, N. Xia, and Z. F. Liu, "SERS titration of 4-mercaptopyridine self-assembled monolayers at aqueous buffer/gold interfaces," *Analytical Chemistry*, vol. 71, no. 7, pp. 1354–1358, 1999.

[51] L. Hovemadsen and D. Bers, "Indo-1 binding to protein in permeabilized ventricular myocytes alters its spectral and Ca binding properties," *Biophysical Journal*, vol. 63, no. 1, pp. 89–97, 1992.

[52] A. Baker, R. Brandes, J. H. M. Schreur, S. A. Camacho, and M. W. Weiner, "Protein and acidosis alter calcium-binding and fluorescence spectra of the calcium indicator indo-1," *Biophysical Journal*, vol. 67, no. 4, pp. 1646–1654, 1994.

[53] M. Marsh, G. Griffiths, and G. E. Dean, "Three-dimensional structure of endosomes in BHK-21 cells," *Proceedings of the National Academy of Sciences of the United States of America*, vol. 83, no. 9, pp. 2899–2903, 1986.

[54] D. Yamashiro and F. Maxfield, "Kinetics of endosome acidification in mutant and wild-type Chinese hamster ovary cells," *The Journal of Cell Biology*, vol. 105, no. 6, part 1, pp. 2713–2721, 1987.

[55] H. Barth, G. Pfeifer, F. Hofmann, E. Maier, R. Benz, and K. Aktories, "Low pH-induced formation of ion channels by clostridium difficile toxin B in target cells," *The Journal of Biological Chemistry*, vol. 276, no. 14, pp. 10670–10676, 2001.

[56] A. Bidani and T. Heming, "Effects of bafilomycin A_1 on functional capabilities of LPS-activated alveolar macrophages," *Journal of Leukocyte Biology*, vol. 57, no. 2, pp. 275–281, 1995.

[57] K. Tanigaki, S. Sasaki, and S. Ohkuma, "In bafilomycin A_1-resistant cells, bafilomycin A_1 raised lysosomal pH and both prodigiosins and concanamycin A inhibited growth through apoptosis," *FEBS Letters*, vol. 537, no. 1–3, pp. 79–84, 2003.

[58] A. G. Cabado, M. R. Vieytes, and L. M. Botana, "Amiloride-dependent transport is the main mechanism implicated in sodium influx regulation in rat mast cells," *Journal of Cellular Physiology*, vol. 156, no. 3, pp. 567–570, 1993.

[59] T. Kleyman and E. Cragoe, "Amiloride and its analogs as tools in the study of ion transport," *The Journal of Membrane Biology*, vol. 105, no. 1, pp. 1–21, 1988.

[60] K. Taouil, J. C. Féray, J. Brunet, M. O. Christen, R. P. Garay, and P. Hannaert, "Inhibition by xipamide of amiloride-induced acidification in cultured rat cardiocytes," *European Journal of Pharmacology*, vol. 324, no. 2-3, pp. 289–294, 1997.

[61] M. J. Clague, S. Urbe, F. Aniento, and J. Gruenberg, "Vacuolar ATPase activity is required for endosomal carrier vesicle formation," *The Journal of Biological Chemistry*, vol. 269, no. 1, pp. 21–24, 1994.

[62] A. W. M. van Weert, K. W. Dunn, H. J. Geuze, F. R. Maxfield, and W. Stoorvogel, "Transport from late endosomes to lysosomes, but not sorting of integral membrane proteins in endosomes, depends on the vacuolar proton pump," *The Journal of Cell Biology*, vol. 130, no. 4, pp. 821–834, 1995.
[63] L. S. Johnson, K. W. Dunn, B. Pytowski, and T. E. McGraw, "Endosome acidification and receptor trafficking: bafilomycin A1 slows receptor externalization by a mechanism involving the receptor's internalization motif," *Molecular Biology of the Cell*, vol. 4, no. 12, pp. 1251–1266, 1993.
[64] R. D. Klausner, J. Van Renswoude, and C. Kempf, "Failure to release iron from transferrin in a Chinese hamster ovary cell mutant pleiotropically defective in endocytosis," *The Journal of Cell Biology*, vol. 98, no. 3, pp. 1098–1101, 1984.
[65] R. D. Klausner, J. Van Renswoude, and G. Ashwell, "Receptor-mediated endocytosis of transferrin in K562 cells," *The Journal of Biological Chemistry*, vol. 258, no. 8, pp. 4715–4724, 1983.

16

Concentration of Inorganic Elements Content in Benthic Seaweeds of Fernando de Noronha Archipelago by Synchrotron Radiation Total Reflection X-Ray Fluorescence Analysis (SRTXRF)

Leandro De Santis Ferreira,[1] Rosana Peporine Lopes,[2] Mabel Norma Costas Ulbrich,[2] Thais Guaratini,[3] Pio Colepicolo,[4] Norberto Peporine Lopes,[1] Ricardo Clapis Garla,[5] Eurico Cabral Oliveira Filho,[6] Adrian Martin Pohlit,[7] and Orghêda Luiza Araújo Domingues Zucchi[1]

[1] *Departamento de Física e Química, Faculdade de Ciências Farmacêuticas de Ribeirão Preto, Universidade de São Paulo, 14040-903 Ribeirão Preto, SP, Brazil*
[2] *Instituto de Geociências, Universidade de São Paulo, 05508-080 São Paulo, SP, Brazil*
[3] *Lychnoflora Pesquisa e Desenvolvimento em Produtos Naturais LTDA, Incubadora SUPERA, Campus da USP, 14040-900 Ribeirão Preto, SP, Brazil*
[4] *Instituto de Química, Universidade de São Paulo, 05513-970 São Paulo, SP, Brazil*
[5] *Departamento de Botânica, Ecologia e Zoologia, Centro de Biociências, Universidade Federal do Rio Grande do Norte, 59072-970 Natal, RN, Brazil*
[6] *Instituto de Biociências, Universidade de São Paulo, 05508-090 São Paulo, SP, Brazil*
[7] *Coordenação de Pesquisa em Produtos Naturais, Instituto Nacional de Pesquisa da Amazônia, 69060-001 Manaus, AM, Brazil*

Correspondence should be addressed to Leandro De Santis Ferreira, leansf@fcfrp.usp.br

Academic Editor: Ricardo Vessecchi

SRTXRF was used to determine As, Ba, Br, Ca, Co, Cr, Cs, Cu, Dy, Fe, K, Mn, Mo, Ni, Pb, Rb, Sr, Ti, V, and Zn in eleven seaweed species commonly found in Fernando de Noronha: *Caulerpa verticillata* (J. Agardh) (Chlorophyta), *Asparagopsis taxiformis* (Delile), *Dictyurus occidentalis* (J. Agardh), *Galaxaura rugosa* (J. Ellis & Solander) J. V. Lamouroux, *G. obtusata* (J. Ellis & Solander) J. V. Lamouroux, *G. marginata* (J. Ellis & Solander) J. V. Lamouroux (Rhodophyta), *Dictyota cervicornis* (Kützing), *Dictyopteris justii* (J. V. Lamouroux), *Dictyopteris plagiogramma* (Montagne) Vickers, *Padina gymnospora* (Kützing) Sonder, and a *Sargassum* sp. (Phaeophyta). Data obtained were compared to those from the analysis of other parts of the world seaweeds using different analytical techniques and were found to be in general agreement in terms of major and minor elemental components. Results provide baseline information about the absorption and accumulation of these elements by macroalgae in the area.

1. Introduction

In the South Atlantic Ocean is located Fernando de Noronha archipelago around 540 km of the northeastern Brazilian coast. This archipelago is composed by one large island and 20 small adjacent islets that represent a mountain chain top developed along an east-west fracture zone of the ocean floor and was built up by volcanic and subvolcanic essentially alkaline and subsaturated rocks [1].

The marine flora of Fernando de Noronha was first studied by Dickie [2]. Most of the investigations carried out since then were taxonomic studies [3–5]. Also, the families Dictyotaceae and Sargassaceae of brown algae, the green algae *Caulerpa verticillata*, and the red algae *Galaxaura* spp. are among the most abundant macroalgae on the rocky and reef shores of the archipelago [5, 6]. Their predominance is probably related to the production of secondary metabolites that inhibit herbivore predation [7].

Seaweeds require various mineral ions for photosynthesis and growth. Also, it has long been established that marine and estuarine macroalgae accumulate metals to levels many times those found in the surrounding waters [7], and several algae have been used for monitoring concentrations of elements [8–12]. This study provides baseline information for further investigations of the absorption and accumulation of 20 elements by eleven macroalgae species commonly found in Fernando de Noronha archipelago. The concentrations of the elements in the seaweeds were determined using Synchrotron Radiation Total Reflection X-Ray Fluorescence Analysis (SRTXRF).

2. Experimental

2.1. Chemicals Reagents and Solutions. All the reagents were purchased from Merck (Darmstadt, HE, Germany) and Synth (Diadema, SP, Brazil). The multielementary solution was prepared using monoelementary solutions purchased from Acros Organics (Geel, ANT, Belgium and New Jersey, NJ, USA), and ultrapure (deionized) water was obtained using a deionizer from Microtec (Ribeirão Preto, SP, Brazil).

2.2. Sampling. Eleven species of seaweeds commonly found in Fernando de Noronha archipelago were studied: *Caulerpa verticillata* (Chlorophyta), *Asparagopsis taxiformis, Dictyurus occidentalis, Galaxaura rugosa, G. obtusata, G. marginata* (Rhodophyta), *Dictyota cervicornis, Dictyopteris justii, Dictyopteris plagiogramma, Padina gymnospora,* and a *Sargassum* sp. (Phaeophyta) Samples were collected in February and March, 2006 at Caieiras Beach (3°50′18.8″S, 32°23′57.3″W) and Sueste Bay (3°52′1.2″S, 32°25′19.7″W) which are on the main island (Figure 1; Table 1). The IBAMA authorization to collect algae was registered with the number 050/2006. Seaweed specimens were collected randomly, that is, some individuals in a population were collect without a rule or defined sequence. Whole plants were uprooted and placed in labeled plastic bags. Seaweed samples were frozen and sent to the laboratory where they arrived 48 h after harvesting. The algae were identified by Prof. Dr. Eurico Cabral de Oliveira Filho. Residual sediment, epiphytes, and animals were removed, and the algae were washed with distilled water to remove seawater and air dried in a circulating air oven at 40°C for 48 h. After drying, around 5 g of each seaweed species was powdered by a triturating process in a grail after freezing the samples with liquid nitrogen. The powdered seaweeds were kept in a freezer until analysis was performed.

2.3. Sample and Calibration Solution Preparation. Samples of 250 mg of each algae species were placed in pyrex test tubes and digested according to a procedure described by Ward et al. [13] . Briefly, 6.0 mL of nitric acid (65%) and hydrogen peroxide (30%) were added to each test tube and homogenized. Test tubes were then placed on a digestion block overnight (*ca.* 12 hours) and heated at 130 ± 5°C until a translucid, particle-free, and fully digested solution was obtained. 5.00 mL of ultrapure water were transferred to the digested (sample) solution using a pyrex volumetric pipette, and the resulting solution was homogenized. The blank was a mixture of nitric acid, hydrogen peroxide, and deionized water and was done using the same procedure performed to the samples. Then 1.00 mL of this solution was removed using a pyrex volumetric pipette and to this aliquot was added 10 μL of Ga (1.0 μL mL^{-1}) as an internal standard [14]. Calibration solutions of multielements that emit X-K and X-L rays were prepared, and Ga element was added as internal standard as above. 5.0 μL of each sample were placed on a perspex support (polished quartz, 28 × 22 mm). The same procedure was done for 5.0 μL of calibration solution. Drying for 1 h under, a 150 W infrared lamp (Phillips model 7, Amsterdam, NH, Netherlands) gave rise to a thin layer of approximately 5 mm diameter. Sample and calibration solutions were irradiated as described below.

Table 1: Classification and sampling for eleven macroalgae studied in Fernando de Noronha archipelago, of northeastern Brazil.

Divisions and species	Site
Chlorophyta	
Caulerpa verticillata (J. Agardh, 1847)	S
Rhodophyta	
Asparagopsis taxiformis (Delile) Trevisan de Saint-Léon, 1845	S
Dictyurus occidentalis (J. Agardh, 1847)	C
Galaxaura rugosa (J. Ellis & Solander) J. V. Lamouroux, 1816	C
Galaxaura obtusata (J. Ellis & Solander) J. V. Lamouroux, 1816	C
Galaxaura marginata (J. Ellis & Solander) J. V. Lamouroux, 1816	C
Phaeophyta	
Dictyota cervicornis (Kützing, 1859)	S
Dictyopteris justii (J. V. Lamouroux, 1809)	S
Dictyopteris plagiogramma (Montagne) Vickers, 1905	C
Padina gymnospora (Kützing) Sonder, 1871	C
Sargassum sp. (C. Agardh, 1820)	S

S: Sueste; C: Caieiras.

2.4. Instrumentation and Analysis Conditions. The equipment used was an X-ray fluorescence beamline constructed at the National Synchrotron Light Laboratory—LNLS (Campinas, São Paulo State, Brazil). For the total reflection of radiation, a series of mirrors are adjusted to allow that the synchrotron radiation hit the sample in small angle. The sample and calibration solutions were analyzed (three replicates of each) for 100 s each with a white synchrotron radiation beam using 0.5 mm of Al as absorber, 1.0 mm of Ta as collimator in the detector, a sample-to-detector distance of 1.1 mm, a height of 1 mm under total reflection conditions, and an angle of incidence of 1 mrad. The characteristic X-rays were detected with the aid of a Ge hyperpure semiconductor detector (resolution of 145 eV for energy of 5.9 keV).

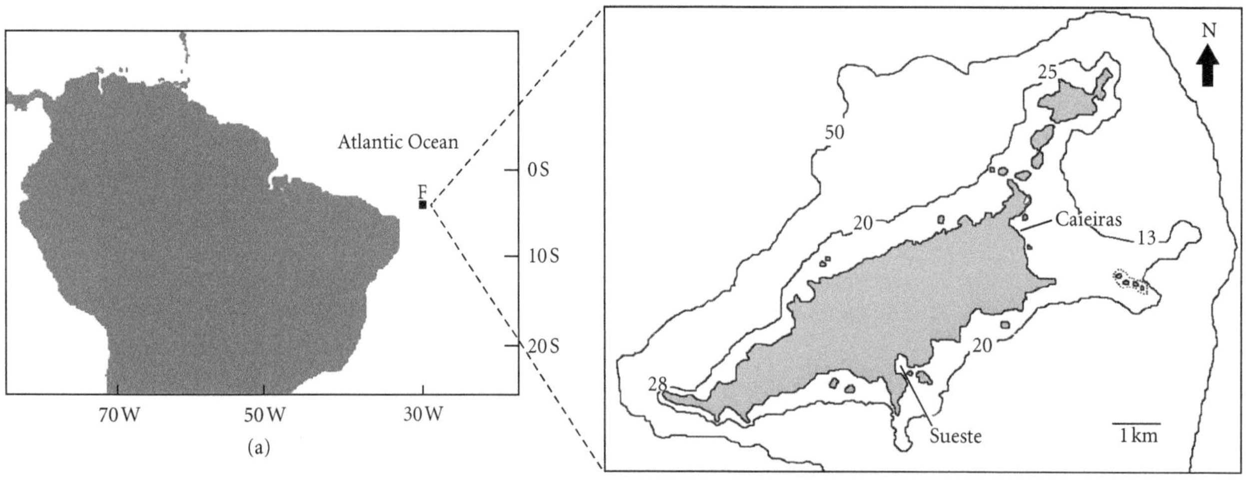

FIGURE 1: Geographical localization of study sites (Caieiras and Sueste) at the main island of Fernando de Noronha archipelago, off northeastern Brazil, where seaweeds were collected.

2.5. Quantitative Analysis for X-Ray Fluorescence. Energy peaks detected in specters of the calibration samples were determined on the spectrometer, and the energy-emitting elements were identified from their X-ray characteristics (analytical lines). The liquid intensities for characteristic X-rays emitted were calculated with a mathematic adjustment in which the contribution of interfering lines on the analytical line (spectral interference) was considered including the escape peak and the addition peak. Mathematical adjustments were calculated with the software AXIL [15].

In quantitative analysis, the fluorescent intensity of the characteristic line is related to the concentration by the expression $I_i = S_i \times C_i \times A_i$, where I_i = fluorescent intensity of element i (cps), S_i = elemental sensitivity of element i (cps μg^{-1} mL), C_i = concentration of element i ($\mu g\,mL^{-1}$), and A = absorption factor.

Given the tiny thickness of the prepared samples, the absorption and/or intensification effects (matrix effect) of the analytical line are negligible. Thus, there is no need to consider the absorption factor [16], and the relation is $I_i = S_i \times C_i$.

To correct like geometry and X-ray flow variation errors during excitation, Ga was used as internal standard. Ga is not present in the macroalgae samples. Referencing to internal standard yields the expression $(I_i/I_{Ga}) = (S_i/S_{Ga}) \times (C_i/C_{Ga})$, where I_{Ga} = fluorescent intensity of Ga (cps), S_{Ga} = elemental sensitivity of Ga (cps μg^{-1} mL), and C_{Ga} = concentration of Ga ($\mu g\,mL^{-1}$).

If we define $S'_i = S_i/S_{Ga}$ and $R_i = C_{Ga} \times (I_i/I_{Ga})$, where S'_i = relative sensitivity for element i (unidimensional) and R_i = product of relative intensity and C_{Ga} ($\mu g\,mL^{-1}$), then

$$R_i = S'_i \times C_i. \quad (1)$$

In the calibration solutions, R_i is directly proportional to C_i; therefore, the angular coefficient of the calibration curve for the element i is its relative sensitivity. If S'_i is known for the elements present in the calibration solutions, then the following function is obtained:

$$\ln S'_i = a + bZ_i + cZ_i^2 + dZ_i^3, \quad (2)$$

where a, b, c, and d are parameters that can be determinated by variance analyses, and Z_i is the atomic number for element i.

The relative sensitivity for any X-K or X-L ray emitting elements present in the samples can thus be calculated. The SANEST program was used to test significance of the parameters at 5% probability for inclusion in the model which was used to determine the above (2) [17]. The concentrations ($\mu g\,mL^{-1}$ or $\mu g\,g^{-1}$) for any inorganic element present in different samples were obtained from (1) after obtaining the experimental limits of detection (LD)(3)

$$LD_i = \frac{3 \cdot \sqrt{BG_i/t} \cdot C_{Ga}}{I_{Ga} \cdot S'_i}, \quad (3)$$

where BG_i = background (cps), t = detection time (s), and other variables are defined above [18, 19].

From the calculation of the experimental LD values, it was established that the values of LD are a polynomial function of the atomic number of the elements present, $LD_i = f(Z_i)$. Thus, using this formula, it is possible to calculate the LD for the elements which are not present in the sample.

3. Results and Discussion

Calibration curves ($\ln S'_i = f(Z_i)$) with significant parameters at 5% level were obtained for all X-K (Table 2, Figure 2) and X-L ray (Table 3, Figure 3) emitting elements through the multielementary calibration solutions. For the SRTXRF technique, the maximum S'_i value obtained is for the internal standard which was the element Ga ($Z = 31$) in this

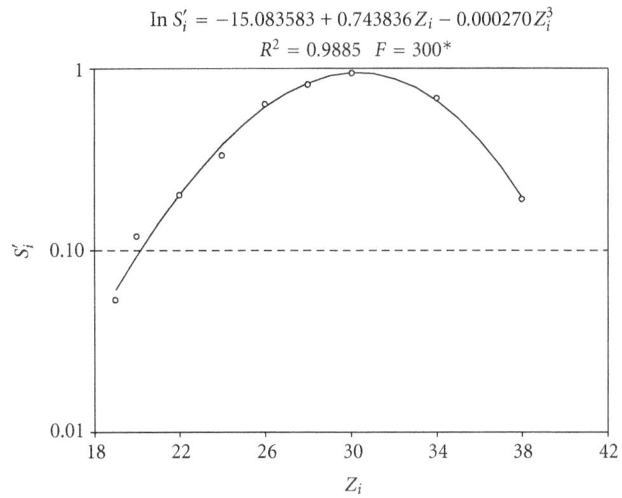

FIGURE 2: Experimental and calculated relative sensitivity for chemical elements emitting X-K rays for $19 \leq Z_i \leq 38$.

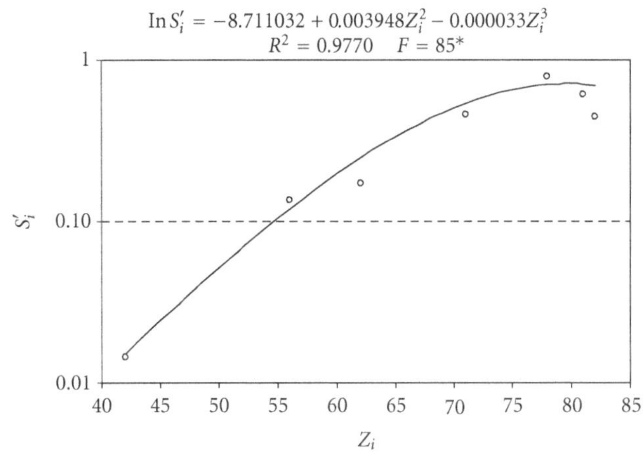

FIGURE 3: Experimental and calculated relative sensitivity for chemical elements emitting X-L rays for $42 \leq Z_i \leq 82$.

experiment. The functions increase for $Z < 31$ until $Z = 31$ and decreasing for $Z > 31$. Theoretically, $S'_{Ga} = 1.0$; however, the experimental value of Ga was 0.93. It is important to correct for this difference in the calculation of the other elements. In this case, if the experimental data were used without experimental correction, an error of 6.6% can be observed for all the elements. Some authors attribute the need to perform these corrections to obtain the values close to true net intensities [20].

The minimum detection limits for X-K and X-L ray emitting elements are presented in Table 4. The lowest recorded detection limit for X-K ray emitting elements was LD = 0.01 ppm for Zn ($Z = 30$) and Ni ($Z = 28$) and the highest detection limit value was LD = 1.72 ppm for K ($Z = 19$). Lowest and highest detection limit values for X-L ray emitting elements were, respectively, 0.01 ppm for Cu ($Z = 29$) and 1.57 ppm for Mo ($Z = 42$). After determining the experimental detection limits, the concentration of each chemical element was estimated (Table 5). The results of the

TABLE 2: Experimental and calculated mean relative sensitivity, S_i', and mean standard deviation for chemical elements emitting X-K rays for $19 \leq Z_i \leq 38$.

Element	Z_i	S_i' (experimental)	$s(m)^a$	S_i'(calculated)
K	19	0.053007	0.004082	0.060650
Ca	20	0.118720	0.006506	0.093774
Ti	22	0.200669	0.015206	0.203081
Cr	24	0.329118	0.017632	0.381365
Fe	26	0.632468	0.027613	0.613014
Ni	28	0.810042	0.030792	0.832589
Zn	30	0.930767	0.025492	0.943175
Se	34	0.665939	0.051780	0.666817
Sr	38	0.190190	0.019045	0.195293

[a] $s(m)$ = standard deviation = S_i' (exprimental)/$\sqrt{n}$, $n = 9$.

TABLE 3: Experimental and calculated mean relative sensitivity, S_i', and mean standard deviation for chemical elements emitting X-L rays for $42 \leq Z_i \leq 82$.

Element	Z_i	S_i' (experimental)	$s(m)^a$	S_i'(calculated)
Mo	42	0.014521	0.001182	0.015121
Ba	56	0.135400	0.008366	0.119381
Sm	62	0.170940	0.007042	0.246660
Lu	71	0.457106	0.014518	0.537822
Pt	78	0.789765	0.060052	0.743550
Tl	81	0.612795	0.047333	0.707681
Pb	82	0.445363	0.012872	0.697799

[a] $s(m)$ = standard deviation = S_i' (exprimental)/$\sqrt{n}$, $n = 9$.

analysis of algae (Table 5) and the chemical composition of the alkaline rocks where the algae were collected (Table 6) were compared.

Essential elements Ca, Fe, K, Mn, and Zn were found in all algae samples as were relevant species CaO, Fe_2O_3, K_2O, MnO, and Zn in the rocks upon which these algae grew. Interestingly, *G. marginata* had little Fe (75.27 ppm) while *D. plagiogramma* had over 150 times that amount (11936.65 ppm). *P. gymnospora* (3028.40 ppm) had the least values of Ca while *G. marginata* and *A. taxiformis* had the highest Ca levels detected (82606.32 and 88908.21 ppm, resp.). Relatively little K was found in *C. verticillata* (504.13 ppm) while the levels of K in *D. occidentalis* were almost 100 times this amount (49523.34 ppm). Also, the *P. gymnospora* had large amounts of Zn (274.44 ppm) whereas five algae species had levels of Zn in the range of 2–7 ppm Also, Sr was found in all rocks analyzed (*ca.* 950–1750 ppm) and was absorbed by all algae species and in relatively large abundance (*ca.* 500–6000 ppm). In contrast, Ba was generally abundant in rock samples (*ca.* 20–1350 ppm) from where the algae were collected, but Ba was only detected in *D. justii*. While Br was detected in several species, it was most concentrated in *A. taxiformis* (257.10 ppm).

The results obtained by SRTXRF analysis of algae are comparable to those obtained for algae from other parts of the world using other analytical methods. For example,

TABLE 4: Minimum detection limits values (ppm) for X-K and X-L ray emitting elements.

Species	As	Ba	Br	Ca	Co	Cr	Cs	Cu	Dy	Fe	K	Mn	Mo	Ni	Pb	Rb	Sr	Ti	V	Zn
Chlorophyta																				
C. verticillata	0.02	—	0.03	0.29	—	0.03	—	—	—	0.02	0.36	0.02	—	—	—	—	0.09	0.05	0.04	0.01
Rhodophyta																				
A. taxiformis	—	—	0.08	0.90	—	0.09	—	—	0.12	0.07	1.12	0.08	—	—	0.06	—	0.21	0.18	0.13	0.03
D. occidentalis	0.02	—	—	0.22	—	0.03	0.12	0.01	0.03	0.02	1.72	0.03	0.43	—	—	0.05	0.08	—	—	0.01
G. rugosa	—	—	0.02	0.25	—	0.02	0.14	—	0.03	0.02	0.37	0.02	0.49	—	0.02	—	0.07	—	—	0.01
G. obtusata	—	—	—	0.40	—	—	0.22	—	0.05	0.03	0.60	0.04	0.86	—	—	—	0.11	—	—	0.01
G. marginata	—	—	—	0.87	—	—	0.46	—	0.12	0.07	1.35	0.08	1.57	—	—	—	0.23	—	—	0.02
Phaeophyta																				
D. cervicornis	—	—	0.07	0.73	—	0.08	—	—	—	0.06	1.26	0.07	—	—	0.06	—	0.21	0.18	0.11	0.03
D. justii	0.02	0.16	0.03	0.30	0.02	—	—	—	—	0.02	0.48	0.03	—	0.01	—	—	0.10	—	0.04	0.01
D. plagiogramma	—	—	—	0.83	—	0.08	0.30	—	0.13	0.08	1.16	0.08	—	0.05	0.07	0.16	0.25	0.11	—	0.05
P. gymnospora	0.01	—	0.02	0.09	0.01	—	0.06	0.01	—	0.01	0.21	0.01	—	0.01	—	0.03	0.04	—	—	0.01
Sargassum sp.	0.03	—	0.03	0.25	0.02	—	—	0.01	—	0.02	0.50	0.03	—	0.01	—	0.06	0.09	—	0.04	0.01

—: not calculated.

TABLE 5: Mean values of elemental composition (ppm) of eleven species of seaweeds collected in Fernando de Noronha Archipelago, off northeastern Brazil and their respective CV, coefficient of variation (%).

Species	As	Ba	Br	Ca	Co	Cr	Cs	Cu	Dy	Fe	K	Mn	Mo	Ni	Pb	Rb	Sr	Ti	V	Zn
Chlorophyta																				
C. verticillata	10.82 (11.26)	—	8.60 (13.63)	20639.78 (0.93)	—	11.91 (12.51)	—	—	—	2559.68 (0.47)	504.13 (5.08)	21.43 (6.40)	—	—	—	—	511.19 (1.80)	108.86 (3.66)	14.06 (8.37)	11.44 (3.83)
Rhodophyta																				
A. taxiformis	—	—	257.10 (2.64)	88908.21 (1.95)	—	25.37 (10.30)	—	—	108.47 (10.46)	4108.09 (1.77)	1988.31 (3.22)	77.14 (9.75)	—	—	12.18 (9.48)	—	2121.10 (1.79)	224.55 (4.35)	39.63 (5.53)	17.35 (1.61)
D. occidentalis	12.03 (8.43)	—	—	15366.59 (0.30)	—	6.26 (7.96)	62.98 (17.10)	1.55 (11.38)	6.12 (15.43)	915.92 (1.15)	49523.34 (1.29)	10.33 (4.35)	465.94 (14.12)	—	—	13.78 (10.85)	718.43 (2.12)	—	—	3.21 (18.63)
G. rugosa	—	—	1.65 (17.96)	22949.59 (1.63)	—	5.28 (2.41)	192.80 (5.05)	—	21.66 (15.51)	513.81 (2.55)	1092.76 (3.99)	15.64 (1.90)	832.84 (12.86)	—	2.00 (9.25)	—	1735.42 (3.63)	—	—	2.36 (12.17)
G. obtusata	—	—	—	41500.55 (6.75)	—	—	358.28 (7.76)	—	39.57 (14.48)	245.12 (5.92)	2611.29 (6.96)	23.58 (23.38)	671.37 (9.99)	—	—	—	3002.25 (9.42)	—	—	3.80 (16.81)
G. marginata	—	—	—	82606.32 (0.88)	—	—	736.05 (3.21)	—	54.34 (9.56)	75.27 (7.73)	15655.44 (0.62)	39.79 (5.66)	836.71 (7.84)	—	—	—	5957.57 (1.53)	—	—	5.70 (18.02)
Phaeophyta																				
D. cervicornis	—	—	20.21 (8.44)	44313.30 (1.04)	—	19.53 (5.43)	—	—	—	3621.88 (0.88)	40527.54 (1.21)	60.85 (4.33)	—	—	27.22 (1.73)	—	2272.40 (0.61)	181.51 (5.68)	44.39 (10.78)	25.54 (2.92)
D. justii	17.35 (2.37)	21.39 (14.88)	3.56 (12.04)	18394.89 (4.26)	3.53 (4.81)	—	—	—	—	1410.02 (4.05)	7370.06 (4.41)	30.33 (3.94)	—	5.35 (16.12)	—	—	3046.38 (4.15)	—	33.87 (3.00)	22.50 (1,67)
D. plagiogramma	—	—	—	56397.66 (2.17)	—	53.64 (7.44)	289.32 (15.38)	—	46.31 (19.97)	11936.65 (2.62)	10593.50 (2.35)	78.34 (6.44)	—	15.15 (5.23)	17.56 (16.64)	47.58 (12.20)	1521.78 (0.91)	537.01 (7.26)	—	24.58 (3.78)
P. gymnospora	15.20 (2.28)	—	8.03 (10.63)	3028.40 (0.55)	1.93 (17.99)	—	148.77 (2.46)	1.10 (11.52)	—	251.77 (0.73)	9910.06 (0.51)	7.04 (6.41)	—	1.91 (13.19)	—	12.75 (17.70)	881.64 (1.21)	—	—	274.44 (0.33)
Sargassum sp.	117.92 (0.10)	—	16.63 (5.92)	14035.01 (0.70)	5.35 (7.23)	—	—	1.61 (9.72)	—	172.12 (1.91)	27196.26 (0.84)	21.53 (5.85)	—	2.68 (8.54)	—	36.43 (9.32)	1453.50 (0.77)	—	14.13 (6.20)	6.15 (5.63)
Average	34.67	21.39	45.11	37103.66	3.60	20.41	298.03	1.42	46.08	2346.40	15179.33	35.09	701.71	6.27	14.74	27.63	2111.06	262.98	29.22	36.10

—: not detected.

Table 6: Representative chemical composition of alkaline rocks from Caieiras (1 to 5) and Sueste (6 to 8).

	Caieiras					Sueste			
	1	2	3	4	5	6	7	8	Average
K_2O^a	1.45	4.85	4.86	2.92	1.83	2.21	5.16	3.67	3.37
CaO^a	11.2	5.80	4.10	6.81	12.5	10.5	0.58	5.73	7.15
TiO_2^a	3.19	2.36	1.18	2.40	3.86	3.37	0.17	2.26	2.35
MnO^a	0.18	0.18	0.15	0.21	0.16	0.21	0.19	0.12	0.18
$Fe_2O_{3total}^a$	12.9	5.99	4.57	7.71	12.5	13.2	2.23	4.79	7.99
V^b	275.1	164.0	71.0	171.8	239.9	257.0	13.0	187.4	172.40
Cr^b	505.1	229.0	—	37.3	303.5	303.0	—	—	275.58
Co^b	55.0	19.1	8.00	15.8	48.2	44.6	—	8.77	28.50
Ni^b	358.9	23.6	17.9	62.4	295.5	309.6	—	19.6	155.36
Cu^b	53.3	7.90	4.20	—	51.3	53.1	—	—	33.96
Zn^b	57.5	97.0	95.3	94.3	122.5	114.1	181.0	80.0	105.21
Rb^b	53.0	142.0	142.5	80.5	47.1	49.1	322.0	96.5	116.59
Sr^b	960.9	820.8	1329	1620	1744	1023	—	1520	1288.24
Mo^b	2.14	—	7.00	6.07	2.57	2.00	—	3.82	3.93
Cs^b	1.20	1.90	2.40	3.67	0.55	1.00	—	4.33	2.15
Ba^b	510.2	946.0	1191	1070	906.4	558.0	19.0	1330	816.33
Dy^b	5.96	7.10	6.60	5.94	8.11	6.90	—	6.90	6.79
Pb^b	—	13.3	14.7	14.5	5.21	7.20	32.0	11.9	14.12

—: not detected; [a]wt %; [b]ppm, —not analyzed . Using the procedure described in Janasi et al. [21], major, minor, and trace elements were determined by inductively coupled plasma-mass spectrometer (ICP-MS).

Hou and Yan [22] studied elements present in 35 species of marine algae from the coast of China by neutron activation analysis in a miniature neutron source reactor (MNSR). They observed that, in brown, red, and green algae, the levels of individual elements were (averages, resp., in parentheses): As (159/<0.36/12.2 ppm), Ba (76.2/109.6/174 ppm), Br (3426/6157/596 ppm), Ca (22.7/29.7/11.2 ppt), Co (0.93/1.15/1.04 ppm), Cr (4.02/4.84/6.33 ppm), Cs (1.11/1.02/0.95 ppm), Fe (1892/2511/3716 ppm), K (67.5/48.4/29.0 ppt), Mn (857/89.4/90.6 ppm), Rb (29.4/21.5/25.8 ppm), Sr (892/313/161 ppm), and Zn (21.7/28.3/23.3 ppm). Besides this study, in another work about 26 marine benthic algae species found in Karachi Coast, Pakistan, the levels (averages for the 26 species are in parentheses) of Ca (26.75 ppt), Co (5.88 ppm), Cr (5.13 ppm), Cu (11.87 ppm), Fe (2.41 ppt), K (69.5 ppt), Pb (13.43 ppm), and Zn (53.28 ppm) were established using flame atomic absorption spectrometry (AAS) [23]. As in the algae in these previous studies, the algae of Fernando de Noronha were found to contain large amounts of Ca, K, and Fe (in parts per thousand, ppt) and small or trace amounts of As, Ba, Co, Cr, Cs, Cu, Pb, Rb, and Zn. Since data on the composition of the surfaces where algae in these previous studies grew was not reported, it is not possible to ascertain the contributions of these surface substrates to the composition of the algae. The similar compositional trends in principle and trace elements present in algae from these previous studies and those of Fernando de Noronha lends support to the usefulness of the SRTXRF technique in the analytical arsenal.

Low numbers of algae species contained Rb (4 algae species, 12–48 ppm), Ti (4 species, 109–537 ppm), and V (5 species, 14–44 ppm) which may be related to geological characteristics of the alkaline rocks present in the study sites. Low concentrations of As (5 species, range 11–118 ppm), Co (3 species, 1.9–5.4 ppm), Cr (6 species, range 5.3–54 ppm), Cu (3 species, range 1.1–1.6 ppm), Ni (4 species, range 1.9–15 ppm), and Pb (4 species, range 2.0–27 ppm) were observed in most of the algae species, except for *Sargassum* sp. which had a higher concentration of As (118 ppm) compared to the other species. These elements may have been absorbed from the seawater through natural weathering and lixiviation of rocks and soil.

An interesting finding was the detection of Dy which is a rare earth metal present in rocks of both collection sites in *A. taxiformis, D. occidentalis, G. rugosa, G. obtusa, G. marginata,* and *D. plagiogramma*. Thus, these algae absorb and store this rare chemical element which is present in rocks in relatively low abundance.

4. Conclusion

The SRTXRF technique proved to be adequate for the determination of 20 chemical elements present in eleven species of common macroalgae of Fernando de Noronha archipelago providing baseline information for the accumulation of metals in two sites. The results indicate a relationship among the metals present in the seaweeds and the rocks present in this area. Besides, the concentrations of common macro- and microelements obtained are comparable to those obtained by other authors using different analytical methods. However, multielement capability in a single analysis, high-sensitivity and precision, short analysis time, and easy sample preparation are some advantages of SRTXRF when compared to other elemental determination techniques such as AAS or ICP-MS.

Acknowledgments

The authors thank the Laboratório Nacional de Luz Síncrotron (LNLS-Campinas, SP) for allowing the use of beam line facilities and for CNPq, FAPESP, and CAPES for financial support. The authors declare that they do not have any conflict of interests.

References

[1] U. G. Cordani, M. N. Ulbrich, E. A. Menor, and R. P. Lopes, "Cenozoic alkaline volcanism of Fernando de Noronha island," in *South American Symposium on Isotope Geology, Field Trip Guide*, pp. 1–24, CBPM/IRD, Salvador, 2003.

[2] G. Dickie, "Enumeration of algae collected at Fernando de Noronha by H. M. Moseley, N. A. naturalist to H. M. S. "Challenger"," *Journal Linnean Society*, vol. 14, pp. 363–365, 1874.

[3] F. Pinheiro-Vieira and M. M. Ferreira-Correia, "Quarta contribuição ao inventário das algas marinhas bentônicas do nordeste brasileiro," *Arquivos de Ciências do Mar*, vol. 10, pp. 189–192, 1970.

[4] E. C. de Oliveira Filho, "An annotated list of the Brazilian seaweeds in Dickie's herbarium," *Botanical Journal of the Linnean Society*, vol. 69, no. 3, pp. 229–238, 1974.

[5] Pedrini A. G., Y. Ugadim, M. R. A. Braga, and S. M. B. Pereira, "Algas marinhas bentônicas do Arquipélago de Fernando de Noronha, Brasil," *Boletim de Botânica da USP*, vol. 13, pp. 93–101, 1992.

[6] V. R. Eston, A. E. Migotto, E. C. O. Filho, S. A. Rodrigues, and J. C. Freitas, "Vertical distribution of benthic marine organisms on rocky coasts of the Fernando de Noronha Archipelago (Brazil)," *Boletim do Instituto Oceanográfico de São Paulo*, vol. 34, pp. 37–53, 1986.

[7] C. S. Lobban and P. J. Harrison, *Seaweed Ecology and Physiology*, Cambridge University, Cambridge, UK, 1997.

[8] G. M. Amado-Filho, L. T. Salgado, M. F. Rebelo, C. E. Rezende, C. S. Karez, and W. C. Pfeiffer, "Heavy metals in benthic organisms from Todos os Santos Bay, Brazil," *Brazilian Journal of Biology*, vol. 68, no. 1, pp. 95–100, 2008.

[9] G. W. Bryan, "The effects of heavy metals (other than mercury) on marine and estuarine organisms," *Proceedings of the Royal Society of London. Series B*, vol. 177, no. 48, pp. 389–410, 1971.

[10] G. W. Bryan and L. G. Hummerstone, "Indicators of heavy metal contamination in the Looe Estuary (Cornwall) with particular regard to silver and lead," *Journal of the Marine Biological Association of the United Kingdom*, vol. 57, pp. 75–92, 1977.

[11] Y. Serfor-Armah, B. J. B. Nyarko, E. K. Osae, D. Carboo, S. Anim-Sampong, and F. Seku, "Rhodophyta seaweed species as bioindicators for monitoring toxic element pollutants in the marine ecosystem of Ghana," *Water, Air, and Soil Pollution*, vol. 127, no. 1-4, pp. 243–253, 2001.

[12] M. Rajfur, A. Kłos, and M. Wacławek, "Sorption properties of algae Spirogyra sp. and their use for determination of heavy metal ions concentrations in surface water," *Bioelectrochemistry*, vol. 80, no. 1, pp. 81–86, 2010.

[13] A. F. Ward, L. F. Marciello, L. Carrara, and V. J. Luciano, "Simultaneous determination of major, minor and trace elements in agricultural and biological samples by inductively coupled argon plasma spectrometry," *Spectroscopy Letters*, vol. 13, pp. 803–831, 1980.

[14] O. L. A. D. Zucchi, S. Moreira, M. J. Salvador, and L. L. Santos, "Multielement analysis of soft drinks by x-ray fluorescence spectrometry," *Journal of Agricultural and Food Chemistry*, vol. 53, no. 20, pp. 7863–7869, 2005.

[15] P. van Espen, H. Nullens, and F. Adams, "A computer analysis of X-ray fluorescence spectra," *Nuclear Instruments and Methods*, vol. 142, no. 1-2, pp. 243–250, 1977.

[16] R. Klockenkämper and A. von bohlen, "Determination of the critical thickness and the sensitivity for thin-film analysis by total reflection X-ray fluorescence spectrometry," *Spectrochimica Acta Part B*, vol. 44, no. 5, pp. 461–469, 1989.

[17] E. P. Zonta and A. Machado, *SANEST: Sistema de Análise Estatística para Microcomputadores*, CIAGRI-ESALQ-USP, São Paulo, Brazil, 1993.

[18] L. A. Currie, "Limits for qualitative detection and quantitative determination: application to radiochemistry," *Analytical Chemistry*, vol. 40, no. 3, pp. 586–593, 1968.

[19] W. Ladisich, R. Rieder, P. Wobrauschek, and H. Aiginger, "Total reflection X-ray fluorescence analysis with monoenergetic excitation and full spectrum excitation using rotating anode X-ray tubes," *Nuclear Instruments and Methods in Physics Research*, vol. 330, no. 3, pp. 501–506, 1993.

[20] M. J. Salvador, D. A. Dias, S. Moreira, and O. L. A. D. Zucchi, "Analysis of medicinal plants and crude extracts by synchrotron radiation total reflection X-ray fluorescence," *Journal of Trace and Microprobe Techniques*, vol. 21, no. 2, pp. 377–388, 2003.

[21] V. A. Janasi, S. Andrade, and H. H. G. J. Ulbrich, "A correção do drift instrumental em ICP-AES com espectrômetro seqüencial e a análise de elementos maiores, menores e traços em rochas," *Boletim do Instituto de Geociências da USP: Série Científica*, vol. 26, pp. 45–58, 1995.

[22] X. Hou and X. Yan, "Study on the concentration and seasonal variation of inorganic elements in 35 species of marine algae," *Science of the Total Environment*, vol. 222, no. 3, pp. 141–156, 1998.

[23] M. A. Rizvi and M. Shameel, "Pharmaceutical biology of seaweeds from the Karachi coast of Pakistan," *Pharmaceutical Biology*, vol. 43, no. 2, pp. 97–107, 2005.

17

Interaction of Avelox with Bovine Serum Albumin and Effect of the Coexistent Drugs on the Reaction

Baosheng Liu, Chao Yang, Xiaona Yan, Jing Wang, and Yunkai Lv

Key Laboratory of Medical Chemistry and Molecular Diagnosis, College of Chemistry & Environmental Science, Hebei University, Ministry of Education, Baoding 071002, China

Correspondence should be addressed to Baosheng Liu, lbs@hbu.edu.cn

Academic Editor: Stig Pedersen-Bjergaard

The interaction between Avelox and bovine serum albumin (BSA) was investigated at different temperatures by fluorescence spectroscopy. Results showed that Avelox could quench the intrinsic fluorescence of BSA strongly, and the quenching mechanism was a static quenching process with Förester spectroscopy energy transfer. The electrostatic force played an important role on the conjugation reaction between BSA and Avelox. The order of magnitude of binding constants (K_a) was 10^4, and the number of binding site (n) in the binary system was approximately equal to 1. The binding distance (r) was less than 3 nm and the primary binding site for Avelox was located in subdomain IIA of BSA. Synchronous fluorescence spectra clearly revealed that the microenvironment of amino acid residues and the conformation of BSA were changed during the binding reaction. In addition, the effect of some antibiotics on the binding constant of Avelox with BSA was also studied.

1. Introduction

Most drugs are able to bind to plasma protein when they entrance in blood plasma system of organism, and serum albumin is the most abundant protein in blood plasma and serves as a depot protein and transport protein for numerous endogenous and exogenous compounds [1]. Generally speaking, drugs could bind with serum albumin mostly through the formation of noncovalent complexes reversibly. The drug-protein complex can be regarded as a form of drug in the biology temporary storage, it can effectively avoid drug elimination from metabolism so quickly that it can maintain the total concentration and effective concentrations of blood medicine in plasma. In addition, binding of drugs to plasma proteins controls their free, active concentrations and provides a reservoir for a longer action, the binding of drugs is responsible for the protective role of albumin. Therefore, interaction of a drug with, and competition for, the binding sites on plasma proteins might strongly affect its distribution, elimination, as well as its pharmacodynamics and toxic properties [2]. Competition between two drugs for their binding to plasma protein can strongly affect the drug disposition of both drugs, with possible serious physiological consequences. Binding parameters are indeed fundamental factors in determining the overall pharmacological activity of a drug, and in this context the determination of the binding parameters of drugs to albumin has become essential to understand their pharmacokinetic, pharmacodynamic, and toxicological profile.

Avelox is a new generation of fluoroquinolone antibacterial agents, which is used to treat a wide variety of bacterial infections including pneumonia, sinusitis, and worsening of chronic bronchitis (the structure is shown in Figure 1). This medicine gradually played a significant role in clinical cure. At present, the molecular interactions between BSA and many drugs have been investigated successfully in biomedical domain [3]. However, the interaction between Avelox and BSA has not been investigated, especially the effect of the common antibiotics on the binding of drugs to BSA. In this paper, we provided investigations on the interaction of Avelox with BSA by fluorescence spectroscopy under physiological pH 7.40 as well as the effect of the coexistent drugs. This study is expected to provide important insight into the essence, potential toxicity between drugs and protein in real terms, and can also provide a useful clinical reference for future combination therapy.

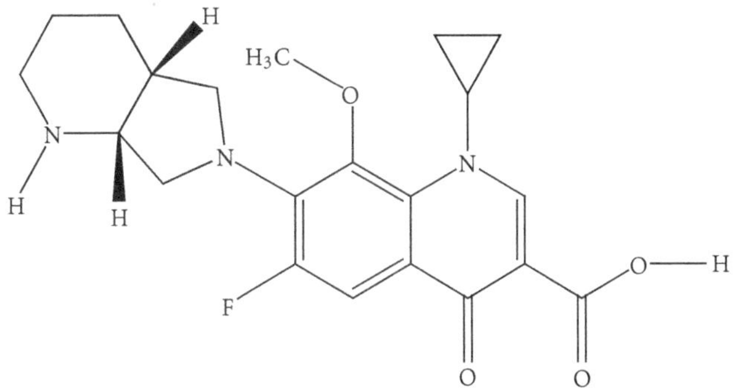

FIGURE 1: Chemical structure of Avelox.

2. Experimental

2.1. Apparatus. All fluorescence spectra were recorded with a Shimadzu RF-540 spectrofluorophotometer and a Hitachi F-4500 spectrofluorophotometer. Absorption was measured with an UV-vis recording spectrophotometer (UV-265 Shimadzu, Japan). All pH measurements were made with a pHS-3C precision acidity meter (Leici, Shanghai). All temperatures were controlled by a CS501 superheated water bath (Nantong Science Instrument Factory).

2.2. Materials. Avelox (no less than 99.9% pure) was obtained from monitor of Chinese veterinary medicine. Bovine serum albumin (BSA, no less than 99% pure) was purchased from Sigma Company. Warfarin (WF), Chlorpheniramine (CP), and digitoxin (DG) were all obtained from Chinese Institute of Drug and Biological Products. Stock solutions of BSA (1.0×10^{-4} M) and Avelox (2.0×10^{-3} M) were prepared. And all the stock solutions were further diluted as working solutions prior to use. All the common antibiotics were diluted to 1.0×10^{-3} M. *Tris*-HCl buffer solution containing NaCl (0.15 M) was used to keep the pH of the solution at 7.40. NaCl solution was used to maintain the ionic strength of the solution. All other reagents were of analytical grade, and all aqueous solutions were prepared with newly double-distilled water and stored at 277 K.

The fluorescence intensities were corrected for the absorption of excitation light and reabsorption of emitted light to decrease the inner filter using the following relationship [4]:

$$F_{cor} = F_{obs} \times e^{(A_{ex}+A_{em})/2}, \quad (1)$$

where F_{cor} and F_{obs} are the corrected and observed fluorescence intensities, respectively. A_{ex} and A_{em} are the absorbance values of Avelox at excitation and emission wavelengths, respectively. The fluorescence intensity used in this paper was corrected.

2.3. Procedures

2.3.1. Fluorescence Experiments. In a typical fluorescence measurement, 1.0 mL of pH = 7.40 *Tris*-HCl, 1.0 mL of 1.0×10^{-5} M BSA solution, and different concentrations of Avelox were added into a 10 mL colorimetric tube, successively. The samples were diluted to scaled volume with water, mixed thoroughly by shaking, and kept static for 15 min at different temperatures (293, 303, and 313 K). The excitation wavelength for BSA was 280 nm (or 295 nm), with the excitation and emission slit widths set at 5 nm. The solution was subsequently scanned on the fluorophotometer and determined the fluorescent intensity at 343 nm. The fluorescence intensity of BSA without Avelox was used as F_0. Meanwhile, the fluorescence intensity of compound without single Avelox ($F_{BSA-Avelox} - F_{Avelox}$) was used as F because Avelox has slight fluorescence emission at 343 nm. The fluorescence intensities in this paper were corrected by (1). Otherwise, we recorded the fluorescence spectra when the $\Delta\lambda$ value between the excitation and emission wavelengths was stabilized at 15 and 60 nm, respectively.

2.3.2. Probe Displace Experiments. At 293 K, different concentrations of probe drug WF (or CP, DG) were respectively added to the mixture of Avelox-BSA systems. The molar ratio of Avelox and BSA was kept at 1, both of their concentrations are 1.0×10^{-6} M in BSA-Avelox systems. The mixtures were diluted to scaled volume with water, mixed thoroughly by shaking, and kept static for 30 min. The excitation wavelength for BSA was 280 nm, with the excitation and emission slit widths set at 5 nm. The fluorescence intensity of BSA only with Avelox was used as F_0 to study the quenching effect of probe drugs on Avelox-BSA system.

3. Results and Discussion

3.1. Interaction between Avelox and BSA. Based on Section 2.3.1, we have drawn the fluorescence spectra of BSA as well as different Avelox added into BSA, respectively. When the excitation wavelengths were at 280 nm and 295 nm, the emission peaks for BSA were both located at 343 nm. In this part, we also found that Avelox had fluorescence at 462 nm obviously, but it had weak fluorescence at 343 nm. According to this phenomenon, we scanned the solution on the fluorophotometer with the range 300–450 nm and recorded the fluorescence spectra of BSA-Avelox without Avelox which were shown in Figure 2. As can be seen from Figure 2, the fluorescence intensity of BSA decreased regularly with the addition of Avelox, and the maximum emission wavelength had 9 nm red shifts from 343 nm to 352 nm. This result implied that Avelox could quench the intrinsic fluorescence of BSA strongly, and the interaction between Avelox and BSA resulted from the formation of a ground-state complex between this drug and protein. In addition, the UV absorption intensity of Avelox which located at 336 nm and 308 nm decreased with increased concentrations of BSA (as seen in Figure 3), it could also reveal that the formation of new complex, in addition suggested that the quenching was a static process.

3.2. Fluorescence Quenching Mechanism of BSA-Avelox System. Fluorescence quenching can occur by different mechanisms, it may be dynamic quenching, resulting from

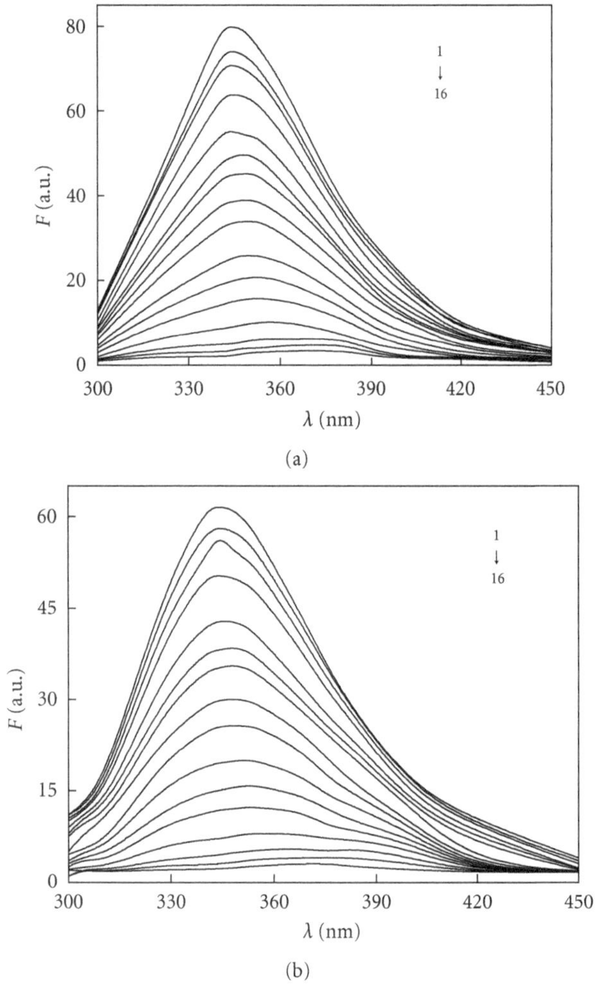

FIGURE 2: Fluorescence spectra of BSA-Avelox system. $C_{BSA} = 1.0 \times 10^{-6}$ M, 1~16; $C_{Avelox} = (0.0, 1.0, 2.0, 4.0, 6.0, 8.0, 10.0, 12.5, 15.0, 20.0, 25.0, 30.0, 40.0, 50.0, 60.0, 70.0) \times 10^{-6}$ M; (a) $\lambda_{ex} = 280$ nm; (b) $\lambda_{ex} = 295$ nm; $T = 293$ K.

the collisional encounter between the drug and protein, or static quenching, resulting from the formation of a ground-state complex between the drug and protein. Higher temperature would result in faster diffusion and typically the dissociation of weekly bound complexes, leading to larger amount of dynamic quenching and smaller amounts of static quenching, respectively [5]. In order to confirm the quenching mechanism, the fluorescence quenching data are analyzed by the Stern-Volmer equation [6]:

$$\frac{F_0}{F} = 1 + K_q \tau_0 [L] = 1 + K_{sv}[L], \tag{2}$$

where F_0 and F are the fluorescence intensities in the absence and presence of ligand, respectively. τ_0 is the average lifetime of fluorescence without ligand, which is about 10^{-8} s. K_{sv} is the Stern-quenching constant. K_q is the quenching rate constant of biomolecule, and $[L]$ is the concentration of the ligand. Based on the linear fit plot of F_0/F versus $[L]$, the K_q values can be obtained. The calculated results were shown in Table 1. It appeared that the values of K_{sv} decreased with

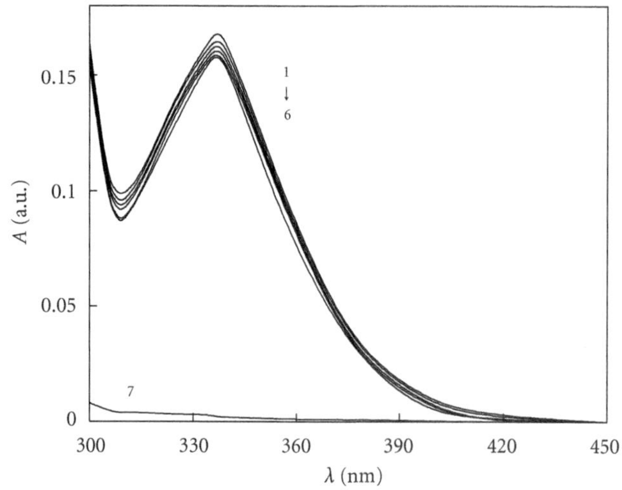

FIGURE 3: Absorption spectra of Avelox-BSA system From (1) to (6), $C_{Avelox} = 1.0 \times 10^{-5}$ M, $C_{BSA} = (0, 0.05, 0.1, 0.5, 0.8, 1.0) \times 10^{-6}$ M, (7): $C_{BSA} = 1.0 \times 10^{-6}$ M.

the increase in temperature for all systems, which indicated that the probable quenching mechanism of the interaction between BSA and Avelox was initiated by complex formation rather than by dynamic collision [7]. In addition, all the values of K_q were much greater than the maximum scatter collision quenching constant of various quenchers ($2 \times 10^{10}\ M^{-1}\ s^{-1}$) [8]. This also suggested that the quenching was a static process [9].

For static quenching process, the relationship between the fluorescence intensity and the concentration of quencher can be usually described by deriving (3) [10] to obtain the binding constant (K_a) and the number of binging sites (n) in most of paper:

$$\log\left(\frac{F_0 - F}{F}\right) = n \log K_a + n \log\left\{[D_t] - n\frac{F_0 - F}{F_0}[B_t]\right\}, \tag{3}$$

where $[D_t]$ and $[B_t]$ are the total concentrations of Avelox and protein, respectively. On the assumption that n in the bracket is equal to 1, the curve of $\log(F_0 - F)/F$ versus $\log\{[D_t] - [B_t](F_0 - F)/F_0\}$ is drawn and fitted linearly, then the value of n can be obtained from the slope of the plot. If the n value obtained is not equal to 1, then it is substituted into the bracket and the curve of $\log(F_0 - F)/F$ versus $\log\{[D_t] - n[B_t](F_0 - F)/F_0\}$ is drawn again. The above process is repeated again and again till n obtained is only a single value or a circulating value. Based on the n obtained, the binding constant K_a can be also obtained. In the work, a calculation program was developed. The calculation process can be finished with calculator based on the simple program, and the calculating results can be obtained by inputting F, $[D_t]$ and $[B_t]$. The calculated results were shown in Table 1.

As seen in Table 1, the values of n were approximately equal to 1 at different temperatures, which indicated the existence of just one main binding site in BSA for Avelox. The order of magnitude of binding constants (K_a) was 10^4 indicating the existence of strong interaction between BSA and Avelox. Meanwhile, the values of K_a decreased

TABLE 1: Quenching reactive parameters of BSA and Avelox at different temperatures.

λ_{ex}	T/K	K_q/(M^{-1} s^{-1})	r_1	K_a/(M^{-1})	n	r_2
	293	5.81×10^{12}	0.9943	5.89×10^4	1.26	0.9971
280	303	5.49×10^{12}	0.9938	5.75×10^4	1.23	0.9903
	313	4.86×10^{12}	0.9951	5.62×10^4	1.24	0.9913
	293	6.35×10^{12}	0.9956	6.02×10^4	1.18	0.9992
295	303	6.12×10^{12}	0.9941	5.89×10^4	1.18	0.9937
	313	5.54×10^{12}	0.9960	5.75×10^4	1.43	0.9954

r_1 is the linear relative coefficient of $F_0/F \sim [L]$; r_2 is the linear relative coefficient of $\log(F_0 - F)/F \sim \log\{[D_t] - n[B_t](F_0 - F)/F_0\}$.
K_q is the quenching rate constant; K_a is the binding constant; n is the number of binding site.

with the increasing temperature, further suggested that the quenching was a static process [11], hence it led to the reduced of the stability of binary systems. On the other hand, comparing the data of 280 nm with 295 nm, values of K_a and n had less difference at different temperatures, it explained that Avelox would mainly quench the fluorescence of tryptophanyl residues from BSA.

3.3. Type of Interaction Force of the Binary Systems. Generally, the interaction forces between the small drug molecule and biological macromolecule include hydrogen bond, Van der Waals force, electrostatic interactions, and hydrophobic force [12]. Ross and Subramanian [13] have characterized the sign and magnitude of the thermodynamic parameter, enthalpy change (ΔH), free energy (ΔG), and entropy change (ΔS) of reaction, associated with various individual kinds of interaction. The reaction ΔH can be regarded as constant if the temperature changes little. Negative ΔH and positive ΔS indicate that electrostatic interaction plays a major role in the binding reaction. Positive ΔH and ΔS are generally considered as the evidence for typical hydrophobic interactions. In addition, Van der Waals force and hydrogen bonding formation in low-dielectric media are characterized by negative ΔH and ΔS [14]. The thermodynamic parameters can be calculated on the basis of the following equations:

$$R \ln K = \Delta S - \frac{\Delta H}{T}, \quad (4)$$

$$\Delta G = \Delta H - T\Delta S. \quad (5)$$

According to the binding constants K_a of Avelox to BSA at different temperatures above (Table 1), the thermodynamic parameters were obtained conveniently. Therefore, the values of ΔH, ΔS, and ΔG were -1.79 kJ mol^{-1}, 85.2 J mol^{-1} K^{-1}, and -26.8 kJ mol^{-1} (T = 293 K), respectively. The negative value of ΔG clarified an automatic reaction between Avelox and BSA. The negative value of ΔH and positive value of ΔS showed that Avelox mainly bound to BSA by the electrostatic attraction.

3.4. Identification of the Binding Sites of Avelox on the BSA. Participation of tyrosine (Tyr) and tryptophan (Trp) groups in drug-serum albumin complexes is assessed using different excitation wavelengths. At 280 nm wavelength, the Trp and Tyr residues in serum albumin are excited, whereas the 295 nm wavelength excites only Trp residues. In the subdomains of BSA, IIA subunit (Trp and Tyr) and IIIA (Tyr) are also considered as the main binding sites for small-drug molecule [15]. Based on the Stern-Volmer equation, comparing the fluorescence quenching of protein excited at 280 nm and 295 nm allows to estimate the participation of Trp and Tyr groups in the complex [16]. As seen in Figure 4, the quenching curves of BSA excited at 280 nm and 295 nm in the presence of Avelox overlap below the molar ratio Avelox : BSA = 80 : 1. This means that if there are fewer than 80 Avelox molecules for 1 BSA molecules, only Trp residues take part in the interaction of Avelox with BSA, and protein fluorescence was quenched by almost 80% at the molar ratio 80 : 1. The quenching of BSA fluorescence exited at 280 nm above this molar ratio was slightly higher than that excited at 295 nm which means that Tyr residues began to take part in the interaction. According to this conclusion, it could be inferred that Avelox molecules only take interaction with Trp residues of BSA at low concentration, whereas both Trp and Tyr residues at high concentration, this conclusion can also

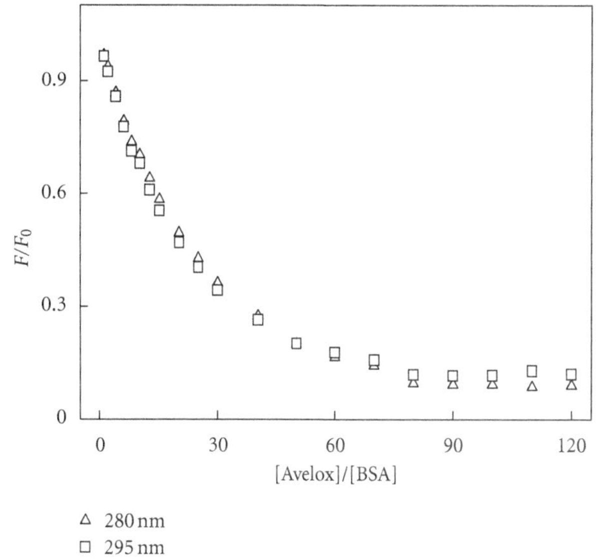

FIGURE 4: Quenching curves of BSA-Avelox system at λ_{ex} = 280 nm and 295 nm $C_{BSA} = 1.0 \times 10^{-6}$ M, $C_{Avelox} = 1.0 \times 10^{-6} \sim 1.2 \times 10^{-4}$ M; T = 293 K.

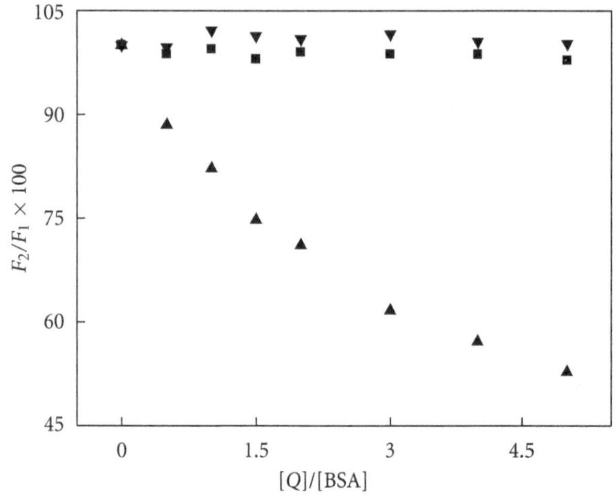

FIGURE 5: Effect of site maker probe on the fluorescence intensities of BSA-Avelox system $C_{Avelox} = C_{BSA} = 1.0 \times 10^{-6}$ M; $C_Q = (0.5, 1.0, 1.5, 2.0, 3.0, 4.0, 5.0) \times 10^{-6}$ M. [Q]: ▲ WF; ■ CP; ▼ DG; λ_{ex} = 280 nm; T = 293 K.

expound why the value of n is slightly more than 1, which has been shown in Table 1.

BSA has two Trp moieties (Trp-134 and Trp-212) located in subdomains IB and IIA, respectively, and only Trp-212 is located within a hydrophobic binding pocket of the protein which usually bind many small ligands, especially heterocyclic ligands of average size and small aromatic carboxylic acids [17]. In this way, it could confirm that the primary binding site for Avelox was located in subdomain IIA of BSA.

Site marker fluorescent probes are useful for rapid identification of drug binding sites [18, 19]. In order to estimate the drug binding sites on BSA, Sudlow et al. have proposed the percentage of probe replacement method using the following equation [20]:

$$\text{probe displacement}(\%) = \frac{F_2}{F_1} \times 100. \quad (6)$$

The concentrations of BSA and Avelox were kept constant (1×10^{-6} M), while that different concentrations of various probe drugs each were added into it. In the study of the substitution reaction, F_1 and F_2 are the fluorescence intensities in the absence and presence probe drug. In this paper, we have considered WF, CP, and DG as probe durgs which bind specifically to sites I, II, and III, respectively. The processed results were shown in Figure 5 by (6).

As seen in Figure 5, relative fluorescence intensities decreased obviously with increasing concentration of WF, whereas it almost kept constant with the adding of CP and DG. It indicated that Avelox was pronouncedly displaced by WF rather than CP or DG, hence it could be inferred that the binding site for Avelox was site I which located in subdomain IIA of BSA.

At room temperature (293 K), control the concentrations of BSA and various probe drugs each at 1×10^{-6} M. According to the method described in Section 2.3.1, successively adding different concentrations of Avelox. The experimental data was calculated by (3). The values of binding constants K_a were 4.68×10^4, 5.50×10^4, and $5.37 \times 10^4\,M^{-1}$ in presence of WF, CP, and DG, respectively. Compared with the binding constant of Avelox-BSA without any probe drug ($5.89 \times 10^4\,M^{-1}$), it can be seen that the binding constant between Avelox and BSA decreased most obviously in the presence of WF, this result indicated that binding of Avelox with BSA was affected by adding WF, which compete the same binding site of BSA. Hence, it also can be proposed that the binding of Avelox primarily occurred in sub-domain IIA (site I).

TABLE 2: Hill's coefficients n_H of BSA-Avelox at different temperatures.

T/K	λ_{ex}/280 nm		λ_{ex}/295 nm	
	n_H	r_3	n_H	r_3
293	0.89	0.9938	0.86	0.9932
303	0.74	0.9959	0.72	0.9953
313	0.70	0.9913	0.69	0.9906

r_3 is the linear relative coefficient of $\lg[Y/(1-Y)] \sim \lg[L]$.

3.5. Hill's Coefficient of BSA-Avelox System. In biochemistry, the binding of a ligand molecule at one site of a macromolecule often influences the affinity for other ligand molecules at additional sites. This is known as cooperative binding. It is classified into positive cooperativity, negative cooperativity, and noncooperativity according to the promotion or inhibition to the affinity for other ligand molecules. Hill's coefficient provides a way to quantify this effect and is calculated graphically on the basis of the following equation [21]:

$$\lg\frac{Y}{1-Y} = \lg K + n_H \lg[L], \quad (7)$$

where Y is the fractional binding saturation; K is the binding constant and n_H is the Hill's coefficient. Hill's coefficient is greater than one, which exhibits positive cooperativity. Conversely, Hill's coefficient is less than one, which exhibits negative cooperativity. A coefficient of 1 indicates noncooperative reaction.

For fluorescence measurement:

$$\frac{Y}{(1-Y)} = \frac{Q}{(Q_m - Q)}, \quad (8)$$

where $Q = (F_0 - F)/F_0$; $1/Q_m$ = intercept of the plot $1/Q$ versus $1/[L]$. Hill's coefficients were presented in Table 2. The values of n_H were slightly less than 1 in the systems, both at excitation wavelengths 280 nm and 295 nm at different temperatures, which indicated negative cooperativeness in the interaction of Avelox with BSA, but they were weak. In addition, the values of n_H were inversely correlated with increasing temperature illustrating the existence of negative cooperativity between Avelox and BSA, that was, the ability of drug bounding to BSA has decreased with the previous ligand (Avelox) bounding to BSA gradually. Furthermore, this negative cooperativity became more powerful with

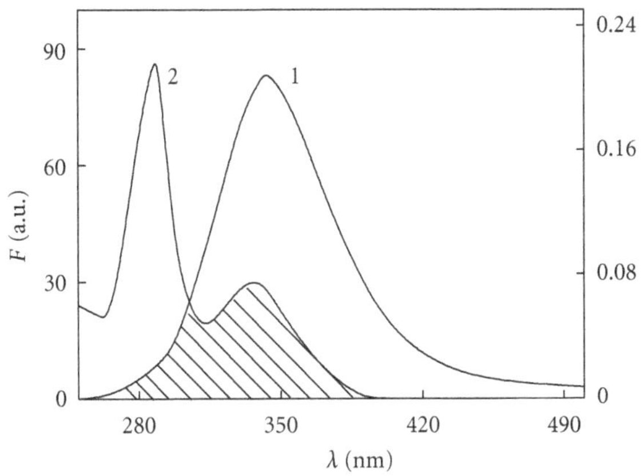

FIGURE 6: Fluorescence emission spectra for BSA (1) and UV absorbance spectra for Avelox (2). $C_{Avelox} = C_{BSA} = 6.0 \times 10^{-6}$ M; $T = 293$ K.

increasing temperature, reducing the amount of Avelox that could be bound to BSA. It was also one of the reasons which led to the reduced K_a with increasing temperature.

3.6. Binding Distances between Avelox and BSA. Fluorescence energy transfer is categorized as radiation energy transfer and nonradiation energy transfer. The fluorescence spectrum will deform if radiation energy transfer occurred. The fluorescence spectrum in Figure 2 has no deformation, thus the energy transfer of BSA and Avelox should be nonradiation energy transfer.

According to Förster's nonradiative energy transfer theory, energy efficiency E, critical energy-transfer distance $R_0(E = 50\%)$, the energy donor, the energy acceptor distance r, the overlap integral, between the fluorescence emission spectrum of the donor, and the absorption spectrum of the acceptor J can be calculated by the formulas [22]:

$$E = 1 - \frac{F}{F_0} = \frac{R_0^6}{\left(R_0^6 + r^6\right)}, \quad (9)$$

$$R_0^6 = 8.78 \times 10^{-25} K^2 \Phi N^{-4} J, \quad (10)$$

$$J = \frac{\sum F(\lambda)\varepsilon(\lambda)\lambda^4 \Delta\lambda}{\sum F(\lambda)\Delta\lambda}, \quad (11)$$

where K^2 is the orientation factor, Φ is the fluorescence quantum yield of the donor, N is a refractive index of the medium, $F(\lambda)$ is the fluorescence intensity of the fluorescence donor at wavelength λ, and $\varepsilon(\lambda)$ is the molar absorption coefficient of the acceptor at this wavelength. The overlap of UV-vis absorption spectra of Avelox and the fluorescence emission spectra of BSA ($\lambda_{ex} = 280$ nm) were shown in Figure 6. Under these experimental conditions, it has been reported that $K^2 = 2/3$, $N = 1.336$, and $\Phi = 0.118$ [23]. Thus J, E, R_0, and r were calculated and shown in Table 3. The donor-to-acceptor distance $r < 7$ nm indicated that the energy transfer from BSA to Avelox occured with high possibility [24]. Furthermore, the value of r was greater than R_0 in this study which suggested that Avelox could strongly quench the intrinsic fluorescence of BSA by a static quenching mechanism [25]. Moreover, the distance r increased, and the energy efficiency E decreased with increasing temperature (Table 3), which resulted in the reduced stability of the binary systems and the values of K_a.

TABLE 3: Parameters of E, J, r, and R_0 between Avelox and BSA at different temperatures.

T/K	E (%)	J/($M^{-1}cm^3$)	R_0/nm	r/nm
293	28.04	1.07×10^{-14}	2.48	2.90
303	26.66	1.04×10^{-14}	2.47	2.92
313	25.60	1.02×10^{-14}	2.46	2.95

R_0 is the critical distance when E is 50%; r is the distance between acceptor and donor; J is the overlap integral between the fluorescence emission spectrum of donor and the absorption spectrum of the acceptor.

3.7. Conformation Investigation of BSA. Synchronous fluorescence spectra are used to investigate the protein conformational change, as it has been shown to give narrow and simple spectra. For the synchronous fluorescence spectra of protein, when the $\Delta\lambda$ value between the excitation and emission wavelengths is stabilized at either 15 or 60 nm, the synchronous fluorescence gives characteristic information for Tyr residues or Trp residues [26]. Because of the red shifts of maximum emission wavelengths of both Tyr and Trp with the less hydrophobic environment, blue shifts of maximun emission wavelengths with the more hydrophobic environment ocurr. These red or blue shifts indicated that the conformation of BSA has been changed [27]. In order to further study the effect on the conformation of BSA, the synchronous fluorescence spectra were measured when the $\Delta\lambda = 15$ nm and $\Delta\lambda = 60$ nm as shown in Figure 7.

It can be seen from Figure 7 that the fluorescence intensities of Tyr and Trp residues decreased regularly with increasing concentration of Avelox. At the same time, the maximum emission wavelength had 4 nm blue shifts at the investigated concentrations range when $\Delta\lambda = 60$ nm, while the maximum emission wavelength had 7 nm red shifts when $\Delta\lambda = 15$ nm. This revealed that the aminoacids microenvironment changed due to the reaction of Avelox and BSA, making the hydrophobicity of Trp residues strengthened, whereas, the hydrophobic environment of Tyr residues more polar. High concentration of drugs makes protein molecules extend, thus reducing energy transfer between aminoacids, reducing the fluorescence intensity [28].

3.8. Effect of the Coexistent Drugs on the Binding of Avelox and BSA. According to this method about the percentage of probe replacement method described in Section 3.4, we have taken assays to study the effects of some common antibiotics on the binding of BSA-Avelox system. The concentrations of BSA and Avelox were kept constant (1×10^{-6} M), while that different concentrations of various antibiotics each were added into it. In the study of the substitution reaction, F_1 and F_2 are the fluorescence intensities in the absence and

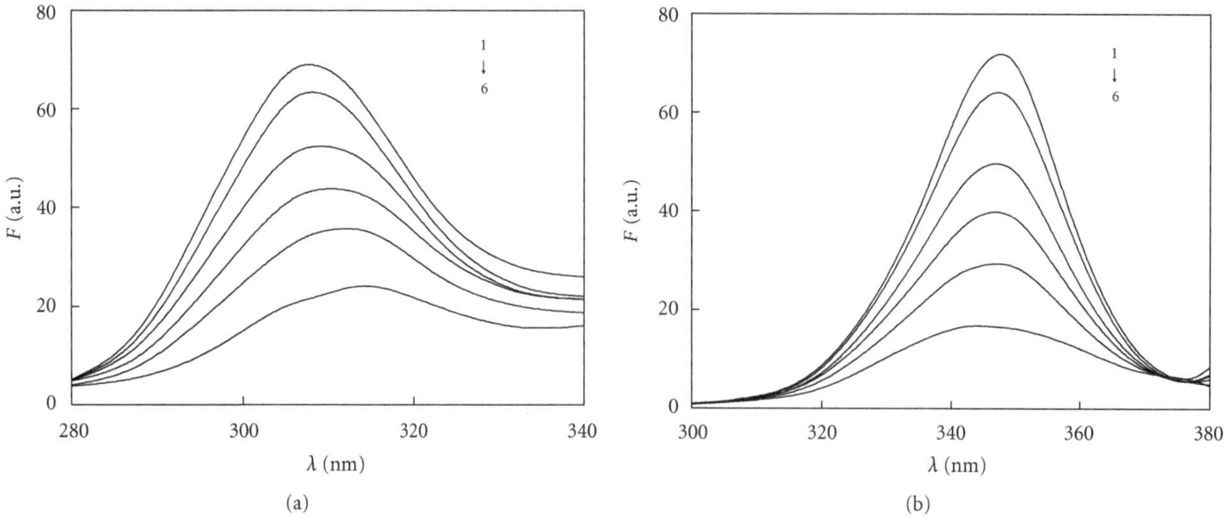

FIGURE 7: Synchronous fluorescence spectra of BSA-Avelox system $C_{BSA} = 1.0 \times 10^{-6}$ M, $1 \sim 6$ $C_{Avelox} = (0.0, 2.0, 6.0, 10.0, 15.0, 25.0) \times 10^{-6}$ M; (a) $\Delta\lambda = 15$ nm, (b) $\Delta\lambda = 60$ nm; $T = 293$ K.

presence of antibiotice. It supposed that there was no effect on the binding of BSA-Avelox system when the value of $(F_2 - F_1)/F_1$ was less than ±5% by (6), that was, this antibiotice could coexist with Avelox. The results showed that it allowed 500 times of gentamicin, thiamphenicol, 300 times of sulfamethoxazole, erythromycin and streptomycin, 100 times of ampicillin, neomycin and kanamycin, 20 times of penicillin G sodium salt and acetyl spiramycin, and 10 times of cefotaxime sodium, cefoperazone sodium, ceftriaxone, and chloramphenicol to coexist with Avelox, but a bit of ciprofloxacin, and norfloxacin could disturb the stability of BSA-Avelox system, making the fluorescence intensity change a lot. It might result that Avelox and ciprofloxacin, norfloxacin all belong to a class of drugs called quinolone antibiotics, having much similar structures, hence, there was certain competition among them for the protein. The competition would lead to binding constant K_a change spontaneously. It could be inferred that there were possible effects on the stayed-time from the blood, as well as plasma concentration and effective concentration, having effects on the efficacy of drugs when the antibiotics such as ciprofloxacin and norfloxacin coexist with Avelox on the binding reaction of BSA. However, aminoglycosides, macrolides, sulfonamides, and penicillin antibiotics have no effects on the binary system, that is, those antibiotics will not affect the transportation of Avelox though BSA when they coexist with Avelox and also have no effects on the efficacy of Avelox.

4. Conclusions

In this paper, the interaction of Avelox with BSA was studied at different temperatures by fluorescence spectroscopic methods. The experimental results indicated that Avelox molecular had embedded into the hydrophobic cavity through electrostatic force and was mainly bound to Trp-212 in BSA. The donor-to-acceptor distance r was less than 7 nm, indicating nonradiation energy transfer. The negative cooperativity existed in BSA-Avelox system for subsequent ligand. From the synchronous fluorescence spectra, it could be shown that the quaternary conformational change of BSA was induced by the interaction of Avelox with the Trp microregion of BSA molecules. The method-like percentage of probe replacement clearly revealed the existence of effects of common antibiotics on the binding of BSA-Avelox system. The study will extend the use of fluorescence spectroscopy; meanwhile, it provides important insights into the future clinical medicine.

Acknowledgments

The authors gratefully acknowledge the financial support of National Science Foundation of China (Grant no. 20675024) and Hebei Provincial Key Basic Research Program (Grant no. 10967126D).

References

[1] K. H. Ulrich, "Molecular aspects of ligand binding to serum albumin," *Pharmacological Reviews*, vol. 33, no. 1, pp. 17–53, 1981.

[2] C. Bertucci and E. Domenici, "Reversible and covalent binding of drugs to human serum albumin: methodological approaches and physiological relevance," *Current Medicinal Chemistry*, vol. 9, no. 15, pp. 1463–1481, 2002.

[3] S. Y. Bi, Y. T. Sun, C. Qiao, H. Y. Zhang, and C. Liu, "Binding of several anti-tumor drugs to bovine serum albumin: fluorescence study," *Journal of Luminescence*, vol. 129, no. 5, pp. 541–547, 2009.

[4] G. Lupidi, A. Scire, E. Camaioni et al., "Thymoquinone, a potential therapeutic agent of Nigella sativa, binds to site I of human serum albumin," *Phytomedicine*, vol. 17, no. 10, pp. 714–720, 2010.

[5] F. Ding, G. Q. Zhao, S. M. Chen, F. Liu, Y. Sun, and L. Zhang, "Chloramphenicol binding to human serum albumin:

determination of binding constants and binding sites by steady-state fluorescence," *Journal of Molecular Structure*, vol. 929, no. 1–3, pp. 159–166, 2009.
[6] J. Q. Liu, J. N. Tian, J. Y. Zhang, Z. D. Hu, and X. G. Chen, "Interaction of magnolol with bovine serum albumin: a fluorescence-quenching study," *Analytical and Bioanalytical Chemistry*, vol. 376, no. 6, pp. 864–867, 2003.
[7] G. J. Zhang, B. Keita, J. C. Brochon et al., "Molecular interaction and energy transfer between human serum albumin and poly-oxometalates," *Journal of Physical Chemistry B*, vol. 111, no. 7, pp. 1809–1814, 2007.
[8] K. Zhu and S. Y. Tong, "A study on the reaction between fluorescein and protein," *Chemical Journal of Chinese Universities*, vol. 17, no. 4, pp. 539–542, 1996.
[9] F. Ding, N. Li, B. Y. Han, F. Liu, L. Zhang, and Y. Sun, "The binding of C.I. Acid Red 2 to human serum albumin: determination of binding mechanism and binding site using fluorescence spectroscopy," *Dyes and Pigments*, vol. 83, no. 2, pp. 249–257, 2009.
[10] S. Y. Bi, C. Y. Qiao, D. Q. Song et al., "Study of interactions of flavonoids with DNA using acridine orange as a fluorescence probe," *Sensors and Actuators, B*, vol. 119, no. 1, pp. 199–208, 2006.
[11] N. Shahabadi and M. Maghsudi, "Binding studies of a new copper (II) complex containing mixed aliphatic and aromatic dinitrogen ligands with bovine serum albumin using different instrumental methods," *Journal of Molecular Structure*, vol. 929, no. 1–3, pp. 193–199, 2009.
[12] N. Wang, L. Ye, F. F. Yan, and R. Xu, "Spectroscopic studies on the interaction of azelnidipine with bovine serum albumin," *International Journal of Pharmaceutics*, vol. 351, no. 1-2, pp. 55–60, 2008.
[13] P. D. Ross and S. Subramanian, "Thermodynamics of protein association reactions: forces contributing to stability," *Biochemistry*, vol. 20, no. 11, pp. 3096–3102, 1981.
[14] M. Gharagozlou and D. M. Boghaei, "Interaction of water-soluble amino acid Schiff base complexes with bovine serum albumin: fluorescence and circular dichroism studies," *Spectrochimica Acta, Part A*, vol. 71, no. 4, pp. 1617–1622, 2008.
[15] A. Sułkowska, M. M. Jurczyk, B. Bojko et al., "Competitive binding of phenylbutazone and colchicine to serum albumin in multidrug therapy: a spectroscopic study," *Journal of Molecular Structure*, vol. 881, no. 1–3, pp. 97–106, 2008.
[16] M. M. Jurczyk, A. Sułkowska, B. Bojko, J. Równicka, and W. W. Sułkowski, "Fluorescence analysis of competition of phenylbutazone and methotrexate in binding to serum albumin in combination treatment in rheumatology," *Journal of Molecular Structure*, vol. 924-926, pp. 378–384, 2009.
[17] Q. L. Guo, R. Li, F. L. Jiang, J. C. Tu, L. W. Li, and Y. Liu, "Characterization of the interactions between itraconazole and human and bovine serum albumins by a spectroscopic method," *Acta Physico-Chimica Sinica*, vol. 25, no. 10, pp. 2147–2154, 2009.
[18] G. Sudlow, D. J. Birkett, and D. N. Wade, "The characterization of two specific drug binding sites on human serum albumin," *Molecular Pharmacology*, vol. 11, no. 6, pp. 824–832, 1975.
[19] G. Sudlow, D. J. Birkett, and D. N. Wade, "Further characterization of specific drug binding sites on human serum albumin," *Molecular Pharmacology*, vol. 12, no. 6, pp. 1052–1061, 1976.
[20] K. Yamasaki, T. Maruyama, U. Kragh-Hansen, and M. Otagiri, "Characterization of site I on human serum albumin: concept about the structure of a drug binding site," *Biochimica et Biophysica Acta*, vol. 1295, no. 2, pp. 147–157, 1996.
[21] B. Bojko, A. Sułkowska, M. Maciążek-Jurczyk, J. Równicka, and W. W. Sułkowski, "The influence of dietary habits and pathological conditions on the binding of theophylline to serum albumin," *Journal of Pharmaceutical and Biomedical Analysis*, vol. 52, no. 3, pp. 384–390, 2010.
[22] Y. Y. Yue, X. G. Chen, J. Qin, and X. J. Yao, "Characterization of the mangiferin-human serum albumin complex by spectroscopic and molecular modeling approaches," *Journal of Pharmaceutical and Biomedical Analysis*, vol. 49, no. 3, pp. 753–759, 2009.
[23] Y. Y. Yue, X. G. Chen, J. Qin, and X. J. Yao, "A study of the binding of C.I. Direct Yellow 9 to human serum albumin using optical spectroscopy and molecular modeling," *Dyes and Pigments*, vol. 79, no. 2, pp. 176–182, 2008.
[24] Y. J. Hu, Y. Liu, L. X. Zhang, R. M. Zhao, and S. S. Qu, "Studies of interaction between colchicine and bovine serum albumin by fluorescence quenching method," *Journal of Molecular Structure*, vol. 750, no. 1–3, pp. 174–178, 2005.
[25] J. B. Xiao, X. Q. Chen, X. Y. Jiang, and M. Hilczer, "Probing the interaction of trans-resveratrol with bovine serum albumin: a fluorescence quenching study with tachiya model," *Journal of Fluorescence*, vol. 18, no. 3-4, pp. 671–678, 2008.
[26] Y. J. Hu, Y. Liu, R. M. Zhao, J. X. Dong, and S. S. Qu, "Spectroscopic studies on the interaction between methylene blue and bovine serum albumin," *Journal of Photochemistry and Photobiology*, vol. 179, no. 3, pp. 324–329, 2006.
[27] B. Huang, G. L. Zou, and T. M. Yang, "Studies on the interaction between adriamycin and bovine serum albumin," *Acta Chimica Sinica*, vol. 60, no. 10, pp. 1867–1871, 2002.
[28] X. H. Wu, J. H. Zhou, X. T. Gu et al., "Study on interaction between hypocrellin A and hemoglobin or myoglobin using synchronous fluorescence spectra," *Spectroscopy and Spectral Analysis*, vol. 26, no. 12, pp. 2287–2290, 2006.

Investigation of Luminescence Characteristics of Osmium(II) Complexes in the Presence of Heparin Polyanions

Yixi Xie, Yu Lei, Shalini Shah, Hao Wu, Jian Wu, Elise Megehee, and Enju Wang

Department of Chemistry, St. John's University, Queens, NY 11439, USA

Correspondence should be addressed to Enju Wang; wange@stjohns.edu

Academic Editor: John P. Hart

The luminescence characteristics of six osmium carbonyl complexes with phenanthroline (phen) or bipyridine (bpy) and pyridine (py), 4-phenylpyridine (4-phpy), or triphenylphosphine (PPh_3) complexes in the presence of polyanion heparin were studied in both ethanol and aqueous solutions. The influence of heparin on the luminescence of the complexes is heavily dependent on the type of ligands in the complexes and the solvent used. In the ethanol solutions, the heparin solution enhanced the luminescence of the five osmium complexes, with the strongest enhancement to the 4-phenylpyridine complexes; linear curves were obtained in the luminescence enhancement ratio (F/F_0) versus the heparin concentration range of 1–40 μg/mL. In aqueous solutions, heparin quenching of the complexes was more significant; a linear quenching curve was obtained with $[Os(phen)_2CO(PPh_3)](PF_6)_2$ in the lower concentration range of 1–12 μg/mL. The interaction of these complexes with heparin in the solutions is discussed. The complexes are shown to be successful in the fast and sensitive detection of heparin in commercial injectable samples.

1. Introduction

Heparina, a highly sulfated glycosaminoglycan, has the highest negative charge density of any known natural biological molecule [1]. Heparin has been extensively used as an injectable anticoagulant in many clinical procedures for the prevention of blood clotting, especially during open heart surgery [2, 3]. Pharmaceutical-grade heparin is derived from mucosal tissues of slaughtered meat animals such as porcine intestine and bovine lung [4]. Native heparin is a polymer with a molecular weight ranging from 3,000 to 50,000, and the average molecular weight of most commercially prepared heparin is in the range of 12,000 to 15,000. Under physiological conditions, the sulfate groups are deprotonated and attract positively charged counterions (usually sodium) to form a heparin salt. It is in this form that heparin is usually administered as an anticoagulant [5, 6]. At the end of the medical procedure, the heparin has to be rendered inactive to prevent possible bleeding problems. Therefore, clinicians need to know the exact concentration of heparin remaining when they attempt to neutralize the anticoagulant activity of heparin via precise doses of protamine. The current heparin detection method used in hospital is based on the blood clotting time [7, 8] and varies widely due to variation in the blood content of the different individuals. Thus, the discovery of accurate, rapid, reliable safe, and low-cost methods of heparin detection remains an important area of sensor research. For example, electrochemical sensors [9, 10] and optical sensors [11, 12] were intensively studies in the 1990s, and more recent studies have focused on fluorescent spectroscopy [13, 14] and UV/Vis spectroscopy using gold nanoparticles [15].

Luminescence spectroscopy is an analytical method that studies the luminescence emission from a chromophore after their electrons are excited from the ground state to a higher energy level. One of the most attractive features of luminescence spectroscopy as an analytical method is its inherent high sensitivity, wide linear response range, and fast detection speed [16]. However, the luminescence intensity of the chromophore (or excited state complex) can be seriously reduced by collision with other chemical species or enhanced by chemical interactions. Through studying the effect of these chemical species on the luminescence intensity of the chromophore, the quantitative determinations of a variety of important inorganic and organic species in trace amounts have been realized by photoluminescence methods [16].

Recently, a series of Os(II) carbonyl (CO) complexes with two bipyridine or phenanthroline ligands and one pyridyl or triphenylphosphine ligand as the five other ligands have been synthesized in our lab. The osmium centers in these complexes are all in the +2 oxidation state with an overall +2 charge of the complex ion, and exhibit high emission intensity in the visible region [17, 18]. This research focuses on studying the effect of the heparin polyanion on the photoluminescence of these polypyridyl osmium(II) complexes, for the purpose of finding potential luminescence markers for the detection of the optically nonactive polyanion heparin. To our surprise, polyanions show unique influences on the luminescence spectra and intensity of these osmium complexes. The 1,10-phenanthroline (phen) complexes were compared with the 2,2′-bipyridine (bpy) complexes. For each of these chromophoric ligands, three analogous sets were compared using pyridine (py), triphenylphosphine (PPh_3), or 4-phenylpyridine (4-phpy) as the sixth ligand. These ligands showed a significant impact on the luminescence characteristics of the osmium complex in the absence as well as in the presence of polyanions. The results of the studies could lead to a new, potential effective detection method for heparin monitoring.

2. Experimental

2.1. Material and Reagents. The six new Os(II) complexes are $[Os(phen)_2CO(PPh_3)](PF_6)_2$, $[Os(phen)_2CO(py)](PF_6)_2$, $[Os(phen)_2CO(4\text{-}phpy)](PF_6)_2$, $[Os(bpy)_2CO(PPh_3)](PF_6)_2$, $[Os(bpy)_2CO(py)](PF_6)_2$, and $[Os(bpy)_2CO(4\text{-}phpy)](PF_6)_2$. The structure of each complex is shown in Figure 1. The synthesis of six osmium complexes was described in [17, 18]. The purity is >98%.

Heparin sodium sulfate (>180 USP units/mg), ethanol (99.5+%), reagent or analytical grade $NaClO_4$, $MgSO_4$, CH_3COONa, NaCl, KBr, $NaNO_3$, NaOH, $NaNO_2$, KSCN, and KI were all obtained from Sigma Aldrich. Injectable heparin solutions, 100 units/mL and 40 units/mL with 5% dextrose, were purchased from B. Braun Medical Inc. (Irving, CA, USA).

2.2. Apparatus. Luminescence measurements were made on a Perkin Elmer luminescence spectrometer LS 50 B equipped with a high-intensity xenon flash lamp. A Branson 1210 sonicator was used to help dissolve the osmium complex in solvent when needed. Thermo Scientific digital pipets were used for the addition of all solutions into the quartz cuvette.

2.3. Stock Solutions. The stock solutions of the osmium complexes were prepared by dissolving 0.6–2.8 mg of each osmium complex in ethanol (ethanol solution) or distilled water (aqueous solution) in a 100 mL volumetric flask. The 1.0 mg/mL heparin aqueous solution was prepared by dissolving 5.0 mg of heparin in 5.00 mL of distilled water. The 0.10 mg/mL and 0.010 mg/mL heparin aqueous solutions were diluted from the 1.0 mg/mL stock solution. For small anions solutions, a 0.100 M solution of each salts was first made and then diluted to 0.0100 M and 0.00100 M.

2.4. Luminescence Measurements. For the luminescence measurements of each osmium complex solution (ethanol, aqueous), 2 mL stock solution was transferred to a quartz cuvette (1 cm × 1 cm × 4.4 cm) placed in the luminescent spectrometer cuvette holder. The emission and excitation slits were both set at 10 nm for the luminescence measurements. A prescan was taken first to obtain the wavelengths of the maximum excitation and emission intensities; then the emission spectrum was recorded by exciting the sample at the appropriate excitation wavelength. In order to study the anion quenching or enhancement of the luminescence of each osmium complex solution, 2.00 mL of the stock solution was placed in the cuvette, and the emission spectrum was measured; then a small amount (10 μL or 20 μL) of the anion solution (0.010~1.0 mg/mL for heparin and 0.10 to 0.0010 M for small anions) was added to the cuvette and the emission spectrum was recorded after the solution was well mixed. The anion solution was continuously added to the cuvette, and the emission spectrum was recorded after each addition. The values of the luminescence intensity were taken at a fixed wavelength at or near the emission maximum. All the luminescence measurements were taken under ambient conditions (~23°C).

Detection of injectable heparin was performed using standard addition method. First, 10 or 20 μL of injectable heparin solution was added to 2 mL of the aqueous $[Os(phen)_2CO(4\text{-}phpy)](PF_6)_2$ solution. Next three portions of 5 μL of 1 mg/mL standard heparin solution were added, and the emission spectrum of the solution was measured before and after each addition. The concentrations of the heparin in the injectable solution were calculated using the standard addition curve.

3. Results and Discussion

3.1. Excitation and Emission Spectra of Osmium Complexes. The excitation and emission spectra of each complex in their ethanol solution (except for $[Os(bpy)_2CO(py)](PF_6)_2$ which could not dissolve in ethanol) and their aqueous solution were first recorded at room temperature under ambient air. All the spectra exhibited broad emission and excitation bands up to 200 nm in width. Figure 2 shows the emission spectra in ethanol solution of the two complexes: $[Os(phen)_2CO(PPh_3)](PF_6)_2$, and $[Os(phen)_2CO(py)](PF_6)_2$, which exibit the largest difference in luminescence intensity.

As shown in Table 1 and Figure 2, the two Os-PPh_3 complexes have much higher luminescence intensities and shorter emission wavelengths than the two Os-(4-phpy) complexes and the two Os-py complexes (520 nm versus 560 nm). The shorter wavelength (higher energy) of the emission is caused by the π-backbonding ability of the phosphine which stabilizes the HOMO of the molecule relative to pyridine. The higher luminescence intensities are attributed to the larger steric effect of the PPh_3 ligand which shields the complex from colliding with the solvent. This slows down

Table 1: Excitation and emission maxima wavelengths (nm) of the six osmium complexes.

Complex	λ_{Em} in EtOH	λ_{Em} in H_2O	λ_{Ex} in EtOH	λ_{Ex} in H_2O
$[Os(phen)_2CO(PPh_3)](PF_6)_2$	520	520	396	396
$[Os(bpy)_2CO(PPh_3)](PF_6)_2$	520	520	396	396
$[Os(phen)_2CO(4\text{-}phpy)](PF_6)_2$	560	560	420	420
$[Os(bpy)_2CO(4\text{-}phpy)](PF_6)_2$	560	560	430	430
$[Os(phen)_2CO(py)](PF_6)_2$	560	560	420	420
$[Os(bpy)_2CO(py)](PF_6)_2$	N/A	560	N/A	420

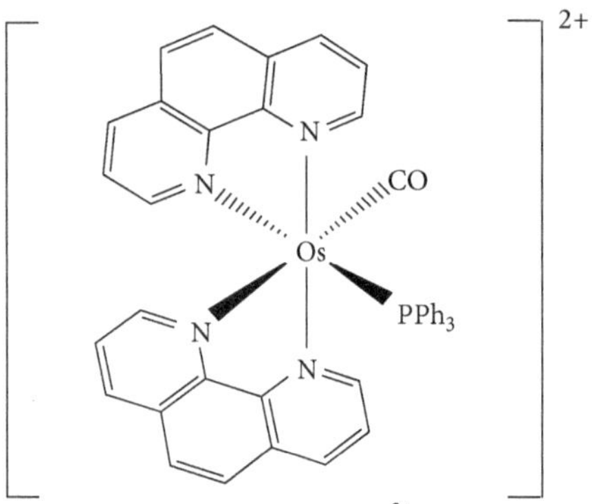

$[Os(phen)_2CO(PPh_3)]^{2+}$

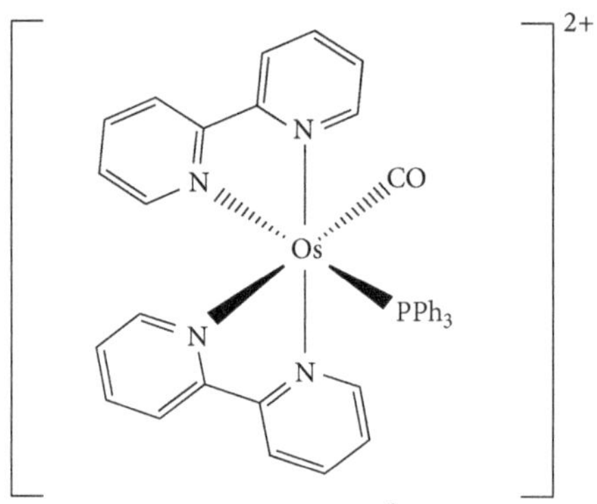

$[Os(bpy)_2CO(PPh_3)]^{2+}$

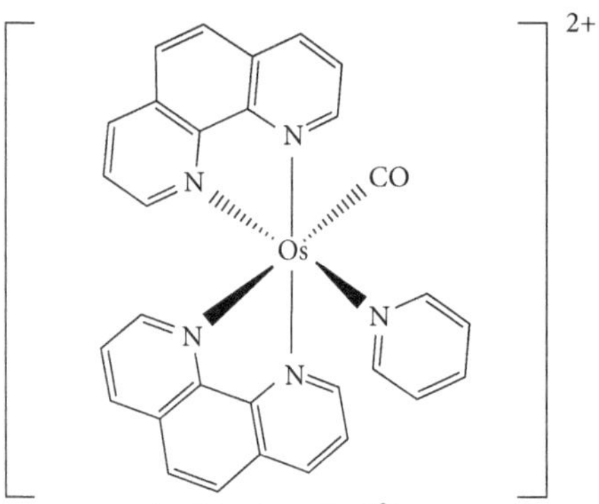

$[Os(phen)_2CO(py)]^{2+}$

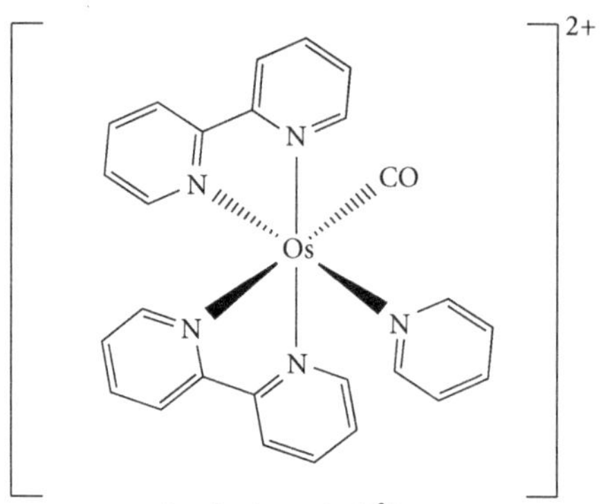

$[Os(bpy)_2CO(py)]^{2+}$

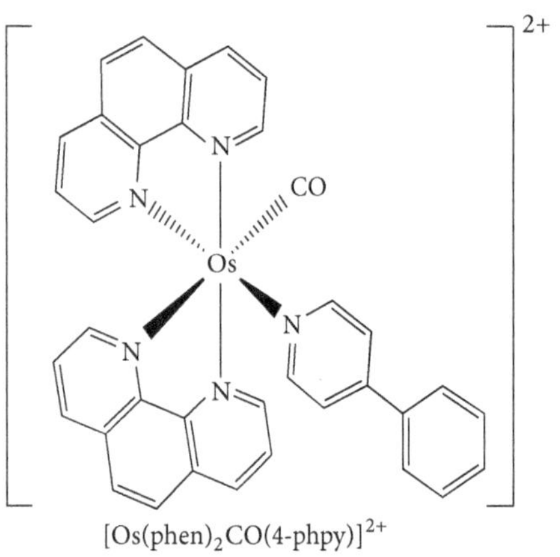

$[Os(phen)_2CO(4\text{-}phpy)]^{2+}$

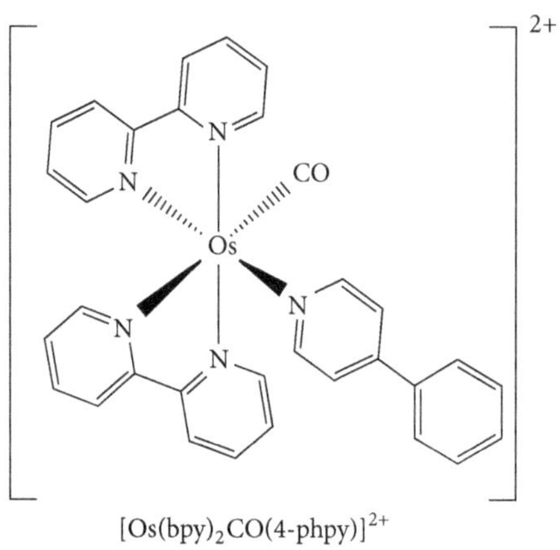

$[Os(bpy)_2CO(4\text{-}phpy)]^{2+}$

Figure 1: Structures of the six osmium complexes.

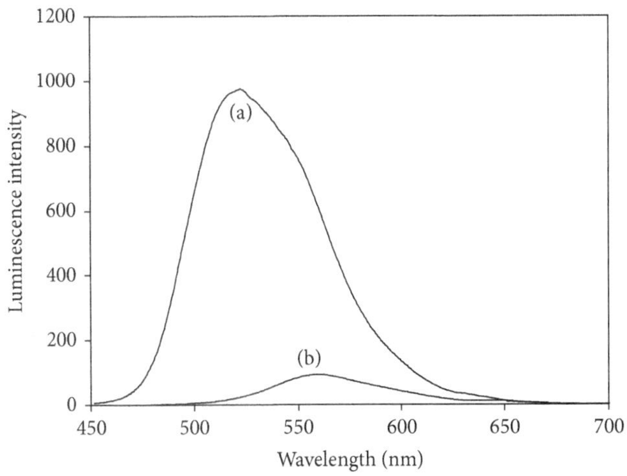

FIGURE 2: Luminescence spectra of (a) $[Os(phen)_2CO(PPh_3)](PF_6)_2$ and (b) $[Os(phen)_2CO(py)](PF_6)_2$ in ethanol solution (2.4 mg/100 mL).

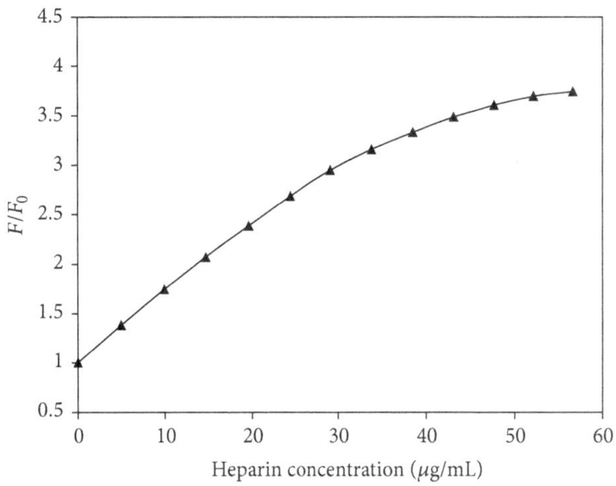

FIGURE 4: Plot of the luminescence enhancement ratio (F/F_0) of $[Os(bpy)_2CO(4\text{-}phpy)](PF_6)_2$ at 560 nm in ethanol solution (2.8 mg/100 mL) versus heparin concentration.

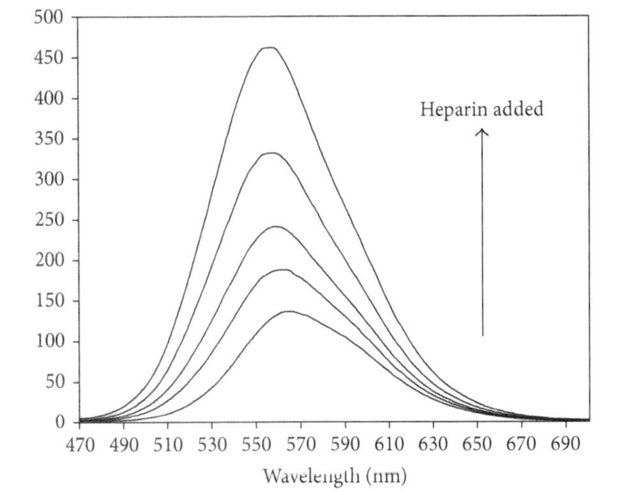

FIGURE 3: Luminescence spectra of $[Os(bpy)_2CO(4\text{-}phpy)](PF_6)_2$ in ethanol solution (2.8 mg/100 mL) in the presence of varing concentrations of heparin. From bottom to top, heparin concentration: 0, 9.9, 19.6, 29.1, and 38.5 μg/mL.

the vibrational relaxation to the ground vibrational level of the excited state that must occur prior to emission of a photon of light that occurs when the electron returns to the ground electronic state. All these six complexes exhibit higher luminescence intensities in aqueous solution than in ethanol solution due to the hydrophobicity of the ligands surrounding the osmium center.

The spectral changes that occur before and after the addition of small inorganic anion solutions suggest that those anions ClO_4^-, SO_4^{2-}, CH_3COO^-, Cl^-, Br^- and NO_3^- exhibit very little quenching of the luminescence of the osmium complex with anion concentrations up to 0.01 M, while NO_2^-, SCN^- and I^- anions can dramatically quench the luminescence of the osmium complexes with concentrations at 0.01 mM. However, OH^- exhibits a small enhancement of the luminescence of the osmium complex; luminescence intensity slightly increased as the solution pH increased from 6 to 9 in aqueous solution.

3.2. Influence of the Luminescence Spectra of Osmium Complexes by Heparin Polyanions. It was expected that heparin would quench the luminescence of the osmium complexes due to its high negative charges. Surprisingly, however, it was found that heparin actually enhanced the luminescence intensity of the osmium complexes in ethanol solutions. The luminescence of each complex in the ethanol and aqueous solution changed quite differently after heparin was added.

3.2.1. Ethanol Solution. In ethanol solution, each complex exhibited an increase in luminescence intensity after heparin was added in concentrations greater than 0.5 μg/mL. However, the amount of increase varied significantly for each complex. The solutions of $[Os(bpy)_2CO(4\text{-}phpy)](PF_6)_2$, $[Os(phen)_2CO(4\text{-}phpy)](PF_6)_2$, and $[Os(phen)_2CO(py)](PF_6)_2$ exhibited a very large increase in luminescence intensity while the solutions of $[Os(phen)_2CO(PPh_3)](PF_6)_2$ and $[Os(bpy)_2CO(PPh_3)](PF_6)_2$ exhibited a small luminescence increase.

Figure 3 shows the luminescence spectra of the ethanol solution of $[Os(bpy)_2CO(4\text{-}phpy)](PF_6)_2$ in the absence and presence of heparin at different concentrations. Addition of heparin to the solution significantly enhanced the luminescence. The luminescence intensity was increased from a relative emission intensity of 136 in the absence of heparin to 462 by addition of heparin in a concentration range of 5.0–38.5 μg/mL. Further, there is an 8 nm blue shift in wavelength of the emission maxima with high heparin concentrations. A plot showing the wider range of heparin concentrations on the enhancement ratio (F/F_0) is shown in Figure 4. Here, F and F_0 are the luminescence (fluorescence) intensity of the osmium complexes solutions at 560 nm with and without heparin, respectively. The luminescence intensity

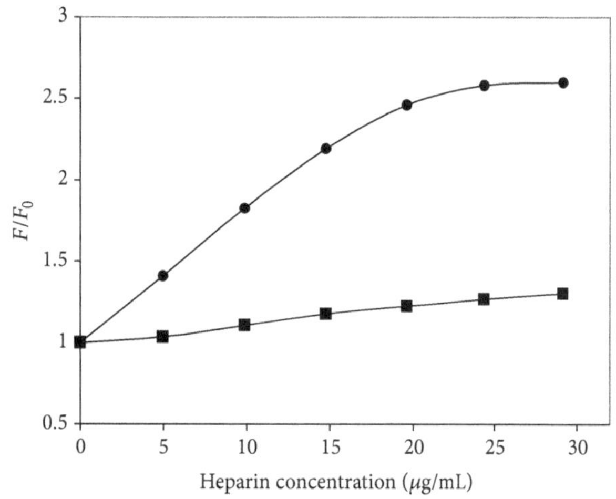

FIGURE 5: Plot of the luminescence enhancement ratio (F/F_0) of (•) $[Os(phen)_2CO(py)](PF_6)_2$ and (■) $[Os(phen)_2CO(PPh_3)](PF_6)_2$ (Ex slit width: 5 nm) both at 2.4 mg/100 mL versus heparin concentration.

increased steadily up to a heparin concentration of 56 μg/mL. In the lower concentration range (up to 30 μg/mL), there is a linear relationship between the F/F_0 ratio and the concentration with an R^2 of 0.99. The heparin enhancement of the $[Os(phen)_2CO(4\text{-phpy})](PF_6)_2$ (2.4 mg/100 mL EtOH) luminescence has a similar pattern to that seen with $[Os(bpy)_2CO(4\text{-phpy})](PF_6)_2$. In this case, the luminescence was enhanced by three and half times when the added heparin concentration was increased to 50 μg/mL. Once again, there is an 8 nm blue shift of the emission maxima at higher heparin concentrations. In the concentration range of 1 to 40 μg/mL, there is a linear relationship between the F/F_0 ratio and the heparin concentration with an R^2 of 0.99.

The heparin enhancement of the luminescence of $[Os(phen)_2CO(py)](PF_6)_2$ (2.4 mg/100 mL EtOH) is smaller than that seen for the 4-phpy complexes. The luminescence intensity was enhanced by two and half times when the added heparin concentration was increased to 20–30 μg/mL, and then the luminescence started to decrease at heparin concentrations in excess of 30 μm/mL. In this case, there is also an 8 nm blue shift of the emission maxima at higher heparin concentrations.

The heparin enhancement of the luminescence of the $[Os(phen)_2CO(PPh_3)](PF_6)_2$ (2.4 mg/100 mL EtOH) and the $[Os(bpy)_2CO(PPh_3)](PF_6)_2$ (2.1 mg/100 mL EtOH) solutions is very small compared to the Os-(4-phpy) and Os-py complexes. Here, the shift in the emission maxima is insignificant (less than 2 nm).

Figure 5 shows the plot of F/F_0 versus heparin concentration for the two phenanthroline complexes with pyridine and PPh_3 as the sixth ligand. From Figures 4 and 5, it can be seen that the 4-phpy complexes, which have the lowest solution luminescence in the absence of polyanions, showed the highest enhancement in the concentration range of 5–40 μg/mL added heparin concentrations. By contrast, the PPh_3 complexes showed the smallest enhancement in the same concentration range.

The osmium complexes luminescence enhancement by heparin is likely contributed by the close contact of heparin and the complexes, which forms a rigid structure that reduced the solvent collision and quenching. Heparin has a long linear structure with anionic sites on its branches when it is dissolved in aqueous solution [1]. When its aqueous solution is added to the ethanol solution of the osmium complex, the anion sites on its branches attract the double positively charged osmium complex cations. As a result, the osmium complex could electrostatically bind to heparin and become surrounded by the heparin molecules (Figure 6). In this case, solvent molecules can no longer get close enough to the osmium complex to help relax the excited state, resulting in the emission maximum moving to higher energy. This effect would also explain the increased emission intensity since the rate of nonradiative decay would be greatly reduced in the absence of favorable interaction with solvent molecules.

The sixth ligand in these complexes plays an important role in the interaction of these charged sites. The complexes $[Os(bpy)_2CO(4\text{-phpy})]^{2+}$, $[Os(phen)_2CO(4\text{-phpy})]^{2+}$, or $[Os(phen)_2CO(py)]^{2+}$ have small narrow 4-phenylpyridine or pyridine ligand. When heparin is added to their ethanol solution, these osmium complexes could readily approach the heparin polyanion and bind up to two heparin anion sites. These two sites may well form two heparin polyanions branches, thus forming ion pairs between the polyanion and osmium complex. Therefore, heparin can cause significant enhancement to the luminescence of the Os-(4-phpy) and Os-py complexes. Further, the luminescence only stops increasing at very high heparin concentrations when all the osmium complex cations are bound to heparin anion sites. By contrast, the $[Os(phen)_2CO(PPh_3)]^{2+}$ and $[Os(bpy)_2CO(PPh_3)]^{2+}$ complexes have a sterically bulky PPh_3 group, which prevents them from approaching the heparin polyanion closely. Therefore, the luminescence intensity of the Os-PPh_3 complexes could not be significantly enhanced by heparin.

3.2.2. Aqueous Solution. The effect of heparin on the emission of these osmium complexes in their aqueous solution is quite different from that of their ethanol solution. The aqueous solution of osmium complexes with a PPh_3 or pyridine group exhibits a decrease in luminescence intensity at low added heparin concentrations; then the luminescence increases at heparin concentration above 15 to 19 μg/mL. However, complexes with the 4-phpy group exhibit a decrease of luminescence at heparin concentrations below 0.5 μg/mL then an increase in luminescence intensity is observed with the addition of heparin in the range of 1 to 15 μg/mL. This shows that as a polyanion, when not binding to the osmium complexes, heparin is an efficient luminescence quencher in aqueous solutions. This is especially true with the PPh_3 complexes, where binding only occurs at very high (>15 μg/mL) heparin concentrations. By contrast, the 4-phpy complexes bind more strongly to heparin, and so less of a quenching effect is observed.

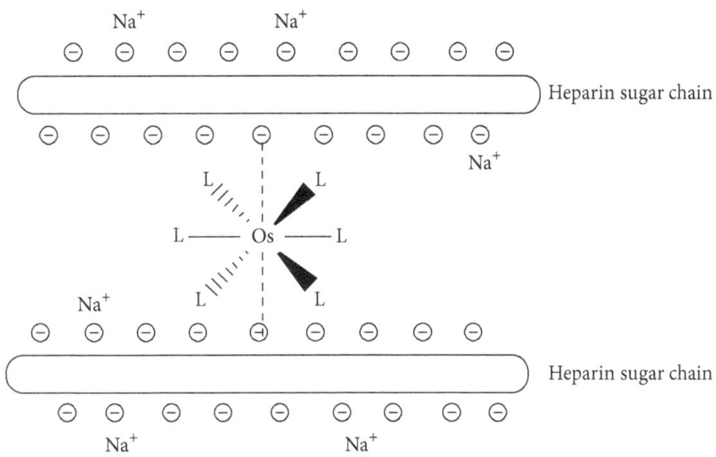

FIGURE 6: Schematic drawing of the interaction of heparin with the double positive-charged osmium complexes.

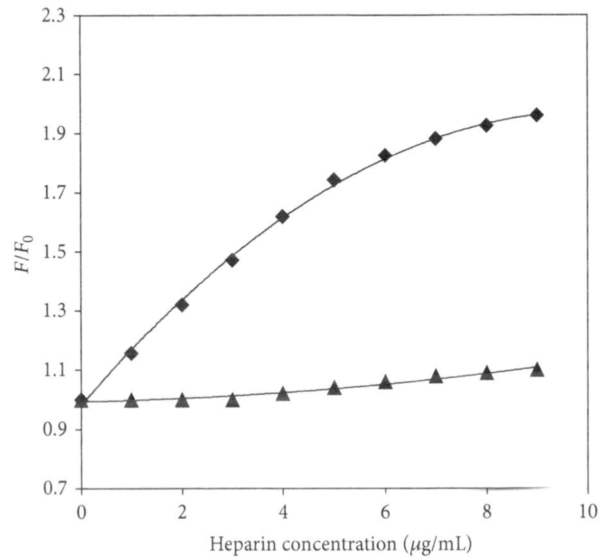

FIGURE 7: Plots of the luminescence enhancement ratio (F/F_0) of (◆) $[Os(phen)_2CO(4\text{-}phpy)](PF_6)_2$, and [▲]$[Os(bpy)_2CO(4\text{-}phpy)](PF_6)_2$ at 560 nm in aqueous solution (1.2 mg/100 mL) versus heparin concentration.

Figure 7 shows the plots of the luminescence enhancement ratio (F/F_0) as a function of heparin concentration for the two Os-(4-phpy) complexes in aqueous solution. The high luminescence enhancement ratio of the 1,10-phenanthroline complexes by heparin shown in Figure 7 suggests that the $[Os(phen)_2CO(4\text{-}phpy)]^{2+}$ complex binds more strongly to the heparin polyanion than the analogous 2,2′-bipyridine complex complexes. Comparing Figures 4, 5, and 7, the enhancement of heparin to the $[Os\text{-}CO(bpy)2(4\text{-}phpy)]^{2+}$ complex is much smaller in aqueous solutions, which indicate a weaker binding to heparin, possibly due to the higher solubility of heparin and the complexes in aqueous solution.

In the aqueous solutions of $[Os(phen)_2CO(PPh_3)](PF_6)_2$ or $[Os(bpy)_2CO(PPh_3)](PF_6)_2$, it is even harder for the osmium complex to bind to heparin because of the high solubility of heparin in aqueous solution and the hindrance of the bulky PPh_3 group. In fact, at low heparin concentration, heparin quenches the luminescence of these two complexes heavily as polyanion. The decrease of the luminescence intensity in the aqueous solution of these two complexes fits the Stern-Volmer equation

$$\frac{F_0}{F} = 1 + K_{sv}[Q]. \quad (1)$$

Here F_0 and F are the luminescence signals in the absence and in the presence of a quencher, respectively; K_{sv} is the Stern-Volmer quenching constant; and $[Q]$ is the concentration of the quencher.

Stern-Volmer plot is shown in Figure 8 for the $[Os(phen)_2CO(PPh_3)]^{2+}$ complex. The high linearity of the Stern-Volmer curve, shown at the concentration range of 1 to 12 μg/mL in the insert, confirms that heparin exhibits dynamic quenching to the Os-PPh_3 complexes. The nonlinearity at higher concentrations may indicate that osmium complex-heparin ion-pairs have formed. Further addition of heparin above 12 μg/mL caused the luminescence of these complexes to increase slightly.

Table 2 summarizes the heparin concentration ranges that induced significant luminescence changes to the complexes studied in ethanol or aqueous solutions. In ethanol solution, $[Os(bpy)_2CO(4\text{-}phpy)](PF_6)_2$ exhibits the greatest enhancement in the presence of heparin as measured by the highest slope of the F/F_0 versus heparin concentration linear curve. This complex exhibits the longest response range and the highest sensitivity to heparin concentration. In aqueous solutions, the $[Os(phen)_2CO(4\text{-}phpy)](PF_6)_2$ and both phosphine complexes showed significant luminescence changes at low heparin concentrations.

Table 2: Linear response ranges and slopes obtained from the F/F_0 curves to heparin of the three osmium complexes.

Complex	Solution	Linear range (μg/mL)	Slope
$[Os(bpy)_2CO(4\text{-}phpy)](PF_6)_2$	Ethanol solution	1–38	0.061
$[Os(phen)_2CO(4\text{-}phpy)](PF_6)_2$	Ethanol solution	1–40	0.072
$[Os(phen)_2CO(py)](PF_6)_2$	Ethanol solution	1–24	0.074
$[Os(phen)_2CO(4\text{-}phpy)](PF_6)_2$	Aqueous solution	1–10	0.116
$[Os(phen)_2CO(PPh_3)](PF_6)_2$	Aqueous solution	1–10	−0.036*

*Value from the Stern-Volmer equation.

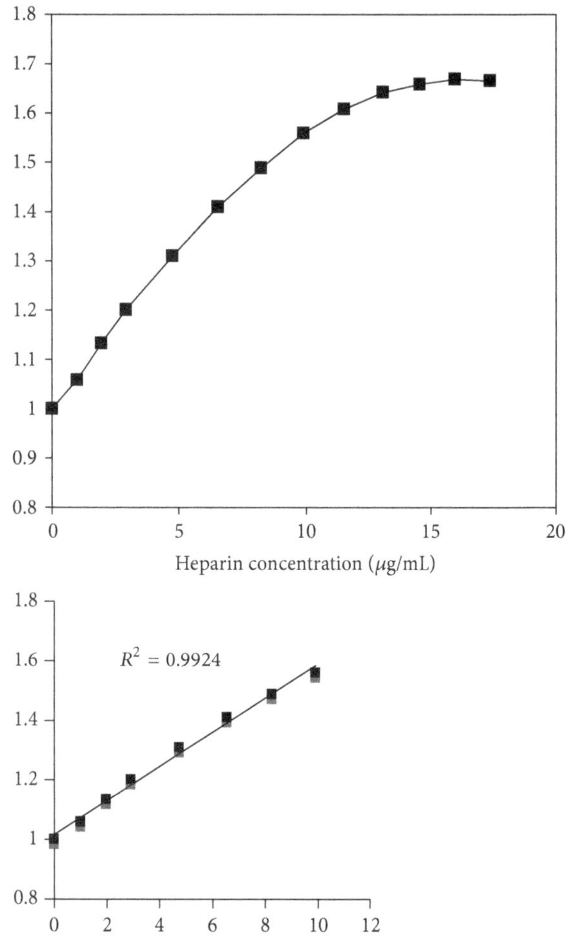

Figure 8: Stern-Volmer plot for the heparin quenching of the luminescence of $[Os(phen)_2CO(PPh_3)](PF_6)_2$ aqueous solution (1.2 mg/100 mL in H_2O).

3.3. Detection of Heparin in Medicinal Solution. The detection of heparin was then performed with commercial medicinal solutions to evaluate the feasibility. Two commercial heparin sodium injection solutions (40 and 100 units/mL in 5% dextrose, in 100 mL plastic bags) from B. Braun Medical Inc. were tested. With the $[Os(phen)_2CO(4\text{-}phpy)]^{2+}$ complex in aqueous solution, the emission intensities of the osmium solution before and after adding 20 μL of these two heparin solutions were measured. The ratios (F/F_0) obtained were 1.30 and 1.97 ($n = 5$), respectively, showing that the osmium complex indeed had a greater response to higher heparin concentration preparations. However, these medical preparation responses are much higher than the pure heparin solution response. When response curve of heparin in the presence of 5% dextrose was performed, an enhanced osmium complex response to heparin concentration was observed. This enhanced response is likely due to the hydrophobic environment that the polysaccharide, dextrose, created around the osmium complex in aqueous solution, thus reducing the extent of nonradiative decay of the excited states and increasing the luminescence intensity of the osmium complex.

To eliminate the background effect, the standard addition method was used. First, 10 or 20 μL of the commercial heparin sample solutions was added to the aqueous osmium complex solution; then three portions of 5 μL of the 1.0 mg/mL heparin standard solution were added. The luminescence values of the complex solution after each addition of the sample and standard heparin solutions were measured at 560 nm. With the standard addition curves and using the conversion of 180 units/mg of sodium heparin, the average heparin concentration in the two sample bags was calculated to be 46 ± 3 and 111 ± 3 units/mL. The relative standard deviation, RSD for the two corresponding samples are 5.6 ($n = 3$) and 2.3 ($n = 6$) respectively, indicating that the precision of the method is acceptable considering the volume added is very small, although they are 11% higher than the labeled value. Further tests, with other commercial sources having different backgrounds, are currently under study.

4. Conclusion

The luminescence of osmium carbonyl complexes is stable at room temperature in ethanol and aqueous solutions, though the intensity and wavelength of this emission are heavily dependent on the other five ligands. The effect of heparin on the luminescence of the osmium complexes depends heavily on both the ligand and solvent used. In ethanol solutions, heparin could enhance the luminescence of five osmium complexes at low heparin concentrations, with the greatest enhancement occurring with the Os-(4-phpy) complexes. This increased emission intensity may be due to binding of heparin to the double charged complexes which changed the solvent environment of the complexes. In aqueous solutions, heparin quenching the osmium complex emission was more significant as heparin is more soluble in water and behaves as an anion quencher. These osmium complexes are shown to be useful for the fast and sensitive detection of heparin in commercially injectable samples.

Acknowledgment

The authors are thankful to St. John's University for the financial support throughout this research.

References

[1] M. Cox and D. Nelson, *Lehninger Principles of Biochemistry*, Freeman, 4th edition, 2004.

[2] B. Casu, "Structure of heparin and heparin fragments," in *Heparin and Related Polysaccharides, Structure and Activities*, F. A. Ofosu, I. Danishefsky, and J. Hirsh, Eds., p. 556, The New York Academy of Sciences, New York, NY, USA, 1989.

[3] J. Hirsh, "Heparin induced bleeding," *Nouvelle Revue Française d'Hématologie*, vol. 26, no. 4, pp. 261–266, 1984.

[4] R. J. Linhardt and N. S. Gunay, "Production and chemical processing of low molecular weight heparins," *Seminars in Thrombosis and Hemostasis*, vol. 25, supplement 3, pp. 5–16, 1999.

[5] I. Bjork and U. Lindahl, "Mechanism of the anticoagulant action of heparin," *Molecular and Cellular Biochemistry*, vol. 48, no. 3, pp. 161–182, 1982.

[6] L. B. Jaques, "Heparins—anionic polyelectrolyte drugs," *Pharmacological Reviews*, vol. 31, no. 2, pp. 99–166, 1979.

[7] B. S. Bull, W. M. Huse, F. S. Brauer, and R. A. Korpman, "Heparin therapy during extracorporeal circulation. II. The use of a dose response curve to individualize heparin and protamine dosage," *Journal of Thoracic and Cardiovascular Surgery*, vol. 69, no. 5, pp. 685–689, 1975.

[8] J. C. van Kerkhof, P. Bergveld, and R. B. M. Schasfoort, "The ISFET based heparin sensor with a monolayer of protamine as affinity ligand," *Biosensors and Bioelectronics*, vol. 10, no. 3-4, pp. 269–282, 1995.

[9] S. C. Ma, V. C. Yang, B. Fu, and M. E. Meyerhoff, "Electrochemical sensor for heparin: further characterization and bioanalytical applications," *Analytical Chemistry*, vol. 65, no. 15, pp. 2078–2084, 1993.

[10] M. E. Meyerhoff, V. C. Yang, J. A. Wahr et al., "Potentiometric polyion sensors: new measurement technology for monitoring blood heparin concentrations during open heart surgery," *Clinical Chemistry*, vol. 41, no. 9, pp. 1355–1356, 1995.

[11] E. Wang, M. E. Meyerhoff, and V. C. Yang, "Optical detection of macromolecular heparin via selective coextraction into thin polymeric films," *Analytical Chemistry*, vol. 67, no. 3, pp. 522–527, 1995.

[12] E. Wang, G. Wang, L. Ma, C. M. Stivanello, S. Lam, and H. Patel, "Optical films for protamine detection with lipophilic dichlorofluorescein derivatives," *Analytica Chimica Acta*, vol. 334, no. 1-2, pp. 139–147, 1996.

[13] J. Li, J. Liu, X. Zhu, Q. Peng, and C. Jiang, *Lumi*, vol. 113, p. 305, 2005.

[14] W. Wei, H. Wang, and C. Jiang, "Spectrofluorimetric determination of trace heparin using lomefloxacin-terbium probe," *Spectrochimica Acta A*, vol. 63, no. 2, pp. 241–246, 2006.

[15] B. K. Jena and C. R. Raj, "Optical sensing of biomedically important polyionic drugs using nano-sized gold particles," *Biosensors and Bioelectronics*, vol. 23, no. 8, pp. 1285–1290, 2008.

[16] J. R. Lakowicz, *Principles of Fluorescence Spectroscopy*, Plenum Press, New York, NY, USA, 1983.

[17] P. Menhaji, *[M.S. thesis]*, Saint John's University, New York, NY, USA, 2006.

[18] C. Leung, *[M.S. thesis]*, St. John's University, New York, NY, USA, 2010.

19

Apolar Compounds in Seaweeds from Fernando de Noronha Archipelago (Northeastern Coast of Brazil)

Leandro De Santis Ferreira,[1] Izabel Cristina Casanova Turatti,[1] Norberto Peporine Lopes,[1] Thais Guaratini,[2] Pio Colepicolo,[3] Eurico Cabral Oliveira Filho,[4] and Ricardo Clapis Garla[5]

[1] *Departamentos de Física e Química, Faculdade de Ciências Farmacêuticas de Ribeirão Preto, Universidade de São Paulo, 14040-903 Ribeirão Preto, SP, Brazil*
[2] *Lychnoflora Pesquisa e Desenvolvimento em Produtos Naturais LTDA, Incubadora SUPERA, Campus da USP, 14040-900 Ribeirão Preto, SP, Brazil*
[3] *Instituto de Química, Universidade de São Paulo, 05513-970 São Paulo, SP, Brazil*
[4] *Instituto de Biociências, Universidade de São Paulo, P.O. Box 11461, 05508-090 São Paulo, SP, Brazil*
[5] *Departamento de Oceanografia e Limnologia, Centro de Biociências, Universidade Federal do Rio Grande do Norte, 59072-970 Natal, RN, Brazil*

Correspondence should be addressed to Ricardo Clapis Garla, rgarla@hotmail.com

Academic Editor: Ricardo Vessecchi

Hyphenated techniques of gas chromatography coupled to mass spectrometer were used to determine fatty acids in eleven species of seaweeds from Fernando de Noronha archipelago. The main compounds detected in all studied species were the alcohol phytol and the fatty acids 14 : 0; 15 : 0; 16 : 0; 18 : 0; 18 : 1 n^9; 18 : 2 $\Delta^{9,12}$; 20 : 4; 20 : 5. These fatty acids are commonly found in seaweeds present in warm regions. Thus, we found no specificity in the presence of a particular set of fatty acids and the studied species indicating that they are not useful as taxonomic indicators. However, they could be used in a comparative study with algae found in polluted area because many of the studied seaweeds are widespread and Fernando de Noronha has low human influence.

1. Introduction

Seaweeds are key ecological factors in shallow marine areas forming the base of the trophic web and structuring ecosystems especially on consolidated substrata. They are known by the production of many bioactive compounds. Hence, the pharmaceutical and the cosmetics industries have a special interest in algae as sources of specific molecules [1–4]. Also, several works have investigated their value in human and animal nutrition [5–8]. Among the algae-derived compounds, polyunsaturated fatty acids are especially important as they act as antioxidant agents involved in many physiological processes [9, 10].

The studies to determine the fatty acids profiles started some years before and many of them focused on the use of these compounds as biomarkers for chemotaxonomy, though their concentration may be susceptible to environmental interference [11–14]. Some of the main important factors that influence the algae fatty acid concentration are the temperature, [15, 16], types of habitat [15, 17], and presence of metals and pollutants [12, 13, 18]. Although less common, other approaches using fatty acids have explored the effects of industrial effluents and environmental variables on the amount and quality of fatty acids produced [19, 20] and their role in food assimilation by herbivorous invertebrates [21].

Fernando de Noronha archipelago is an isolated group of islands formed by relatively recent volcanic processes. The islands are located approximately 350 km off the northeastern Brazilian coast and are part of Pernambuco State [22]. The archipelago was established as a marine protected area since 1988, and its marine flora is composed of 128 taxa, including

Table 1: Classification for the eleven species of seaweeds from Fernando de Noronha archipelago in northeastern Brazil analyzed in this study.

Divisions and species
Chlorophyta
Caulerpa verticillata J. Agardh, 1847
Rhodophyta
Asparagopsis taxiformis (Delile) Trevisan de Saint-Léon, 1845
Dictyurus occidentalis J. Agardh, 1847
Dichotomaria marginata (J. Ellis & Solander) Lamark, 1816
Dichotomaria obtusata (J. Ellis & Solander) Lamarck, 1816
Galaxaura rugosa (J. Ellis & Solander) J. V. Lamouroux, 1816
Ochrophyta
Dictyota cervicornis Kützing, 1859
Dictyopteris justii J. V. Lamouroux, 1809
Dictyopteris plagiogramma (Montagne) Vickers, 1905
Padina gymnospora (Kützing) Sonder, 1871
Sargassum sp.

Table 2: Saturated fatty acids present in the eleven seaweeds studied.

Fatty Acids	At	Gr	Db	Dm	Do	Cv	Dc	Dj	S	Pg	Dp
12:0	−	−	−	−	−	+	+	+	−	−	−
14:0	+	+	+	+	+	+	+	+	+	+	+
15:0	+	+	+	+	+	+	+	+	+	+	+
16:0	+	+	+	+	+	+	+	+	+	+	+
17:0	+	+	−	+	+	+	+	+	+	+	+
18:0	+	+	+	+	+	+	+	+	+	+	+
20:0	+	+	+	+	+	−	+	+	+	+	+
22:0	+	+	+	+	−	−	−	+	+	+	+
24:0	+	+	+	+	−	+	−	+	+	−	−

+ presence; − absence. At: *Asparagopsis taxiformis*; Gr: *Galaxaura rugosa*; Db: *Dichotomaria obtusata*; Dm: *Dichotomaria marginata*; Do: *Dictyurus occidentalis*; Cv: *Caulerpa verticillata*; Dc: *Dictyota cervicornis*; Dj: *Dictyopteris justii*; S: *Sargassum* sp.; Pg: *Padina gymnospora*; Dp: *Dictyopteris plagiogramma*.

44 species of Chlorophyta, 62 of Rhodophyta, and 22 of Phaeophyta [23]. The most abundant benthic algae groups found in the archipelago are represented by the families Dictyotaceae and Sargassaceae. Other seaweeds also commonly found are the green algae *Caulerpa verticillata* and red algae *Galaxaura* spp. [23, 24]. As part of a broader research project aimed to study chemical compounds of seaweeds from Fernando de Noronha, herein is provided baseline information on the fatty acids produced by eleven species commonly found in the archipelago.

2. Experimental

2.1. Field Collection. A list of the studied species is presented in Table 1. Samples were collected between February and March 2006 at two sites of the main island, Caieiras Beach (3°50′18.8″S, 32°23′57.3″W) and Sueste Bay (3°52′1.2″S, 32°25′19.7″W). Research permit from the Brazilian Environmental Agency (IBAMA) to collect algae was registered under number 050/2006. Seaweeds were randomly collected by hand by uprooting the whole plant, which were placed in labeled plastic bags, frozen, and sent to the laboratory. The algae were identified, cleaned from epiphytes, animals, and sediment, washed with distilled water, and in an oven at 40°C dried for seven days.

2.2. Preparation of Samples. Ten milliliters (10 mL) of dichloromethane (J. T. Baker, Phillipsburg, NJ, USA) were added to 1 g (dried weight) of each species of algae and submitted to ultrasonic bath for 30 minutes. This procedure was repeated three times and the total extract was concentrated on nitrogen gas. Subsequently, 2.0 mL of 1.0 M sodium methoxide were added to the extract and shaken occasionally during five minutes at 65°C. After cooling the extract, 1.0 mL of water and three samples of 1.0 mL of chloroform (J. T. Baker, Phillipsburg, NJ, USA) was added to each sample, shaken for one minute, and centrifuged at 3000 rpm, or 1612.8 g, to extract methyl esters of the fatty acids. The chloroform phase (3.0 mL), was removed and nitrogen gas was used to evaporate the solvent. Samples were suspended in 1.0 mL ethyl acetate (J. T. Baker, Phillipsburg, NJ, USA), and sodium sulphate anhydrous (Sigma Inc., St. Louis, MO, USA) was added to remove water. This methodology was adapted from Eder et al. [25, 26].

2.3. GC-MS Analysis. Samples were analyzed through GC-MS (gas chromatography coupled to mass spectrometer detector) in a Shimadzu QP2010 with ionization source of 70 eV and fragmentation by electronic ionization (EI). The volume of 1.0 μL for each sample was injected at 220°C in a DBWAX column (30 m × 0.25 mm × 0.25 μm). The analysis occurred with 1-minute sample time in the splitless mode, a column flow of 1.3 mL min^{-1}, a linear velocity of 41.4 cm s^{-1}, and scan between *m/z* 40 and *m/z* 500. Oven's temperature started with 50°C, increasing to 20°C min^{-1} until 200°C, kept in this temperature for 5 minutes, then increased to 5°C min^{-1} until 230°C and kept in this temperature for 30 minutes. For identification of compounds, the peaks were compared with some standards and always consulting the libraries WILEY version no. 7 and NIST version nos. 12 and 62.

3. Results and Discussion

The GC-MS analysis had excellent resolution, and the characteristics of the detected unsaturated and saturated fatty acids are given in Tables 2 and 3. The eleven species have myristic acid (14:0), pentadecanoic acid (15:0), palmitic acid (16:0), stearic acid (18:0), oleic acid (18:1, n^9), linoleic acid (18:2 $\Delta^{9,12}$), arachidonic acid (20:4), and eicosapentaenoic acid (20:5). The major compounds were hexadecanoic acid methyl ester and phytol, which were detected in all samples analyzed. Other apolar compounds,

Table 3: Unsaturated fatty acids founded in the eleven seaweeds studied.

Fatty Acids	At	Gr	Db	Dm	Do	Cv	Dc	Dj	S	Pg	Dp
$14:1\ n^5$	−	−	−	−	−	+	−	−	−	−	−
$16:1\ n^7$	+	+	+	−	+	+	+	+	+	−	−
$16:1\ n^9$	+	+	−	−	+	+	+	+	+	+	−
$16:1\ n^9$ ETHYL	−	−	−	−	+	−	−	−	−	−	−
$16:1\ n^{11\text{ or }13}$	+	−	−	−	+	+	+	+	+	+	−
$16:2\ \Delta^{7,10\text{ or }9,12}$	+	−	−	−	−	+	−	−	−	−	−
$16:3\ \Delta^{4,7,10}$	−	−	−	−	−	−	−	+	+	+	+
$18:1\ n^9$	+	+	+	+	+	+	+	+	+	+	+
$18:1\ n^{10\text{ or }11\text{ or }12}$	+	+	+	+	+	+	+	+	+	−	+
$18:1\ n^{13\text{ or }14\text{ or }16}$	+	−	−	−	+	+	+	+	−	+	−
$18:2\ \Delta^{8,11}$	−	−	−	−	−	−	−	+	−	−	−
$18:2\ \Delta^{9,12}$(E,E) or (Z,Z)	+	+	+	+	+	+	+	+	+	+	+
$18:3\ \Delta^{6,9,12}$	+	−	−	−	+	+	−	+	+	+	−
$18:3\ \Delta^{9,12,15}$	+	+	+	−	+	+	+	+	+	+	−
$20:1\ n^{9\text{ or }11}$	−	+	−	−	−	−	−	−	+	+	−
$20:2\ \Delta^{11,14}$	+	+	−	−	−	+	−	+	+	+	−
$20:3\ \Delta^{7,10,13\text{ or }8,11,14}$	+	−	−	−	+	+	+	+	+	+	−
$20:4$	+	+	+	+	+	+	+	+	+	+	+
$20:5$	+	+	+	+	+	+	+	+	+	+	+
$22:1\ n^{13}$	+	−	−	−	−	−	−	−	+	−	−

+ presence; − absence. At: *Asparagopsis taxiformis*; Gr: *Galaxaura rugosa*; Db: *Dichotomaria obtusata*; Dm: *Dichotomaria marginata*; Do: *Dictyurus occidentalis*; Cv: *Caulerpa verticillata*; Dc: *Dictyota cervicornis*; Dj: *Dictyopteris justii*; S: *Sargassum sp.*; Pg: *Padina gymnospora*; Dp: *Dictyopteris plagiogramma*.

Table 4: Other apolar compounds found in seaweed species from Fernando de Noronha.

Compound	At	Gr	Db	Dm	Do	Cv	Dc	Dj	S	Pg	Dp
1-Tetradecanol	−	−	−	+	−	−	−	−	−	−	−
1-Hexadecanol	+	+	+	+	−	−	−	−	+	−	+
1-Hexadecanol-2-methyl	−	−	+	−	−	−	−	−	−	−	−
1-Pentadecanol	−	−	−	+	−	−	−	−	−	−	−
1-Octadecanol	+	+	+	+	−	−	−	−	+	−	+
Phytol	+	+	+	+	+	+	+	+	+	+	+
Hexadecane, 1-iodo	+	−	−	−	−	−	−	−	−	−	−
Benzeneacetamide, N-aminocarbonyl	+	−	−	−	−	−	−	−	−	−	−
3-Buten-2-one, 4-(4-hydroxy-2,2,6-trimethyl-7-oxabicyclo[4.1.0]hept-1-yl)	−	+	+	+	−	−	−	−	+	−	+
Tetrapentacontan,1,54-dibromo	−	−	−	−	−	−	−	−	+	−	−
3-Octyl oxiraneoctanoic acid methyl ester	−	−	−	−	−	−	−	+	−	−	−
1-Docosanol	+	−	−	−	−	−	−	−	−	−	−
Hexadecanamide	+	−	−	−	−	−	−	−	−	−	−
Loliolide	+	+	+	+	−	−	−	−	+	−	+
Isololiolide	+	+	+	+	−	−	−	−	+	−	+

+ presence; − absence. At: *Asparagopsis taxiformis*; Gr: *Galaxaura rugosa*; Db: *Dichotomaria obtusata*; Dm: *Dichotomaria marginata*; Do: *Dictyurus occidentalis*; Cv: *Caulerpa verticillata*; Dc: *Dictyota cervicornis*; Dj: *Dictyopteris justii*; S: *Sargassum* sp.; Pg: *Padina gymnospora*; Dp: *Dictyopteris plagiogramma*.

mainly alcohols, were also identified and are provided in Table 4.

The only seaweed in which 20 : 0 fatty acid was absent was the green algae, *C. verticillata*, and the acid 17 : 0 was absent in only one red algae. There were problems in isomer identification while conducting tentative taxonomic analysis, especially in double and triple bonds of C16 and C18 fatty acids, due to great similarity in retention time and in the molecular fragmentation in mass spectrum.

All compounds detected are common in macroalgae [14]. Few differences were observed in the production of fatty acids by brown, green, and red algae. Also, as only one species of green seaweed was analyzed, this prevents further taxonomic comparisons with brown and red algae. In spite of the fact that GC-MS is a highly efficient and sensitive method of analysis, it was unable to precisely determine which of the polyunsaturated fatty acids with 16 or 18 atoms of carbon was present in each species. Other apolar compounds detected had a more restrict distribution among species, and although most of them were detected in only one species, it was not possible to use them as chemotaxonomy markers, that is, phytol or loliolide. Moreover, the fatty acids referred to as chemotaxonomic agents in previous investigations [14, 17] did not allow discerning brown, green, and red algae in the present study.

Fernando de Noronha is an area subjected to low levels of pollution because it is a marine protected area where anthropic activities are controlled. Hence, it would be interesting to compare the fatty acid profiles obtained in this study with the ones from coastal areas of the main land near of the archipelago region that has different degrees of pollution. For example, *Sargassum* spp. fatty acid profiles or other algae species, which are widely distributed and resistant to environmental changes [20], might be compared to the profiles of specimens collected in areas subjected to agrotoxics, solvents, and metals, as already done by Tewari et al. [19] with mercury. In addition, the baseline information obtained may also be applicable to a long-term ongoing research focusing on the sea urchins, which are important grazers in the archipelago [27], in order to determine possible relationships between the algae species consumed and their assimilation efficiency by those invertebrates.

4. Conclusion

The methodology used allowed determining the fatty acid profile of eleven seaweed species from Fernando de Noronha archipelago. As no specificity was found in the presence of a particular set of fatty acids, these compounds could not be used as taxonomic indicators in this case.

Conflict of Interests

The authors declare that do not have any conflict of interests.

Acknowledgment

The authors thank FAPESP, CAPES, and CNPq for financial support and the Brazilian Environmental Agency (IBAMA) and the Administration of Fernando de Noronha (ADEFN) for providing research permits.

References

[1] R. Martí, M. J. Uriz, and X. Turon, "Seasonal and spatial variation of species toxicity in Mediterranean seaweed communities: correlation to biotic and abiotic factors," *Marine Ecology Progress Series*, vol. 282, pp. 73–85, 2004.

[2] E. Schefuß, G. J. M. Versteegh, J. H. F. Jansen, and J. S. Sinninghe Damsté, "Lipid biomarkers as major source and preservation indicators in SE Atlantic surface sediments," *Deep-Sea Research I*, vol. 51, no. 9, pp. 1199–1228, 2004.

[3] F. Song, X. Fan, X. Xu, J. Zhao, Y. Yang, and J. Shi, "Cadinane sesquiterpenes from the brown Alga Dictyopteris divaricata," *Journal of Natural Products*, vol. 67, no. 10, pp. 1644–1649, 2004.

[4] Q. Xian, H. Chen, H. Liu, H. Zou, and D. Yin, "Isolation and identification of antialgal compounds from the leaves of *Vallisneria spiralis* L. by activity-guided fractionation," *Environmental Science and Pollution Research*, vol. 13, no. 4, pp. 233–237, 2006.

[5] M. P. Viera, J. L. Gómez Pinchetti, G. Courtois de Vicose et al., "Suitability of three red macroalgae as a feed for the abalone *Haliotis tuberculata* coccinea Reeve," *Aquaculture*, vol. 248, no. 1–4, pp. 75–82, 2005.

[6] E. Marinho-Soriano, P. C. Fonseca, M. A. A. Carneiro, and W. S. C. Moreira, "Seasonal variation in the chemical composition of two tropical seaweeds," *Bioresource Technology*, vol. 97, no. 18, pp. 2402–2406, 2006.

[7] L. M. P. Valente, A. Gouveia, P. Rema, J. Matos, E. F. Gomes, and I. S. Pinto, "Evaluation of three seaweeds Gracilaria bursa-pastoris, Ulva rigida and Gracilaria cornea as dietary ingredients in European sea bass (*Dicentrarchus labrax*) *juveniles*," *Aquaculture*, vol. 252, no. 1, pp. 85–91, 2006.

[8] C. Denis, M. Morançais, M. Li et al., "Study of the chemical composition of edible red macroalgae *Grateloupia turuturu* from Brittany (France)," *Food Chemistry*, vol. 119, no. 3, pp. 913–917, 2010.

[9] K. Bouarab, F. Adas, E. Gaquerel, B. Kloareg, J. P. Salaün, and P. Potin, "The innate immunity of a marine red alga involves oxylipins from both the eicosanoid and octadecanoid pathways," *Plant Physiology*, vol. 135, no. 3, pp. 1838–1848, 2004.

[10] Z. B. Aoun, R. B. Said, and F. Farhat, "Anti-inflammatory, antioxidant and antimicrobial activities of aqueous and organic extracts from Dictyopteris membranacea," *Botanica Marina*, vol. 53, no. 3, pp. 259–264, 2010.

[11] K. Uchida and K. Mogi, "Cellular fatty acid spectra of Pediococcus species in relation to their taxonomy," *Journal of General and Applied Microbiology*, vol. 18, no. 2, pp. 109–129, 1972.

[12] J. W. Moore and S. Ramamoorthy, *Heavy Metals in Natural Waters*, Spriger, Berlin, Germany, 1984.

[13] M. Kainz and A. Mazumder, "Effect of algal and bacterial diet on methyl mercury concentrations in zooplankton," *Environmental Science and Technology*, vol. 39, no. 6, pp. 1666–1672, 2005.

[14] S. V. Khotimchenko, V. E. Vaskovsky, and T. V. Titlyanova, "Fatty acids of marine algae from the pacific coast of North California," *Botanica Marina*, vol. 45, no. 1, pp. 17–22, 2002.

[15] V. E. Vaskovsky, S. V. Khotimchenko, B. Xia, and L. Hefang, "Polar lipids and fatty acids of some marine macrophytes from

the yellow sea," *Phytochemistry*, vol. 42, no. 5, pp. 1347–1356, 1996.

[16] L. Iveša, M. Blažina, and M. Najdek, "Seasonal variations in fatty acid composition of *Caulerpa taxifolia* (M. Vahl.) C. Ag. in the northern Adriatic Sea (Malinska, Croatia)," *Botanica Marina*, vol. 47, no. 3, pp. 209–214, 2004.

[17] S. V. Khotimchenko, "Fatty acids of species in the genus *Codium*," *Botanica Marina*, vol. 46, no. 5, pp. 456–460, 2003.

[18] R. Barreiro, L. Picado, and C. Real, "Biomonitoring heavy metals in estuaries: a field comparison of two brown algae species inhabiting upper estuarine reaches," *Environmental Monitoring and Assessment*, vol. 75, no. 2, pp. 121–134, 2002.

[19] A. Tewari, S. Thampan, and H. V. Joshi, "Effect of chlor-alkali industry effluent on the growth and biochemical composition of two marine macroalgae," *Marine Pollution Bulletin*, vol. 21, no. 1, pp. 33–38, 1990.

[20] H. R. Harvey and M. C. Kennicutt II, "Selective alteration of *Sargassum* lipids in anoxic sediments of the Orca Basin," *Organic Geochemistry*, vol. 18, no. 2, pp. 181–187, 1992.

[21] T. E. Cox and S. N. Murray, "Feeding preferences and the relationships between food choice and assimilation efficiency in the herbivorous marine snail *Lithopoma undosum* (Turbinidae)," *Marine Biology*, vol. 148, no. 6, pp. 1295–1306, 2006.

[22] K. M. Knesel, Z. S. Souza, P. M. Vasconcelos, B. E. Cohen, and F. V. Silveira, "Young volcanism in the Borborema Province, NE Brazil, shows no evidence for a trace of the Fernando de Noronha plume on the continent," *Earth and Planetary Science Letters*, vol. 302, pp. 38–50, 2011.

[23] G. Pedrini, Y. Ugadim, M. R. A. Braga, and S. M. B. Pereira, "Algas Marinhas bentônicas do Arquipélago de Fernando de Noronha, Brasil," *Boletim de Botânica da USP*, vol. 13, pp. 93–101, 1992.

[24] V. R. Eston, A. E. Migotto, E. C. Oliveira Filho, S. A. Rodrigues, and J. C. Freitas, "Vertical distribution of benthic marine organisms on rocky coasts of the Fernando de Noronha Archipelago (Brazil)," *Boletim do Instituto Oceanográfico de São Paulo*, vol. 34, pp. 37–53, 1986.

[25] K. Eder, "Gas chromatographic analysis of fatty acid methyl esters," *Journal of Chromatography B*, vol. 671, no. 1-2, pp. 113–131, 1995.

[26] K. Eder, A. M. Reichlmayr-Lais, and M. Kirchgessner, "Studies on the methanolysis of small amounts of purified phospholipids for gas chromatographic analysis of fatty acid methyl esters," *Journal of Chromatography A*, vol. 607, no. 1, pp. 55–67, 1992.

[27] J. S. Eklöf, M. de la Torre-Castro, M. Gullström et al., "Sea urchin overgrazing of seagrasses: a review of current knowledge on causes, consequences, and management," *Estuarine, Coastal and Shelf Science*, vol. 79, pp. 569–580, 2008.

20

Elemental Contents in Hair of Children from Two Regions in Dar Es Salaam

Najat K. Mohammed

Department of Physics, University of Dar es Salaam, P.O. Box 35063, Dar es Salaam, Tanzania

Correspondence should be addressed to Najat K. Mohammed, njkassim@yahoo.com

Academic Editor: A. Denizli

The work presented in this paper is part of the study which aims at determining the levels of elements in hair of children in Tanzania as a bioindicator of their nutrition and health. In this paper, the levels of trace elements in hair from children living in Dar es Salaam have been analysed. The analysis was carried out by long and short irradiation INAA at the reactor centre of the Institute of Nuclear Physics, Rez Czech Republic. 22 samples were collected from children living at Kiwalani about 12 km from Dar es Salaam city and 16 samples from children living at Mlimani, the main campus of University of Dar es Salaam. A total of 34 elements were found in the hair of the children. There were no big differences between the concentration levels of the essential elements in hair samples collected from the children which might indicate the same food consumption habits.

1. Introduction

Trace element concentrations in human body are normally determined by the analysis of blood, serum, plasma, urine, or hair. Hair has an advantage over blood and urine as it accumulates elements in higher concentrations [1] and its collection is fast, painless, and easy [2]. Further more, the analysis of hair gives long-term information on the behaviour of the elements in the body (up to 2-3 years depending on length of hair) in contrast to blood and urine which provide short-term information [3]. Hair is also a stable biological material, which can easily be stored at room temperature for a long time without a change in its composition.

A number of researchers have used hair samples to assess the nutritional and environmental exposure to humans. Othman and Spyrou (1980) reported increased levels of mercury (Hg) in hair samples collected from female adults and children in the Machakos district in Kenya, which are associated with the use of Hg compounds as cosmetics and for lightening the colour of the skin [4]. Airey [5], analysed human hair samples for Hg in relation to fish consumption. The results showed a direct relationship between mean hair Hg and the rate of fish consumption and hence to the diet of the society. Ikingura et al. (1996) reported concentrations of Hg in hair samples which were consistent to the concentration levels found in fish consumed by people living in the Lake Victoria goldfields in Tanzania [6]. Fordyce et al. (2000), reported a clear link between dietary Se intake and its concentrations in hair [7]. Hair from children has also been used in several research projects to assess their health and nutritional status [8–12]. The effect of dietary factor, geographical locations, and environmental exposure on the hair elemental concentration of children from Tanzania has been presented in a previous paper [13].

In this study, hair samples of children from two socioeconomic groups in Dar es Salaam have been analysed.

2. Sample Collection

A total of 34 samples of hair from children of age between 5–15 years (Average 10 years) were collected from two regions of Dar es Salaam representing two socioeconomic classes. 22 samples were collected from children living at Kiwalani about 12 km from Dar es Salaam city and 16 samples from children living at Mlimani, the main campus of University of Dar es Salaam. The children from Kiwalani had parents with low income and primary school education whilst children from Mlimani were from the academic staff of the University of Dar es Salaam. In this paper, the former group will be considered as the low class while the later group as the middle-class children.

About 70 to 500 mg of hair was cut using precleaned stainless steel scissors as near as possible to the scalp from the left side of the nape. All children or the parents (when a child was below 7 years) were interviewed at the time of sampling to obtain some general information on their health. The hair samples were then cut into 1 cm long strands put into plastic bags and kept in desiccators.

3. Sample Preparations

The hair samples were washed according to methods proposed by the IAEA 1977 [14]. The samples were then air dried for a minimum of 48 hours in a clean room laboratory and then inserted into a Polyethylene disc and sent to the Rez reactor centre for irradiation.

4. Sample Analysis

The irradiations were carriedout in the nuclear reactor LVR-15 of the Nuclear Research Institute Řež, operated at 9 MW. For short irradiation, samples and standards were irradiated in the active core of the nuclear reactor. The samples and standards were irradiated separately in polyethylene (PE) rabbits for 1 min, waited for 10 min before counted for 10 min. After irradiation, the capsules containing the samples were surface washed with water and ethyl alcohol and counted using a well-shielded n-type hyper pure germanium detector (HPGe) of relative efficiency 20.8%.

For intermediate and long-lived irradiations, the samples and standards were irradiated in a water-cooled Al can in a channel located at the outskirts of the reactor core with neutron fluence rate 8×10^{13} and 2×10^{13} n cm^{-2} s^{-1} for thermal and fast neutrons, respectively. The samples were irradiated for 2 hours, followed by 4 days of initial decay and 30 min counting time. Further 2 hours counting was carried out after 1 month decay. A total of 40 samples and standards each wrapped in Al foil were placed in one container of 9.5 cm length to form a column and the neutron flux monitors (Fe disc of about 60 mg and 10 mm in diameter) were inserted between each set of 5 samples with the standards. After the irradiation, samples and standards were counted in a shielded p-type HPGe detector of relative efficiency 20.8%.

The concentration levels of the elements were calculated by the comparative method of INAA, using multielement standards (MES) [15]. Quality control was carried out using certified hair reference material GBW 07601 (National Research Centre for Certified Reference Materials, China), NIST Orchid Leaves (SRM 1571), Apple Leaves (SRM 1515), and Urban Particulate (SRM 1648) analysed with the samples. As Table 1 shows, the experimental values were all in good agreement with the certified values within 16% deviation.

5. Results and Discussion

Twenty nine elements which are Ag, Al, As, Au, Ba, Br, Ca, Cd, Ce, Cl, Co, Cr, Cu, Eu, Fe, I, K, Mg, Mn, Na, S, Sb, Sc, Se, Sm, Sr, U, V, Zn were detected in hair samples of children living in two regions of Dar es Salaam (Table 2). The data from the children living at Mlimani, the University of Dar es Salaam (middle class) were compared with the data from the children living in Kiwalani (low class). The Mann-Whitney test indicated that children from Mlimani had significantly ($P < 0.05$) higher concentrations of Al, K, Sr, and Sc whilst children from Kiwalani had higher concentrations of I. The essential elements Zn in hair from both groups were similar (~105 μg/g) whilst Fe and Cu were slightly but not significantly (19% and 27%, resp.) higher in children from Mlimani than children of the other group.

The mean concentrations for Zn and Cu in both groups were lower than the mean concentrations of these elements in children of the same age group reported in Kazakhstan, Hong Kong, and Italy [10–12]. Hambidge et al. (1976), reported Zn concentrations in the range of <105 μg/g in the hair of low-income children who had their hair Zn concentration increased by 71% after 3 month of being provided with rich in Zn meals [8]. In this study, 41% and 58% of the samples from Kiwalani and Mlimani, respectively, had concentration below 105 μg/g. The mean concentration of Cu from both groups were lower than the concentrations reported in Kazakhstan, Hong Kong, and Italy, although these values were within the range for normal concentration of hair Cu in healthy children (6–23 μg/g) reported by Weber et al. [9]. Nevertheless, 14% and 8% of the samples from Kiwalani and Mlimani, respectively, had concentrations of less than 6 μg/g, which Weber et al. (1990) obtained in the malnourished children [9].

There were no statistical differences in the means of the essential elements Fe, Zn, and Cu between the two groups which might indicate similar consumption habits. Most of the Tanzanian diets include cereals (rice and maize) as the main dish and beans and green leaves are taken as sauces. Normally the bioavailability of essential elements such as Fe and Zn is low in cereals and food of plant nature because these foods have high contents of phytic acid as well as fibers and polyphenols, which form insoluble salts with Fe and Zn and thus hindering their absorption [16]. The absorption of Fe and Zn could be enhanced if a meal is taken with a portion of animal protein because these proteins prevent the elements from forming the insoluble phytates with the phytic acid in the foods [17].

Since there were no big differences on essential elemental concentration between the two groups, the observation demonstrated that residents of Dar es Salaam irrespective of their classes have the same food consumption habits. Although the number of samples (22 for low class and 16 for middle class) was not big enough to come to a definite conclusion, there is a need for nutrition awareness and diet

TABLE 1: The concentration levels of elements (μg/g) in four-standard reference materials obtained in this work (Exp.) together with the certified values (Cert.).

	Hair (GBW07601)		Urban particulate (SRM 1648)		Apple leaves (SRM 1515)		Orchid leaves (SRM 1571)	
	Exp.	Cert.	Exp.	Cert.	Exp.	Cert.	Exp.	Cert.
Na	139 ± 5.9	152 ± 10	4028 ± 21	4250 ± 20	29 ± 1	24 ± 1	81 ± 0.7	82 ± 6
K%			1.0 ± 0.05	1.1 ± 0.01			1.4 ± 0.02	1.5 ± 0.07
Al					277 ± 2	286 ± 9		
S%	4.5 ± 0.1	4.3 ± 0.2						
Cl	26 ± 4	35*			564 ± 6	579 ± 23		
V	<0.2	0.03*			0.26 ± 0.03	0.3 ± 0.03		
Mn	6 ± 0.2	6.3 ± 0.5			52 ± 0.2	54 ± 3		
Mg					2583 ± 15	2710 ± 80		
Ca%					1.4 ± 0.02	1.5 ± 0.02		
Cu	10 ± 0.8	10.6 ± 0.7	545 ± 19	609 ± 27	7 ± 0.6	5.6 ± 0.2	12 ± 0.7	12 ± 1
Zn	187 ± 12	190 ± 5	4305 ± 53	4760 ± 140			22 ± 0.6	25 ± 2
Br	0.3 ± 0.01	0.4*	494 ± 3	500 ± 25	1.2 ± 0.1	1.8	9 ± 0.1	10 ± 1
Sr	21 ± 3	24 ± 1		207 ± 15			30 ± 4	37 ± 3
I	0.7 ± 0.1	0.8*						
Ba	17 ± 0.4	17 ± 1	667 ± 92	737 ± 40	44 ± 1	49 ± 2	42 ± 4	44 ± 4
Dy	<0.03	0.017						
U	0.1 ± 0.001	0.09*						
As	0.3 ± 0.001	0.28 ± 0.04	115 ± 1	115 ± 10			11 ± 0.1	10 ± 1
Cd	<4	0.11 ± 0.02	79 ± 9	75 ± 7				0.1 ± 0.02
Sb	0.082 ± 0.004	0.1 ± 0.01	46 ± 2	45 ± 2			3 ± 0.1	3 ± 0.3
La	0.055 ± 0.006	0.05 ± 0.01	36 ± 0.5	44 ± 2			1 ± 0.02	
Sm	0.012 ± 0.002	0.012*	4 ± 0.03	4 ± 0.3			0.1 ± 0.001	
Au	0.003 ± 0.001	0.0025*						
Sc	8 ± 0.1**	8 ± 1**	6 ± 0.2	7 ± 0.1			0.1 ± 0.002	0.1 ± 0.01
Cr	<0.260	0.37 ± 0.05	357 ± 9	403 ± 12			2.2 ± 0.1	2.6 ± 0.3
Fe	52 ± 6	54 ± 6*	3.7 ± 0.1%	3.9 ± 0.1%			252 ± 11	300 ± 28
Co	0.07 ± 0.006	0.07 ± 0.01	16 ± 0.4	18 ± 2			0.14 ± 0.01	0.16 ± 0.04
Se	0.60 ± 0.06	0.60 ± 0.03	25 ± 9	27 ± 1				0.08 ± 0.01
Ag	<0.07	0.029 ± 0.006	6 ± 0.3	6 ± 0.2				
Ce	0.15 ± 0.01	0.12 ± 0.03	56 ± 1	55 ± 4			1 ± 0.1	1 ± 0.1

*Not certified.
**ng/g.

planning for both groups especially as the middle class group could afford to have additional animal protein in their meals to enhance the absorption of Fe and Zn. Education programmes of food and nutrition are an alternative way of promoting nutritional awareness in a population. This method could be effective in any society because malnutrition is not always a result of not having enough food; sometimes it is just on how to turn the foods into nutritious meals.

6. Conclusion

The results in this study showed no big differences between the means of the essential elements Fe, Zn, and Cu in hair samples collected from children in the two socioeconomic classes living in Dar es Salaam. This observation might indicate the same food consumption habits. The mean concentrations for Zn and Cu in both groups were lower than the mean concentrations of these elements in children of the same age group reported in Kazakhstan, Hong Kong, and Italy [10–12]. Although the number of samples was not big enough to come to a definite conclusion, the socioeconomic classes of children in Dar es Salaam did not have much influence on their hair elemental concentrations. Since the concentration of hair Zn has been associated with nutrition, poverty, and/or cultural habits of the people [18], education on nutrition awareness and diet planning is needed for all socio groups in Tanzania especially as the high-income group could afford to have balanced diet.

TABLE 2: The concentration of elements in $\mu g/g$ (mean ± Std. error mean) found in hair children living in Dar es Salaam.

Elements	Mlimani ($n = 16$)	Kiwalani ($n = 22$)
Na	15.93 ± 2.71	31.17 ± 6.14
Mg	74 ± 13	113 ± 26
Al	103 ± 14	72 ± 14
S	46114 ± 577	46827 ± 582
Cl	1305 ± 142	1274 ± 192
K	44.78 ± 6.00	30.51 ± 4.93
Ca	447 ± 36	490 ± 48
V	0.18 ± 0.02	0.15 ± 0.03
Mn	4.88 ± 1.00	7.77 ± 1.24
Cu	10.64 ± 1.54	8.37 ± 0.53
Zn	104 ± 9	107 ± 4.81
Br	6.29 ± 0.87	6.34 ± 0.78
Sr	28.04 ± 2.25	7.42 ± 1.50
I	3.55 ± 1.03	6.74 ± 0.88
Ba	11.62 ± 2.31	6.67 ± 0.51
U	0.062 ± 0.003	0.05 ± 0.01
As	0.22 ± 0.06	0.17 ± 0.02
Sb	0.30 ± 0.06	0.40 ± 0.10
La	0.13 ± 0.03	0.29 ± 0.11
Sm	0.023 ± 0.004	0.013 ± 0.002
Au	0.13 ± 0.08	0.16 ± 0.04
Sc	0.014 ± 0.002	0.009 ± 0.002
Cr	0.40 ± 0.06	0.33 ± 0.05
Fe	59 ± 7	50 ± 9
Co	0.10 ± 0.02	0.19 ± 0.07
Se	0.49 ± 0.02	0.84 ± 0.36
Ag	0.15 ± 0.02	0.24 ± 0.03
Ce	0.31 ± 0.03	0.02 ± 0.17
Eu	0.025 ± 0.001	0.116 ± 0.002

Acknowledgments

This research project has been supported by the European Commission under the 6th Framework Programme through the Key Action: Strengthening the European Research Area, Research Infrastructures, Contract no.: RII3-CT-2003-505925. The authors would also like to acknowledge the Ford Foundation and the International Programme in Physical Sciences of the University of Uppsala (IPPS) for financial assistance and the staff of NPI Rez, Czech Republic for their technical advise and support.

References

[1] V. Iyengar and J. Woittiez, "Trace elements in human clinical specimens: evaluation of literature data to identify reference values," *Clinical Chemistry*, vol. 34, no. 3, pp. 474–481, 1988.

[2] J. Dombovári and L. Papp, "Comparison of sample preparation methods of elemental analysis of human hair," *Microchemical Journal*, vol. 59, no. 2, pp. 187–193, 1998.

[3] I. Rodushkin and M. D. Axelsson, "Application of double focusing sector field ICP-MS for multielemental characterization of human hair and nails. Part II. A study of the inhabitants of northern Sweden," *Science of the Total Environment*, vol. 262, no. 1-2, pp. 21–36, 2000.

[4] I. Othman and N. M. Spyrou, "The abundance of some elements in hair and nail from the Machakos district of Kenya," *Science of the Total Environment*, vol. 16, no. 3, pp. 267–278, 1980.

[5] D. Airey, "Total mercury concentrations in human hair from 13 countries in relation to fish consumption and location," *Science of the Total Environment*, vol. 31, no. 2, pp. 157–180, 1983.

[6] J. R. Ikingura, H. Akagi, J. Mujumba, and C. Messo, "Environmental assessment of mercury dispersion, transformation and bioavailability in the Lake Victoria Goldfields, Tanzania," *Journal of Environmental Management*, vol. 81, no. 2, pp. 167–173, 2006.

[7] F. M. Fordyce, Z. Guangdi, K. Green, and X. P. Liu, "Soil, grain and water chemistry in relation to human selenium-responsive diseases in Enshi District, China," *Applied Geochemistry*, vol. 15, no. 1, pp. 117–132, 2000.

[8] K. M. Hambidge, P. A. Walravens, R. M. Brown et al., "Zinc nutrition of preschool children in the Denver Head Start program," *American Journal of Clinical Nutrition*, vol. 29, no. 7, pp. 734–738, 1976.

[9] C. W. Weber, G. W. Nelson, M. Vazquez De Vaquera, and P. B. Pearson, "Trace elements in the hair of healthy and malnourished children," *Journal of Tropical Pediatrics*, vol. 36, no. 5, pp. 230–234, 1990.

[10] C. K. Man, Y. H. Zheng, and P. K. Mak, "Hair analysis of spastic children in Hong Kong," *Science of the Total Environment*, vol. 191, no. 3, pp. 291–295, 1996.

[11] N. Violante, O. Senofonte, G. Marsili, P. Meli, M. E. Soggiu, and S. Caroli, "Human hair as a marker of pollution by chemical elements emitted by a thermoelectric power plant," *Microchemical Journal*, vol. 67, no. 1–3, pp. 397–405, 2000.

[12] M. Chiba, K. Sera, M. Hashizume et al., "Element concentrations in hair of children living in environmentally degraded districts of the East Aral Sea region," *Journal of Radioanalytical and Nuclear Chemistry*, vol. 259, no. 1, pp. 149–152, 2004.

[13] N. K. Mohammed, J. Mizera, and N. M. Spyrou, "Elemental contents in hair of children from Zanzibar in Tanzania as bioindicator of their nutritional status," *Journal of Radioanalytical and Nuclear Chemistry*, vol. 276, no. 1, pp. 125–128, 2008.

[14] IAEA, Activation Analysis of Hair as an Indicator of Contamination of Man by Environmental Trace Element Pollutants. A cooperative Report on the Co-ordinated Research Programme: nuclear-based methods for Analysis of Pollutants in Human Hair. Report IAEA/RL/50, 1977.

[15] Z. Řanda, M. Vobecký, J. Kuncír, and J. Benada, "Multielement standards in routine reactor neutron activation analysis," *Journal of Radioanalytical Chemistry*, vol. 46, no. 1, pp. 95–107, 1978.

[16] N. Martínez-Navarrete, M. M. Camacho, J. Martínez-Lahuerta, J. Martínez-Monzó, and P. Fito, "Iron deficiency and iron fortified foods—a review," *Food Research International*, vol. 35, no. 2-3, pp. 225–231, 2002.

[17] E. R. Morris, *Trace Elements in Human and Animal Nutrition*, Edt Mertz, London, UK, 5th edition, 1987.

[18] J. Erten, A. Arcasoy, A. O. Cavdar, and S. Cin, "Hair zinc levels in healthy and malnourished children," *American Journal of Clinical Nutrition*, vol. 31, no. 7, pp. 1172–1174, 1978.

21

Natural Radioactivity in Soil and Water from Likuyu Village in the Neighborhood of Mkuju Uranium Deposit

Najat K. Mohammed and Mohamed S. Mazunga

Department of Physics, University of Dar es Salaam, P.O. Box 35063, Dar es Salaam, Tanzania

Correspondence should be addressed to Najat K. Mohammed; njkassim@yahoo.com

Academic Editor: Xiaolin Hou

The discovery of high concentration uranium deposit at Mkuju, southern part of Tanzania, has brought concern about the levels of natural radioactivity at villages in the neighborhood of the deposit. This study determined the radioactivity levels of 30 soil samples and 20 water samples from Likuyu village which is 54 km east of the uranium deposit. The concentrations of the natural radionuclides ^{238}U, ^{232}Th, and ^{40}K were determined using low level gamma spectrometry of the Tanzania Atomic Energy Commission (TAEC) Laboratory in Arusha. The average radioactivity concentrations obtained in soil samples for ^{238}U (51.7 Bq/kg), ^{232}Th (36.4 Bq/kg), and ^{40}K (564.3 Bq/kg) were higher than the worldwide average concentrations value of these radionuclides reported by UNSCEAR, 2000. The average activity concentration value of ^{238}U (2.35 Bq/L) and ^{232}Th (1.85 Bq/L) in water samples was similar and comparable to their mean concentrations in the control sample collected from Nduluma River in Arusha.

1. Introduction

The radioactivity level from the natural radionuclides is termed as background radiation which will depend on the amount of the radioactive materials in the environment. The background radiation can be high if the environment is polluted either from man-made or natural activities. It can also be high in regions with deposit of mineral resources such as uranium ores and phosphate [1]. Materials from the deposit may be brought to the surface soil through processes such as weathering of rocks and soil formation. They can also leach into the groundwater system, contaminate it, and lead to pollution far away from the source.

In Tanzania, high concentrations of uranium deposits of up to 464 ppm have recently been reported at Mkuju in Namtumbo area at Ruvuma region. This has brought concern about the level of natural radioactivity in the soil and drinking water at villages in the neighborhood of the deposit. This is because there are reports in the literature which indicate high radioactivity levels in regions near uranium deposits [2, 3]. Rabesiranana et al. (2008) in Madagascar and Sartandel et al. (2009) in India reported enhanced level of ^{238}U and ^{232}Th in the soil around uranium deposits as compared to the world average [2, 4]. Enkhbat et al. (2008) reported high activity concentration of ^{226}Ra, ^{232}Th, and ^{40}K in soil around Gurvanbulag uranium deposit in the eastern part of Mongolia [3].

The present study has determined concentration levels of the ^{238}U decay series and the ^{232}Th decay series as well as ^{40}K in soil and water from Likuyu village which is 54 km east of the Mkuju uranium deposit in Tanzania. The paper will use ^{238}U and ^{232}Th to indicate the concentrations of these decay series. The data will offer useful and necessary information in the monitoring of environmental contamination which will provide appropriate and better protection guidelines to the public.

2. Material and Methods

2.1. Study Area. The Likuyu-Seka-Maganga village is at Namtumbo district in Ruvuma region southern Tanzania. The village is situated at longitude 36°16′ east and latitude 10°19′ south of the Equator. It is near to the Undendeule Forest Reserve (UFR) at the south and the Selous Game Reserve (SGR) at the north. The Likuyu village is situated at about

54 km east of the Mkuju uranium deposit. The only accessible road to the Mkuju uranium deposit is through Likuyu village.

The ambient temperature at the Likuyu village varies from minimum of 18.8°C to a maximum of 28.2°C. The wind blows from the Mkuju uranium deposit through the Likuyu village in the month of December to February with speed of up to 11.6 m/s. Two rivers, which are Mkuju River and Kilowelo River, flow from the Mkuju uranium deposit to the northern and southern parts of the Likuyu village, respectively. The water from these rivers is domestically used by the people of the Likuyu village. The village has a total population of about 10,000 people settled in eight hamlets, seven of which lie along the road to the Mkuju uranium deposit (Mantra, 2010) [5].

2.2. Sampling Methodology. The village was divided into three sampling zones: the northern, the central, and the southern zones. Ten samples of soil were collected randomly from each sampling area to make a total of 30 samples of soil. The samples were collected at 0–5 cm depth level as reported elsewhere [6]. The collected samples were then placed in labeled polythene bags and transferred to the TAEC laboratory for preparation and analysis.

Twenty water samples were collected from the two rivers passing through the village. The samples were filled in a one liter acid precleaned polyethylene container to avoid wall absorption (IAEA, 1989) and were wrapped using insulation tape to avoid spill during transport of the samples from sampling point to the TAEC laboratory for preparation and analysis [7].

2.3. Sample Preparation. In the laboratory, the soil samples were sieved by a 2 mm mesh to remove larger objects and then ground using mortar and pestle to fine powder in order to have the same matrix as the reference sample. After that, the samples were dried in an oven at a temperature around 45–50°C for several hours until constant weights were attained and then placed in desissators to avoid moisture absorption as described elsewhere [8]. Samples of about 500 g each were packed in Marinelli beaker of about 500 cm^3 volume and sealed using silicon and plastic tapes for air tight. The samples were left for a minimum of 28 days for radioactive secular equilibrium between ^{222}Rn-radon gas and its decay products (^{214}Pb, ^{214}Bi, and ^{226}Ra), from the ^{238}U decay series to be acquired.

The activity of ^{238}U was determined using the gamma-lines of ^{226}Ra (186 keV), ^{214}Pb (295 keV and 351 keV), ^{214}Bi (1238 keV and 1378 keV), ^{228}Ac (338 keV, 911 keV, 965 keV, 969 keV), and ^{208}Tl (860 keV). The gamma-lines of 609.3 keV, 1120 keV (from ^{238}U series), 583.2 keV, 727.3 keV, and 795 keV (^{232}Th series) were omitted from the analyses as sensitivity analysis has shown elsewhere that they are affected by coincidence summing [9]. The ^{40}K was measured from its gamma line energy of 1460.8 keV.

2.4. Gamma Spectrometry. A lead-shielded coaxial high-purity germanium detector (HPGe) of relative efficiency 51% and resolution of 2.1 keV for 1332 keV ^{60}Co gamma-ray source was used for low level counting of the samples. The detector chamber was shielded with three layers of copper, cadmium, and lead of 30 mm, 3 mm, and 100 mm thick, respectively. The γ-ray energy calibration was performed daily using standard radiation sources of ^{133}Ba, ^{109}Cd, ^{54}Mn, ^{57}Co, ^{60}Co ^{65}Zn, ^{22}Na, and ^{137}Cs. Efficiency calibration was determined using In Situ Object Counting-system (ISOCS) of Genie 2000 software.

The activity concentration (in Bq kg^{-1}), A_{Ei} of a radionuclide i and for a peak of energy E, in a samples was calculated using the following formula [10]:

$$A_{Ei} = \frac{N_{Ei}}{\gamma_d \times \varepsilon_E \times m_s \times T}, \quad (1)$$

where N_{Ei} is the net peak area at energy E, ε_E is the detection efficiency at the energy E, T is the counting live time in seconds, γ_d is the gamma ray yield per disintegration of the specific radionuclide for a transition at energy E, and m_s is the mass of the dry weight in kg of the measured sample.

The IAEA soil 375 was used as a standard reference material to evaluate the precision and accuracy of the gamma results. The standard reference material was counted at the same time as used for counting the samples (8 hours), and then its activity at various energies was calculated and compared with the certified value after correction for the decay by considering the date (December 31st, 1991) provided in the data sheet. As shown in Table 1, the experimental concentration values agreed well with the recommended values within 10% accuracy.

3. Results and Discussion

3.1. Soil Samples. The average natural radionuclide activity concentrations in the soil samples collected from the three zones at the Likuyu village are reported in Table 2. Specific activities concentrations of ^{238}U, ^{232}Th, and ^{40}K are reported in Bq kg^{-1} dry weight.

The lowest mean values of all three radionuclides were found in samples from the central part of the village. The northern zone had the highest concentrations of ^{238}U and ^{40}K, which were each 1.2 times higher than their values determined in samples from the central zone. Samples from the southern zone had the highest mean value of ^{232}Th. This value was 1.3 times higher than the mean value obtained in samples collected from the central zone. Rivers Mkuju and Kilowero flow from the Mkuju uranium deposit to the northern and southern zones, respectively. Higher values of radionuclides in samples from these zones compared to the values determined in samples from central zone might indicate the rivers transportation of radionuclides from Mkuju uranium deposit to the village.

The mean activity concentrations of natural radioactivity found in the soil in the present study were compared with the range and average of natural radioactivity concentration levels in soils of eight countries reported by UNSCEAR (2000) (Table 3) [11]. The mean concentration of ^{238}U found in this study was found to be higher than its mean concentrations in soil of 75% of the countries reported by UNSCEAR (2000). The value is also higher than the world average

Table 1: The Standard reference values and experimental activity values of the IAEA reference soil 375 (Bq/kg ± SD).

Radionuclides	Energy (keV)	Experimental activity concentration	Certified reference values
^{226}Ra	186.5 keV	20.54 ± 0.9	20.0 ± 0.9
^{214}Pb	295.2	19.46 ± 0.8	19.96 ± 0.8
	351.92	20.61 ± 1.1	19.96 ± 0.5
^{214}Bi	1764	20.94 ± 0.8	19.96 ± 0.4
	2477.7	22.46 ± 1.2	19.96 ± 0.9
^{40}K	1460	499.93 ± 2.1	423.40 ± 0.2
^{228}Ac	338.5	21.58 ± 1.5	20.46 ± 0.3
^{208}TI	860.4	22.20 ± 1.1	20.46 ± 0.3
	2614.5	22.44 ± 1.7	20.46 ± 0.6
^{228}Ac	910.7	22.29 ± 0.7	20.46 ± 0.5
	968.5	23.69 ± 0.9	20.46 ± 0.5

Table 2: The average activity concentration of radionuclides in the collected soil samples (Bq/kg ± SEM).

Soil samples (N = 10)	Activity concentration		
	^{238}U	^{232}Th	^{40}K
Northern zone	57 ± 5	36 ± 3	616 ± 113
Southern zone	52 ± 4	41 ± 3	558 ± 10
Central zone	46 ± 3	32 ± 3	519 ± 7
Average	52 ± 4	36 ± 3	564 ± 10

Table 3: Comparison of natural radioactivity levels in the soil of Likuyu village with those in other countries as given in UNSCEAR (2000) [11].

Countries	Activity concentration (Bq/kg)		
	^{238}U	^{232}Th	^{40}K
Malaysia	66 (49–86)	82 (63–110)	310 (170–430)
China	32 (2–440)	41 (1–360)	440 (38–760)
Egypt	37 (6–120)	18 (2–96)	320 (29–650)
Algeria	30 (2–110)	25 (2–140)	370 (66–1150)
India	29 (7–81)	64 (14–160)	400 (38–760)
Portugal	49 (26–82)	51 (22–100)	840 (220–1230)
Hong Kong SAR	84 (25–130)	95 (16–200)	530 (80–1100)
United States	35 (4–140)	35 (4–130)	370 (100–700)
World average	30 (16–110)	35 (11–64)	400 (140–850)
Present study	**52**	**36**	**564**

value; however, it lies within the ranges of ^{238}U revealed in soils of all countries in the UNSCEAR report (UNSCEAR, 2000). The mean concentration of ^{232}Th obtained in the study is comparable to the world average value but lower than the mean concentrations in 63% of the countries reported by UNSCEAR (2000) [11].

3.2. Water Samples. The average natural radionuclide activity concentrations in the water samples collected from the two rivers in Likuyu village are reported in Table 4. Specific activities concentrations of ^{238}U, ^{232}Th, and ^{40}K are reported in Bq L^{-1}.

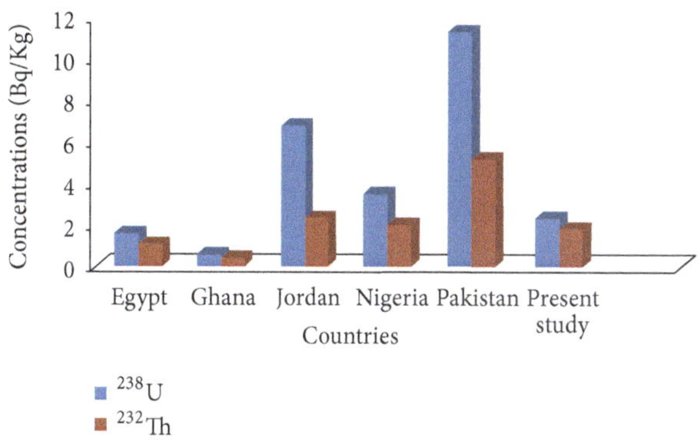

Figure 1: The average radioactivity concentrations values of water samples obtained in the present study compared with studies conducted elsewhere in the world.

Table 4: The average activity concentration of radionuclides in the collected water samples (Bq/L ± SEM).

Water samples	Mean radioactivity concentration		
	^{238}U	^{232}Th	^{40}K
Mkuju River (W1)	2.5 ± 0.4	1.9 ± 0.2	11.0 ± 2.0
Kilowelo River (W2)	2.2 ± 0.3	1.8 ± 0.1	9.2 ± 1.2
Control sample	2.1 ± 0.3	1.7 ± 0.2	2.0 ± 0.8

As Table 4 shows, samples from W1 and W2 had similar concentration of each radionuclide ^{238}U, ^{232}Th, and ^{40}K. The concentration of ^{238}U and ^{232}Th was similar to the concentration in the control sample collected from Nduluma River in Arusha, whilst the mean activity level of ^{40}K was 5 times higher than that of the control sample.

Figure 1 compares the mean concentration of radionuclides from the two rivers in this study with those obtained in rivers of 5 countries reported in the literature.

The mean concentration of ^{238}U from the two rivers in Likuyu was similar and approximately 4 times higher than the activity value reported by Faanu et al. (2011) in a river at Tarkwa Gold mine in Ghana [12]. This value was only 1.5 times higher than the value reported by El Arabi et al. (2006) in natural water resources from Elba protective area in Egypt [13].

The mean value reported in this study was also 3 times lower than the value reported in water samples collected from a river in Jordan. The mean concentration level of ^{232}Th found in this study was lower than the mean activity reported in Pakistan, Jordan, and Nigeria, but slightly higher than the value reported in Ghana and Egypt.

4. Conclusion

This study determined the activity concentrations of natural occurring radionuclides in soil and water samples collected at Likuyu village which is 54 Km east of the Mkuju uranium deposit in southern Tanzania. The measurements were carried out using low level gamma ray spectrometry technique at the TAEC laboratory in Arusha.

The mean concentration of ^{238}U (51.7 Bq/Kg) soil analyzed in this study was found to be higher than its mean concentrations in soil of 75% of the countries reported by UNSCEAR (2000). The value is also higher than the world average value; however, it lies within the ranges of ^{238}U shown in soils of all countries in the UNSCEAR report (UNSCEAR, 2000). The mean concentration of ^{232}Th obtained in the study is comparable to the world average value but lower than the mean concentrations in 63% of the countries reported by UNSCEAR (2000). In this study, the highest mean values of all three radionuclides were found in samples from northern and southern parts of the village, which might be influenced by the flow of rivers Mkuju and Kilowero from the Mkuju uranium deposit to the village.

The results of radioactivity concentrations in the water sample obtained in this study were comparable with the concentration values of natural radioactivity reported in the literature. Moreover, their concentrations were similar to those obtained in the control sample collected from Nduluma River in Arusha. It is recommended to consider seasonal variations in determining radioactivity at Likuyu village. This is because during rainy season rivers might have elevated concentration of radionuclides due to the soil erosion at Mkuju Uranium Deposit.

Conflict of Interests

The authors have no financial or personal interest with Tanzania Atomic Energy Commission (TAEC).

Disclosure

TAEC was mentioned in this paper because the samples were analysed at TAEC Laboratory in Arusha.

Acknowledgments

The authors are greatly acknowledging the Tanzania Atomic Energy Commission for the financial and material support which made this study possible. They are also grateful to Mantra Resources Tanzania Limited for allowing them to perform research in their licensed area and for providing transportation of the samples from Mkuju to Dar es Salaam. And TAEC has also provided extra fund to the researchers for sampling in Mkuju.

References

[1] J. A. Oyedele, "Assessment of the natural radioactivity in the soils of Windhoek city, Namibia, Southern Africa," *Radiation Protection Dosimetry*, vol. 121, no. 3, pp. 337–340, 2006.

[2] N. Rabesiranana, M. Rasolonirina, F. Terina, A. F. Solonjara, and R. Andriambololona, "Top soil radioactivity assessment in a high natural radiation background area: the case of Vinaninkarena, Antsirabe-Madagascar," *Applied Radiation and Isotopes*, vol. 66, no. 11, pp. 1619–1622, 2008.

[3] N. Enkhbat, N. Norov, B. Bat-Erdene, G. Khuukhenkhuu, and B. Otgooloi, "Study of natural background radiation around gurvanbulag uranium deposit area," in *Proceedings of the 1st International Ulaanbaatar Conference on Nuclear Physics and Applications (UBC '08)*, pp. 144–147, Ulaanbaatar, Mongolia, September 2008.

[4] S. J. Sartandel, S. K. Jha, S. V. Bara, R. M. Tripathi, and V. D. Puranik, "Spatial distribution of uranium and thorium in the surface soil around proposed uranium mining site at Lambapur and its vertical profile in the Nagarjuna Sagar Dam," *Journal of Environmental Radioactivity*, vol. 100, no. 10, pp. 831–834, 2009.

[5] E. I. S. Mantra, "Mantra Tanzania Limited Environmental Impact Statement for the Proposed Uranium Mining Project at Mkuju River Project, Namtumbo," Volume 1, Final Report, March 2010.

[6] O. S. Ajayi, "Measurement of activity concentrations of ^{40}K, ^{226}Ra and ^{232}Th for assessment of radiation hazards from soils of the southwestern region of Nigeria," *Radiation and Environmental Biophysics*, vol. 48, no. 3, pp. 323–332, 2009.

[7] IAEA, Radioactive Fallout in Food and Agriculture IAEA-TECDOC-494, IAEA, Vienna, Austria, 1989.

[8] B. Holynsca and J. A. Jasion, "Simultaneous determination of some trace metals in plant material by EDXRF," *Journal of Radioanalytical and Nuclear Chemistry*, vol. 105, no. 2, 1986.

[9] R. T. Newman, R. Lindsay, K. P. Maphoto et al., "Determination of soil, sand and ore primordial radionuclide concentrations by full-spectrum analyses of high-purity germanium detector spectra," *Applied Radiation and Isotopes*, vol. 66, no. 6-7, pp. 855–859, 2008.

[10] M. Tzortzis, H. Tsertos, S. Christofides, and G. Christodoulides, "Gamma-ray measurements of naturally occurring radioactive samples from Cyprus characteristic geological rocks," *Radiation Measurements*, vol. 37, no. 3, pp. 221–229, 2003.

[11] United Nations Scientific Committee on Effects of Atomic Radiation (UNSCEAR) Sources and effects of ionizing radiation, Annex A&B, New York, 2000.

[12] A. Faanu, J. H. Ephraim, and E. O. Darko, "Assessment of public exposure to naturally occurring radioactive materials from mining and mineral processing activities of Tarkwa Goldmine in Ghana," *Environmental Monitoring and Assessment*, vol. 180, no. 1-4, pp. 15–29, 2011.

[13] A. M. El Arabi, N. K. Ahmed, and K. Salahel Din, "Natural radionuclides and dose estimation in natural water resources from Elba protective area, Egypt," *Radiation Protection Dosimetry*, vol. 121, no. 3, pp. 284–292, 2006.

22

Preliminary Assessment of the Chemical Stability of Dried Extracts from *Guazuma ulmifolia* Lam. (Sterculiaceae)

Gisely C. Lopes,[1] Renata Longhini,[1] Paulo Victor P. dos Santos,[2] Adriano A. S. Araújo,[3] Marcos Luciano Bruschi,[1] and João Carlos P. de Mello[1]

[1] *Programa de Pós-Graduação em Ciências Farmacêuticas, Departamento de Farmácia, Universidade Estadual de Maringá, Avenida Colombo, 5790, 87020-900 Maringá, PR, Brazil*
[2] *Student of Pharmacy, Universidade Estadual de Maringá, Avenida Colombo, 5790, 87020-900 Maringá, PR, Brazil*
[3] *Departamento de Fisiologia, Universidade Federal de Sergipe, Avenida Marechal Rondon, s/n, Cidade Universitária, 49100-000 São Cristóvão, SE, Brazil*

Correspondence should be addressed to João Carlos P. de Mello, mello@uem.br

Academic Editor: Ricardo Vessecchi

We report the results of a preliminary estimation of the stability of the dried extract from bark of *Guazuma ulmifolia* Lam. ("Mutamba"), with and without added colloidal silicon dioxide (CSD). The physical and chemical properties and the compatibility of CSD in the extract were evaluated for 21 days of storage under stress conditions of temperature ($45 \pm 2^{\circ}$C) and humidity ($75 \pm 5\%$). Thermogravimetry (TG) was supplemented using selective high-performance liquid chromatography (HPLC) for determination of stability of the characteristic constituents (chemical markers), namely, procyanidin B2 (PB2) and epicatechin (EP). The results showed that PB2 is an appropriate compound to be used as a chemical marker in the quality control of dried extracts of *G. ulmifolia*. The stress study showed that there was no significant difference between the two formulations. However, considering the TG data and the high temperatures involved, the results suggest that CSD increases the stability of the dried extract of *G. ulmifolia*.

1. Introduction

The reasons for the determination of stability of pharmaceuticals are based on concern for public health. The World Health Organization (WHO) defines the stability of drugs and medicines as the ability of a pharmaceutical product to maintain its chemical, physical, microbiological, and biopharmaceutical properties within specified limits throughout the duration of product usage [1].

Several studies reported on the stability of drugs and medicines [2–4]. To the best of our knowledge, the number of stability studies of plant extracts is not the same [5–7]. Measuring the chemical stability of extracts is challenging because of their chemical complexity, which may include hundreds of different compounds. Moreover, the presence of enzymes such as glycosidases, esterases, or oxidases plays an important role in the breakdown of secondary plant metabolites.

Assessment of the chemical stability of plant extracts, many of which are promising candidates for phytomedicines, plays an important role in the process of new drug development. A variety of environmental conditions, such as light, heat, humidity, and the freeze/thaw cycle, can significantly affect the chemical stability of drugs during storage and handling. Identification of stability-affecting factors facilitates the selection of packaging material and the definition of storage and handling conditions [8].

Guazuma ulmifolia Lam. (Sterculiaceae), popularly known as "Mutamba", is a tropical American plant found from Mexico to southern South America. In the popular medicine of several Latin-American countries, it is used for the treatment of burns, diarrhea, inflammations, and alopecia. Polysaccharides, epicatechin (EP), and procyanidin oligomers, such as procyanidins B2 (PB2) and B5, three trimers [procyanidin C1; epicatechin-(4β $\longrightarrow$6)-epicatechin-(4β $\longrightarrow$8)-epicatechin; epicatechin-(4 β $\longrightarrow$8)-epicatechin-(4β $\longrightarrow$6)-epicatechin], and one tetramer [9, 10] have been isolated and identified from its extract. The antidiabetic properties [11, 12], hypotensive and vasorelaxant activity [13, 14], antiulcer [15, 16], antibacterial activities [17, 18], and antiviral activity [19] of the bark, aerial parts, fruits, crude extract, and fractions have been attributed to the presence of proanthocyanidins.

However, there are no studies on the stability of the constituents of *G. ulmifolia* dried extracts. The determination of proanthocyanidins in bark of *G. ulmifolia* was carried out using HPLC, and it was observed that PB2 and EP compounds can be used as chemical markers for routine quality control analysis (Figure 1) [20].

The stability of the constituents in the extract of *G. ulmifolia* is important because the pharmacological properties depend on the chemical viability of the extract. As pointed out above, the procyanidins are pharmacologically active constituents of *G. ulmifolia*; however, they are unstable condensed tannins [21]. Their stability is affected by several factors such as pH, storage, temperature, chemical structure, concentration, light, oxygen, solvents, flavonoids, proteins, metallic ions, and the presence of enzymes [21]. A compatibility study of excipients is essential to develop a stable pharmaceutical dosage form, especially when the active agent is unstable.

The aim of the present study was to evaluate the chemical stability of the dried extract from the bark of *Guazuma ulmifolia* Lam. (Sterculiaceae), with and without an added pharmaceutical excipient.

2. Experimental

2.1. Plant Material. Bark of *Guazuma ulmifolia* Lam., Sterculiaceae, was collected in August 2005 in the city of Jataizinho, state of Paraná, Brazil (S 23°18′26.1″; W 050°58′19.4″; 377 m altitude; Garmin v.2.24). The species was identified by Professor Dr. Cássia Mônica Sakuragui. Voucher specimens are deposited in the herbarium of the Department of Biology of the State University of Maringá under number HUEM 12.051.

2.2. Chemicals and Reagents. All reagents and solvents were of analytical and HPLC grade, including ethyl acetate and trifluoroacetic acid (TFA) (Merck, Darmstadt, Germany). Ultra-pure water obtained by a Milli-Q UF-Plus apparatus (Millipore, Bedford, USA) with conductivity of 18.2 MΩ·cm at 25°C was used in all experiments. Epicatechin (EP) (Sigma, USA) and procyanidin B2 (PB2) (isolated and certified by spectroscopic methods at the Pharmacognosy Laboratory of Maringá State University) of the highest grade (purity > 99.0%) were used as standards. Colloidal silicon dioxide (CSD) was purchased from Degussa (Essen, Germany). All other solvents and chemicals were of analytical grade.

2.3. Preparation of Extracts. Air-dried stem bark (900 g) was exhaustively extracted with 9.0 L of Me_2CO-H_2O (7 : 3) by turbo-extraction (Ultra-Turrax model UTC115KT; IKA; USA) for 20 min at ≤40°C. The extractive dispersion was filtered and evaporated under reduced pressure to 1.0 L and freeze-dried (Christ model Alpha 1-2, Germany), yielding 120 g of crude extract (CE). One gram of CE was dissolved in a mixture of 10 mL water and 400 mg of CSD was added [22]. This mixture (CEA) was freeze-dried under the same conditions described for CE.

2.4. Stability Study. CE and CEA were evaluated for thermal stability under accelerated conditions for 21 days [23]. Samples of the CE and CEA were weighed (200 mg) and packaged in opaque white polyethylene flasks with a capacity of 10 g. The CE and CEA samples were stored in a climate chamber (BINDER, model KBF 240, USA) with a constant relative humidity of 75 ± 5% and maintained at 45 ± 2°C, without direct light. Samples were analyzed at the initial time (t_0) and 2, 7, 14, and 21 days after exposure to the atmospheric conditions described above.

2.5. HPLC Analysis. Accurately weighed 50 mg of CE and 70 mg of CEA were dissolved in 500 μL water, mixed in a tube shaker, and extracted with 500 μL of ethyl acetate, in a microtiter shaker at 1800 rpm (Minishaker, model MS1, IKA, USA) for 3 min (n = 9). Tubes were placed in a refrigerated microcentrifuge (model 5415R, Eppendorf, USA) at 4000 rpm, for 4 min at 5°C, for the total separation of the phases. The ethyl-acetate phase was separated. After evaporation of the solvent and drying under air flow, the residue was reconstituted to 10 mL with methanol : water (1 : 1) (test solution-SS). The sample was filtered through a 0.5 μm membrane filter (Millipore, Bedford, USA).

The analyses were carried out using a HPLC system (Gilson, USA) consisting of a solvent delivery pump (Model 321), a variable wavelength UV/VIS detector (Model 156), a manual injection valve (Rheodyne, USA) with a 20 μL loop, degasser (Model 184), and a thermostated column compartment (Model 831). Data collection and analyses were performed using UniPoint LC System software (Gilson, Villiers-le-Bel, France). A gradient was eluted on a Phenomenex Gemini C-18 column (250 mm × 4.6 mm) (Phenomenex International, USA), 5 μm particle size, Phenomenex SecurityGuard (C-18 cartridge) (20 mm × 4.6 mm). The mobile phase consisted of water (0.05% TFA) as solvent A and acetonitrile (0.05% TFA) as solvent B, and both were degassed and filtered through a 0.45 μm pore size filter (Millipore, Bedford, USA). Separations were affected by a gradient as follows: 0 min 13% B in A; 10 min 17% B; 16 min 18.35% B; 20 min 22.65% B; 23 min 29.81% B; 25 min 65% B; followed by a 7 min reequilibration time. The mobile-phase flow rate was 0.8 mL/min, and the injection volume was 20 μL. The chromatographic runs were carried out at 28°C. UV detection was performed at 210 nm.

The purity of peaks was checked by a Diode Array Detector coupled to a Varian ProStar module (Varian, Palo Alto, CA, USA) with ProStar 210 Solvent Delivery and a

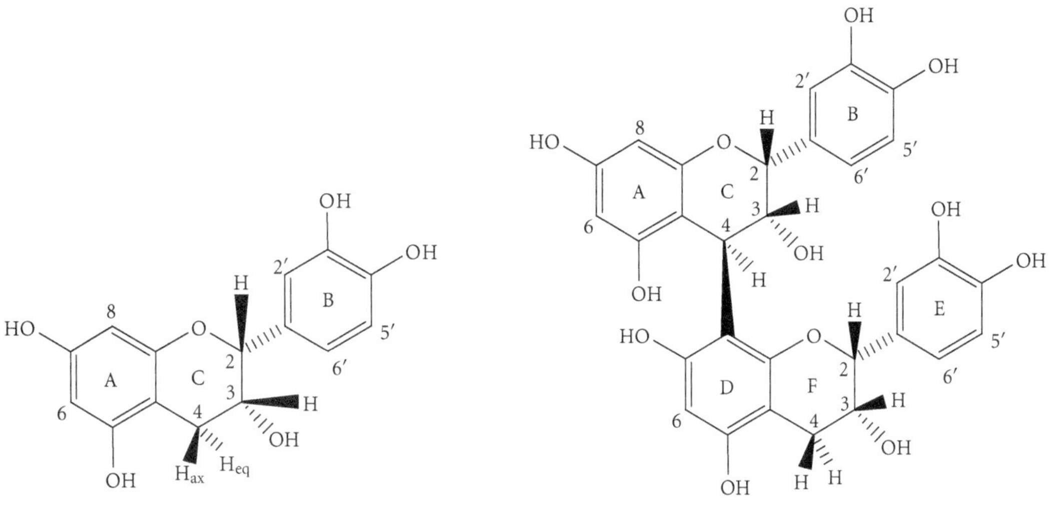

Epicatechin (EP) Epicatechin-(4β ⟶ 8)-epicatechin (PB2)

Figure 1: Chemical markers of the *G. ulmifolia* dried extracts.

ProStar 335 HPLC-DAD, comparing the UV spectra of each peak with those of authentic reference samples.

An EP reference standard stock solution of 400 μg/mL was prepared in methanol : water (1 : 1). Calibration standard solutions at five levels were prepared by serially diluting the stock solution to concentrations of 10.00, 40.00, 70.00, 100.00, and 120.00 μg/mL. A PB2 stock solution of 250 μg/mL was prepared in methanol : water (1 : 1). Calibration standard solutions at seven levels were prepared by serially diluting the stock solution to concentrations of 20.00, 40.00, 50.00, 70.00, 90.00, 120.00, and 150.00 μg/mL. The samples were filtered through a 0.5 μm membrane (Millipore, Bedford, USA) prior to injection. Each analysis was repeated five times, and the calibration curves were fitted by linear regression [20].

2.6. Thermogravimetry (TG). A simultaneous thermal analysis (STA) system (NETZSCH, model STA 409 PG/4/G Luxx, USA) was used for recording the TG curves of the CE and CEA. About 10 mg of sample was weighed accurately using an STA balance. The weighed sample was heated in a closed aluminum pan at a programmed rate of 10°C/min in a temperature range from 30 to 500°C under a nitrogen flow of 50 mL/min. An empty aluminum pan was used as a reference.

2.7. Total Tannins. The percentage of total tannins in CE and CEA at t_0 and day 21 was evaluated using the Folin-Ciocalteau reagent and following a method from the British Pharmacopoeia [24]. Samples of 100 mg and 166 mg of CE and CEA, respectively, were employed. Each analysis was repeated three times.

2.8. Statistical Analysis. Experimental data were analyzed by one-way ANOVA, and the statistical significance of means was determined by the LSD and Tukey's HSD tests. The Dunnett test was employed to compare contents on different days of the analyses. Differences were considered significant at $P < 0.05$.

3. Results and Discussion

Two dried extracts of *G. ulmifolia*, prepared by different techniques (with or without CSD), were evaluated for the stability of their main components (markers): PB2 and EP.

The development of analytical conditions for the analyses herbal drugs and pharmaceutical formulations containing these extracts must necessarily go through a specific validation. In previous work, a rapid and robust LC assay for separation and quantitative analysis of PB2 and EP in extract of *G. ulmifolia* was developed. The method was validated by regulation RE 899/2003 of the National Health Surveillance Agency, Brazil, and the ICH guidelines [20].

Considering that the stability assay was developed with the aim to get an initial response about the chemical stability of the extracts for research and development purposes, the used protocol allowed the extracts to be evaluated under accelerated conditions [25].

Quantification of these markers in the samples in the test of stability was carried out using external standards (PB2 and EP). In the evaluation of linearity, based on $1/x$-weighted linear regression analysis, the responses for both standards in related concentration ranges were linear. The calibration equations were $Y = 818.21x - 2177.9$ ($n = 7$, $R = 0.9990$) for PB2 and $Y = 885.51x + 953.56$ ($n = 5$, $R = 0.9994$) for EP. The RSDs of the slopes were $\leq 5\%$ for both analytes ($n = 5$).

No degradation in the CE and CEA samples under stress conditions was observed. No changes in the chromatographic profile occurred during the period of analysis (Figure 2).

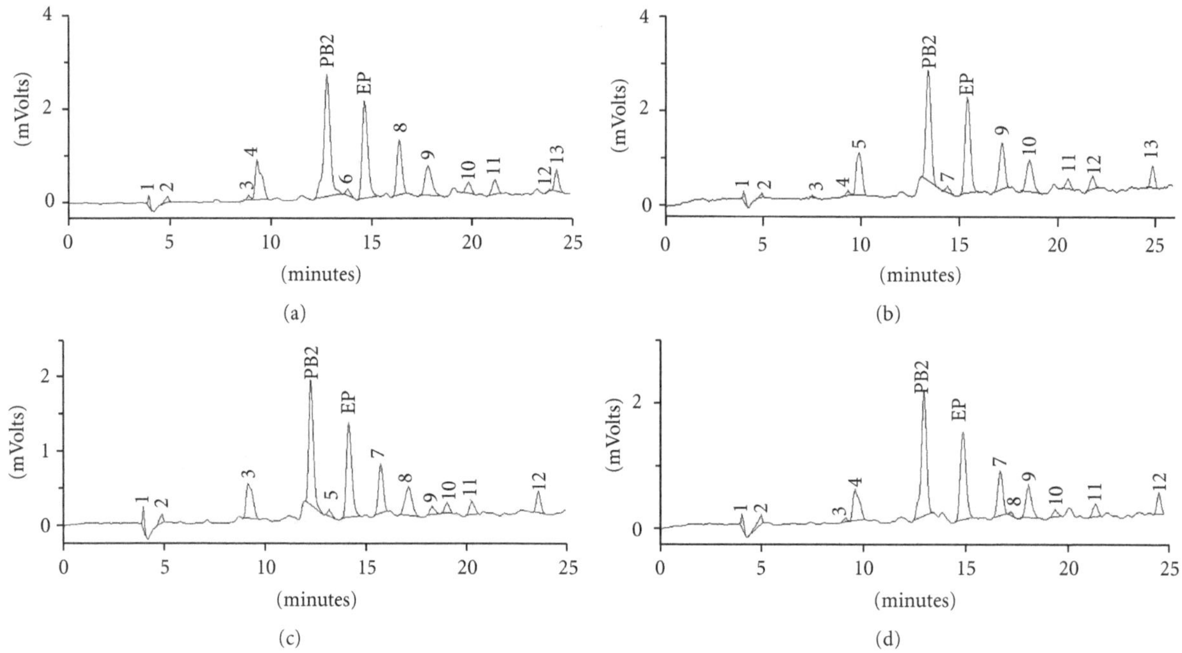

FIGURE 2: Typical HPLC chromatograms of stress test samples of the crude extract (CE) and the crude extract + colloidal silicon dioxide (CEA) of *Guazuma ulmifolia*. (a) CE at time zero; (b) CE at day 21; (c) CEA at time zero; (d) CEA at day 21.

The peak purity test confirmed that the PB2 and EP peaks remained homogeneous and pure throughout the stress test (data analyzed under DAD). The UV spectra of the compounds (PB2 and EP) did not change between the beginning and end of elution of their individual values, confirming the absence of degradation products.

The chemical stability assay of the CE and CEA dried-extract formulations was determined according to the concentration of PB2 and EP at a storage temperature of 45°C and 75% humidity for 21 days. The final concentration was expressed as μg/mL of PB2 and EP in the dried extract (Table 1).

Figure 3 shows the mean values of the PB2 and EP in the CE and CEA samples for each day of storage analyzed.

The PB2 content remained constant after 21 days of storage, in both the CE and CEA. The EP in the CEA showed a significant change ($P < 0.05$) in concentration from t_0 to day 21. However, no significant change in the concentration of EP was observed in the CE stored under the same conditions.

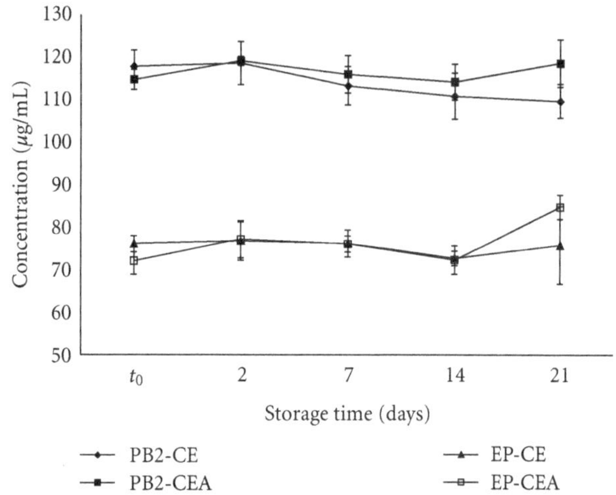

FIGURE 3: The stability of the crude extract (CE) and the crude extract + colloidal silicon dioxide (CEA) of *Guazuma ulmifolia* at 45°C, 75% relative humidity and 21 days.

Figure 4 shows the influence of the presence of the excipient in the dried extract. Apparently, the physical and chemical properties of the CSD can significantly accelerate the increase of EP in the CEA after 21 days. In relation to the concentration of PB2, there was no significant difference between the CE and CEA during the 21 days of analysis.

Proanthocyanidins are commonly composed of monomers of catechin and/or epicatechin with linkages of 4⟶6 and/or 4⟶8. Besides these, other monomers are common: gallocatechin, epigallocatechin, robinetinidin, and fisetinidin [26]. Proanthocyanidins differ structurally according to the number of hydroxyl groups present at aromatic rings and the stereochemistry of the asymmetric carbons of the heterocyclic nucleus. The presence of *O*-methylation, *O*-glycosylation, and *O*-galloylation increase the structural complexity [27].

PB2 is a dimeric proanthocyanidin with chemical linkage of the type $4\beta \longrightarrow 8$. Fletcher et al. [28] showed by NMR studies of the procyanidin peracetate that linkages 4⟶6 and 4⟶8 are found in two energetically protected conformations. Therefore, the linkage between epicatechin monomers forming the PB2 may be physically more stable.

TABLE 1: Stability of the constituents procyanidin B2 (PB2) and epicatechin (EP) of the dried extract of *Guazuma ulmifolia*. Mean ± SD (RSD %).

Days of storage	CE (mg/mg)		CEA (mg/mg)	
	PB2	EP	PB2	EP
Time zero	0.0236 ± 4.5 (3.8)	0.0152 ± 1.4 (1.9)	0.0230 ± 2.7 (2.3)	0.0144 ± 2.3 (3.2)
2	0.0237 ± 6.0 (5.1)	0.0154 ± 3.5 (4.5)	0.0238 ± 1.2 (1.0)	0.0154 ± 3.4 (4.4)
7	0.0226 ± 5.2 (4.6)	0.0152 ± 2.4 (3.2)	0.0232 ± 5.1 (4.4)	0.0152 ± 1.4 (1.9)
14	0.0221 ± 10.5 (9.4)	0.0147 ± 1.2 (1.7)	0.0228 ± 4.8 (4.2)	0.0145 ± 2.4 (3.4)
21	0.0220 ± 9.9 (9.0)	0.0152 ± 6.9 (9.1)	0.0237 ± 7.8 (6.6)	0.0170 ± 2.4 (2.8)

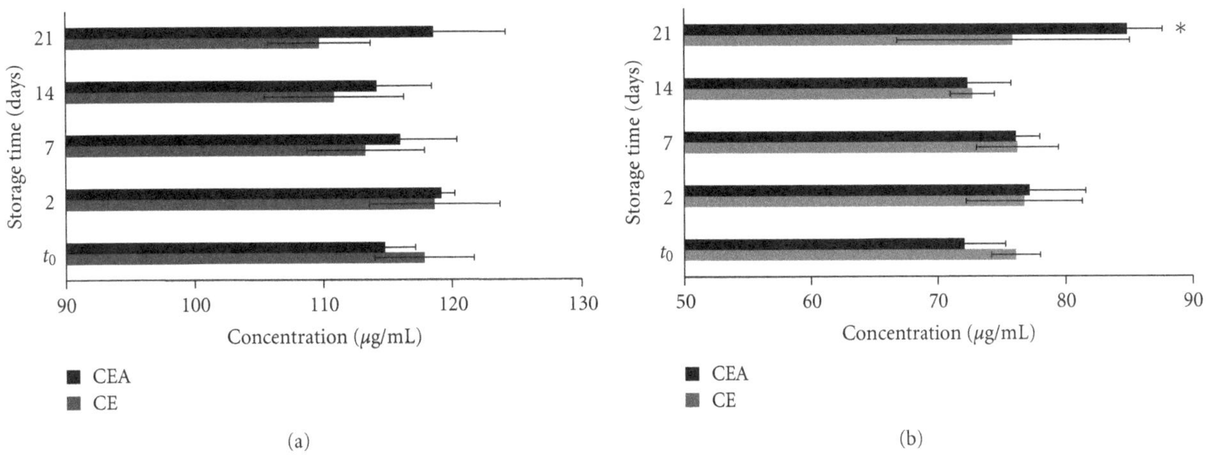

FIGURE 4: Influence of colloidal silicon dioxide on the dried-extract (CE and CEA) of *Guazuma ulmifolia*. (a) procyanidin B2 (PB2) and (b) epicatechin (EP).

The significant change in concentration of EP in the CEA probably occurred by physical interaction of oligomers and/or polymers of condensed tannins in the extract and CSD. This excipient has a large surface area and a high polarity of silanol groups present on its surface, which leads to adsorption of water and formation of hydrogen bonds [29], facilitated by its hygroscopic property [30]. Therefore, CSD is commonly used as a desiccant agent to protect hygroscopic chemicals and drugs from atmospheric moisture [29]. Thus, this excipient is an excellent candidate adjuvant for the stabilization of plant extracts.

Extracts rich in phenolic substances are congruent with this assumption because they are rich in hydroxyls, capable of hydrogen bond interactions. Döner et al. [31] evaluated the bonding between polyvinylpolypyrrolidone (PVP) and different classes of flavonoids. The bonding increases with the number of hydroxyl groups present in the flavonoid nucleus. Compounds that contain 7- and 4′-hydroxyl groups bond most effectively; the same principle can be extrapolated to the CEA. The increase in the concentration of EP in the CEA (Figure 4(b)) may result from an interaction by hydrogen bonding between oligomers and/or polymers of the condensed tannins and the silanol hydroxyl group of CSD.

Gore and Banker [29] observed that silica has the ability to form a monolayer adsorption of water vapor, suggesting that polar water molecules are adsorbed at specific sites on the silica surface. Oligomeric flavonols and polymers of condensed tannins may show the same pattern of connection to the CSD. Bonding of these substances with CSD would weaken the bonds within the compound, releasing monomeric substances. This would explain the statistical difference found at day 21.

However, the analyses of the total tannin content of CE and CEA at time t_0 and day 21 after the stress tests showed no significant differences. The results for CE were 26.1% ± 0.5 (RSD% 2.0) and 26.6% ± 0.8 (RSD% 3.0), and for CEA were 25.0% ± 1.2 (RSD% 4.9) and 25.4% ± 0.7 (RSD% 2.8) at times t_0 and day 21, respectively.

These results suggest that the physical interactions occurred in the extract CEA produced no alterations in the content of proanthocyanidin. However, they show that the quality control of extracts containing high content of phenolic compounds must be accomplished using dimeric compounds, which are more physically stable.

Figure 5 and Table 2 show the TG data in the temperature range from 25 to 500°C for the CE and CEA. TG curves of CE and CEA presented a characteristic profile of elimination of water surface between 35 and 100°C, thermal stability between 100 and 185°C, following thermal decomposition. The thermal decomposition of CEA occurs in two stages, (Δm_2 = 4.80% and DTG_{peak} = 213°C and Δm_3 = 19.79%

Table 2: Thermogravimetry parameters of the crude extract (CE) and the crude extract + colloidal silicon dioxide (CEA) of *Guazuma ulmifolia*.

Sample	Mass loss (%)	Days of storage				
		t_0	2	7	14	21
CE	1	3.68	4.44	4.51	4.85	4.90
	2	9.70	8.44	8.98	8.38	8.27
	3	32.46	33.06	32.29	31.79	30.75
	Total mass loss (%)	45.87	46.22	46.81	45.45	44.85
CEA	1	2.31	3.54	3.11	2.07	2.36
	2	4.80	4.15	5.13	4.85	4.97
	3	19.79	19.39	20.89	18.98	19.21
	Total mass loss (%)	27.33	27.60	29.47	26.03	26.93

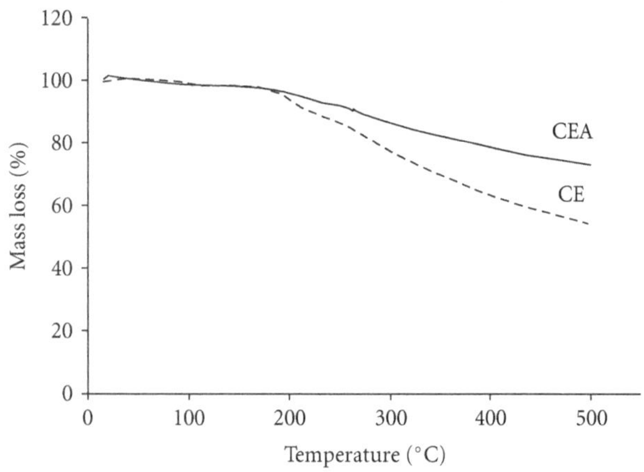

Figure 5: Thermogravimetry curves for the (dashed line) crude extract (CE) and (staked line) crude extract + colloidal silicon dioxide (CEA) of *Guazuma ulmifolia* at time zero.

and DTG_{peak} = 340°C). A similar thermal decomposition was observed for CEA. However, TG curves of CE presented mass losses greater than those of CEA (Δm_2 = 9.70% and DTG_{peak} = 218°C and Δm_3 = 32.46% and DTG_{peak} = 336°C).

The TG analysis indicated no trend in the behavior of CE and CEA, that is, during the days of analysis, the total mass underwent no significant increase or decrease.

The analysis of these data leads us to suppose that CSD conferred some stability, because the CEA lost significantly less total mass. It can be assumed that the CEA was protected from heat to some extent by the CSD, influencing the process of degradation, with smaller percentages of loss of mass from the fusion of chemicals. Thus, we can conclude that the CSD limited the access of water to the extract and/or prevented its degradation [32, 33].

Over the 21 days of the study, there was no significant difference between the two extracts, but considering the TG data and the high temperatures involved, the data suggest that over the long term the CSD would be a good protector for the plant extracts of *G. ulmifolia*. However, further studies should be performed to confirm these results.

As a conclusion, PB2 is an appropriate compound to use as a chemical marker in quality control of the dried extract of *G. ulmifolia*. The stress test showed that the content of total tannins was unchanged. Therefore, in this 21-day screening study, proanthocyanidins in the dried extract of the *G. ulmifolia* showed good compatibility with CSD under stress conditions.

Conflict of Interests

The authors declare that there is no conflict of interests.

Acknowledgments

The authors thank the Brazilian funding agencies CNPq (Conselho Nacional de Desenvolvimento Científico e Tecnológico), Fundacão Araucária, INCT_if, and CAPES (Coordenação de Aperfeiçoamento de Pessoal de Nível Superior). They express gratitude to Professor Dr. Antonio Medina Neto and Professor Dr. Jurandir Hillmann Rohling for technical support with TG, to Dr. Janet W. Reid for revising the English text, and to Professor Dr. Adolf Nahrstedt for contributions.

References

[1] World Health Organization, "International Stability Testing: guidelines for stability testing of pharmaceutical products containing well established drug substances in conventional dosage forms," WHO Technical Report Series 863, 1996, Annex 5.

[2] G. Chawla and A. K. Bansal, "Molecular mobility and physical stability of amorphous irbesartan," *Scientia Pharmaceutica*, vol. 77, no. 3, pp. 695–709, 2009.

[3] A. A. S. Araújo, S. Storpirtis, L. P. Mercuri, F. M. S. Carvalho, M. Santos Filho, and J. R. Matos, "Thermal analysis of the antiretroviral zidovudine (AZT) and evaluation of the compatibility with excipients used in solid dosage forms," *International Journal of Pharmaceutics*, vol. 260, no. 2, pp. 303–314, 2003.

[4] Z. R. Dedania, R. R. Dedania, N. R. Shetha, J. B. Patel, and B. Patel, "Stability indicating HPLC determination of Risperidone in bulk drug and pharmaceutical formulations,"

International Journal of Analytical Chemistry, vol. 2011, Article ID 124917, 6 pages, 2011.

[5] A. K. Shah, B. A. Avery, and C. M. Wyandt, "Content analysis and stability evaluation of selected commercial preparations of St. John's wort," *Drug Development and Industrial Pharmacy*, vol. 31, no. 9, pp. 907–916, 2005.

[6] P. Jin, S. Madieh, and L. L. Augsburger, "The solution and solid state stability and excipient compatibility of parthenolide in feverfew," *AAPS PharmSciTech*, vol. 8, no. 4, article 105, 2007.

[7] Y. Liu and P. A. Murphy, "Alkamide stability in *Echinacea purpurea* extracts with and without phenolic acids in dry films and in solution," *Journal of Agricultural and Food Chemistry*, vol. 55, no. 1, pp. 120–126, 2007.

[8] S. Gafner and C. Bergeron, "The challenges of chemical stability testing of herbal extracts in finished products using state-of-the-art analytical methodologies," *Current Pharmaceutical Analysis*, vol. 1, pp. 203–215, 2005.

[9] M. Hör, M. Heinrich, and H. Rimpler, "Proanthocyanidin polymers with antisecretory activity and proanthocyanidin oligomers from *Guazuma ulmifolia* bark," *Phytochemistry*, vol. 42, no. 1, pp. 109–119, 1996.

[10] J. C. B. Rocha, F. Pedrochi, L. Hernandes, J. C. P. de Mello, and M. L. Baesso, "Ex vivo evaluation of the percutaneous penetration of proanthocyanidin extracts from *Guazuma ulmifolia* using photoacoustic spectroscopy," *Analytica Chimica Acta*, vol. 587, no. 1, pp. 132–136, 2007.

[11] F. J. Alarcon-Aguilara, R. Roman-Ramos, S. Perez-Gutierrez, A. Aguilar-Contreras, C. C. Contreras-Weber, and J. L. Flores-Saenz, "Study of the anti-hyperglycemic effect of plants used as antidiabetics," *Journal of Ethnopharmacology*, vol. 61, no. 2, pp. 101–110, 1998.

[12] A. J. Alonso-Castro and L. A. Salazar-Olivo, "The anti-diabetic properties of *Guazuma ulmifolia* Lam are mediated by the stimulation of glucose uptake in normal and diabetic adipocytes without inducing adipogenesis," *Journal of Ethnopharmacology*, vol. 118, no. 2, pp. 252–256, 2008.

[13] C. Caballero-George, P. M. Vanderheyden, T. De Bruyne et al., "In vitro inhibition of [3H]-angiotensin II binding on the human AT1 receptor by proanthocyanidins from *Guazuma ulmifolia* bark," *Planta Medica*, vol. 68, no. 12, pp. 1066–1071, 2002.

[14] G. A. Magos, J. C. Mateos, E. Páez et al., "Hypotensive and vasorelaxant effects of the procyanidin fraction from *Guazuma ulmifolia* bark in normotensive and hypertensive rats," *Journal of Ethnopharmacology*, vol. 117, no. 1, pp. 58–68, 2008.

[15] M. Hör, H. Rimpler, and M. Heinrich, "Inhibition of intestinal chloride secretion by proanthocyanidins from *Guazuma ulmifolia*," *Planta Medica*, vol. 61, no. 3, pp. 208–212, 1995.

[16] B. Berenguer, C. Trabadela, S. Sánchez-Fidalgo et al., "The aerial parts of *Guazuma ulmifolia* Lam. protect against NSAID-induced gastric lesions," *Journal of Ethnopharmacology*, vol. 114, no. 2, pp. 153–160, 2007.

[17] A. Camporese, M. J. Balick, R. Arvigo et al., "Screening of anti-bacterial activity of medicinal plants from Belize (Central America)," *Journal of Ethnopharmacology*, vol. 87, no. 1, pp. 103–107, 2003.

[18] M. C. Navarro, M. P. Montilla, M. M. Cabo et al., "Antibacterial, antiprotozoal and antioxidant activity of five plants used in izabal for infectious diseases," *Phytotherapy Research*, vol. 17, no. 4, pp. 325–329, 2003.

[19] A. M. M. Felipe, V. P. Rincão, F. J. Benati et al., "Antiviral effect of *Guazuma ulmifolia* and Stryphnodendron adstringens on poliovirus and bovine herpesvirus," *Biological and Pharmaceutical Bulletin*, vol. 29, no. 6, pp. 1092–1095, 2006.

[20] G. C. Lopes, M. L. Bruschi, and J. C. P. de Mello, "RP-LC-UV determination of proanthocyanidins in *Guazuma ulmifolia*," *Chromatographia*, vol. 69, supplement 2, pp. S175–S181, 2009.

[21] A. Castañeda-Ovando, M. D. L. Pacheco-Hernandez, M. E. Páez-Hernández, J. A. Rodríguez, and C. A. Galán-Vidal, "Chemical studies of anthocyanins: a review," *Food Chemistry*, vol. 113, no. 4, pp. 859–871, 2009.

[22] K. C. B. De Souza, P. R. Petrovick, V. L. Bassani, and G. González Ortega, "The adjuvants Aerosil 200 and Gelita-Sol-P influence on the technological characteristics of spray-dried powders from *Passiflora edulis* var. flavicarpa," *Drug Development and Industrial Pharmacy*, vol. 26, no. 3, pp. 331–336, 2000.

[23] M. V. R. Velasco, C. R.M. Maciel, F. D. Sarruf et al., "Desenvolvimento e Teste Preliminar da Estabilidade de formulaç ões cosméticas acrescidas de extrato comercial de *Trichilia catigua* Adr. Juss (e) *Ptychopetalum olacoides* Bentham," *Revista de Ciencias Farmaceuticas Basica e Aplicada*, vol. 29, no. 2, pp. 179–194, 2008.

[24] British Pharmacopoeia Commission, *British Pharmacopoeia*, The Stationary Office, 2008.

[25] H. C. Ansel, N. G. Popovich, and L. V. Allen Jr., *Farmacotécnica. Formas Farmacêuticas e Sistemas de Liberação de Fármacos*, Editorial Premier, 6th edition, 2000.

[26] G. C. Lopes, F. A. V. Machado, C. E. M. Toledo, C. M. Sakuragui, and J. C. P. de Mello, "Chemotaxonomic significance of 5-deoxyproanthocyanidins in *Stryphnodendron* species," *Biochemical Systematics and Ecology*, vol. 36, no. 12, pp. 925–931, 2008.

[27] T. De Bruyne, L. Pieters, H. Deelstra, and A. Vlietinck, "Condensed vegetable tannins: biodiversity in structure and biological activities," *Biochemical Systematics and Ecology*, vol. 27, no. 4, pp. 445–459, 1999.

[28] A. C. Fletcher, L. J. Porter, E. Haslam, and R. K. Gupta, "Plant proanthocyanidins. Part 3. Conformational and configurational studies of natural procyanidins," *Journal of the Chemical Society, Perkin Transactions 1*, no. 14, pp. 1628–1637, 1977.

[29] A. Y. Gore and G. S. Banker, "Surface chemistry of colloidal silica and a possible application to stabilize aspirin in solid matrixes," *Journal of Pharmaceutical Sciences*, vol. 68, no. 2, pp. 197–202, 1979.

[30] A. H. Kibbe, *Handbook of Pharmaceutical Excipients*, London, UK, 3rd edition, 2000.

[31] L. W. Döner, G. Bécard, and P. L. Irwin, "Binding of flavonoids by polyvinylpolypyrrolidone," *Journal of Agricultural and Food Chemistry*, vol. 41, no. 5, pp. 753–757, 1993.

[32] T. F. Moura, D. Gaudy, M. Jacob, A. Terol, B. Pauvert, and A. Chauvet, "Vitamin C spray drying: study of the thermal constraint," *Drug Development and Industrial Pharmacy*, vol. 22, no. 5, pp. 393–400, 1996.

[33] E. A. F. Vasconcelos, M. G. F. Medeiros, F. N. Raffi, and T. F. A. L. Moura, "Influência da temperatura de secagem e da concentração de Aerosil®200 nas características dos extratos secos por aspersão da *Schinus terebinthifolius* Raddi (Anacardiaceae) ," *Revista Brasileira de Farmacognosia*, vol. 15, pp. 243–249, 2005.

Permissions

The contributors of this book come from diverse backgrounds, making this book a truly international effort. This book will bring forth new frontiers with its revolutionizing research information and detailed analysis of the nascent developments around the world.

We would like to thank all the contributing authors for lending their expertise to make the book truly unique. They have played a crucial role in the development of this book. Without their invaluable contributions this book wouldn't have been possible. They have made vital efforts to compile up to date information on the varied aspects of this subject to make this book a valuable addition to the collection of many professionals and students.

This book was conceptualized with the vision of imparting up-to-date information and advanced data in this field. To ensure the same, a matchless editorial board was set up. Every individual on the board went through rigorous rounds of assessment to prove their worth. After which they invested a large part of their time researching and compiling the most relevant data for our readers. Conferences and sessions were held from time to time between the editorial board and the contributing authors to present the data in the most comprehensible form. The editorial team has worked tirelessly to provide valuable and valid information to help people across the globe.

Every chapter published in this book has been scrutinized by our experts. Their significance has been extensively debated. The topics covered herein carry significant findings which will fuel the growth of the discipline. They may even be implemented as practical applications or may be referred to as a beginning point for another development. Chapters in this book were first published by Hindawi Publishing Corporation; hereby published with permission under the Creative Commons Attribution License or equivalent.

The editorial board has been involved in producing this book since its inception. They have spent rigorous hours researching and exploring the diverse topics which have resulted in the successful publishing of this book. They have passed on their knowledge of decades through this book. To expedite this challenging task, the publisher supported the team at every step. A small team of assistant editors was also appointed to further simplify the editing procedure and attain best results for the readers.

Our editorial team has been hand-picked from every corner of the world. Their multi-ethnicity adds dynamic inputs to the discussions which result in innovative outcomes. These outcomes are then further discussed with the researchers and contributors who give their valuable feedback and opinion regarding the same. The feedback is then collaborated with the researches and they are edited in a comprehensive manner to aid the understanding of the subject.

Apart from the editorial board, the designing team has also invested a significant amount of their time in understanding the subject and creating the most relevant covers. They scrutinized every image to scout for the most suitable representation of the subject and create an appropriate cover for the book.

The publishing team has been involved in this book since its early stages. They were actively engaged in every process, be it collecting the data, connecting with the contributors or procuring relevant information. The team has been an ardent support to the editorial, designing and production team. Their endless efforts to recruit the best for this project, has resulted in the accomplishment of this book. They are a veteran in the field of academics and their pool of knowledge is as vast as their experience in printing. Their expertise and guidance has proved useful at every step. Their uncompromising quality standards have made this book an exceptional effort. Their encouragement from time to time has been an inspiration for everyone.

The publisher and the editorial board hope that this book will prove to be a valuable piece of knowledge for researchers, students, practitioners and scholars across the globe.

List of Contributors

Noha N. Atia Salwa R. El-Shabouri and Wesam M. El-Koussi
Department of Pharmaceutical Analytical Chemistry, Faculty of Pharmacy, Assiut University, Assiut 71526, Egypt

Ashraf M. Mahmoud
Department of Pharmaceutical Analytical Chemistry, Faculty of Pharmacy, Assiut University, Assiut 71526, Egypt
Department of Pharmaceutical Chemistry, Faculty of Pharmacy, Najran University, P.O. Box 1988, Najran 61441, Saudi Arabia

Mohsen Zeeb
Agricultural, Medical and Industrial Research School, Nuclear Science and Technology Research Institute, P.O. Box 31485/498, Karaj, Iran
Department of Analytical Chemistry, Faculty of Chemistry, University College of Science, University of Tehran, P.O. Box 14155-6455,Tehran, Iran

Mahdi Sadeghi
Agricultural, Medical and Industrial Research School, Nuclear Science and Technology Research Institute, P.O. Box 31485/498, Karaj, Iran

Marina Franco, Renata Jasionowska and Elisa Salvatore
Dipartimento di Chimica, Sapienza Università, 00185 Roma, Italy

Jin Wang, Yong-de Yue, Hao Jiang and Feng Tang
SFA Key Laboratory of Bamboo and Rattan Science and Technology, International Centre for Bamboo and Rattan, No. 8 Futong Dongdajie, Wangjing, Chaoyang District, Beijing 100102, China

Paul J. Gates
School of Chemistry, University of Bristol, Cantock's Close, Bristol BS8 1TS, UK

Norberto P. Lopes
Faculdade de Ciências Farmacêuticas de Ribeirão Preto, Universidade de São Paulo, Via do Café S/N, 14040-903 Ribeirão Preto, SP, Brazil

Hiroyuki Tanaka
Graduate School of Pharmaceutical Science, Kyushu University, 3-1-1 Maidashi, Higashi-ku, Fukuoka 812-0855, Japan

Osamu Morinaga, Takuhiro Uto, Nguyen Huu Tung and Yukihiro Shoyama
Faculty of Pharmaceutical Science, Nagasaki International University, 2827-7 Huis Ten Bosch, Sasebo 859-3298, Japan

Shunsuke Fuji
Faculty of Health Management, Nagasaki International University, 2827-7 Huis Ten Bosch, Sasebo 859-3298, Japan

Frederick Asare Aboagye
Phytochemistry Department, Centre for Scientific Research into Plant Medicine, University of Ghana, P.O. Box 73, Mampong-Akuapem, Ghana

Xiao Wei Li
State Key Laboratory of Natural and Biomimetic Drugs, School of Pharmaceutical Sciences, Peking University, No.38 Xue-yuan Road, Haidian District, Beijing 100191, China

Waraporn Putalun
Faculty of Pharmaceutical Science, Khon Kaen University, Khon Kaen 40002, Thailand

Shibdas Banerjee and Shyamalava Mazumdar
Department of Chemical Sciences, Tata Institute of Fundamental Research, Homi Bhabha Road, Colaba, Mumbai 400005, India

Ahlam Jameel Abdulghani
Department of Chemistry, College of Science, Baghdad University, Jaderiya, Baghdad, Iraq

Hadi Hassan Jasim and Abbas Shebeeb Hassan
Department of Chemistry, College of Science, Al-Mustansiriyah University, Baghdad, Iraq

M. A. Elbagermi and A. I. Alajtal
Department of Chemistry, Faculty of Science, University of Misurata, P.O. Box 1338, Misurata, Libya

H. G. M. Edwards
Raman Spectroscopy Group, University Analytical Centre, Division of Chemical and Forensic Sciences, University of Bradford, West Yorkshire, BD7 1DP, UK

Chevelle A. Cason, Thomas A. Fabré, Andrew Buhrlage and Heather A. Bullen
Department of Chemistry, Northern Kentucky University, Highland Heights, KY 41099, USA

Kristi L. Haik
Department of Biological Sciences, Northern Kentucky University, Highland Heights, KY 41099, USA

Sandra Ribeiro de Morais
Programa de Pós-Graduação em Biologia, Instituto de Ciências Biológicas, Universidade Federal de Goiás, 74001 970 Goiânia, GO, Brazil
Instituto de Ciências da Saúde, Universidade Paulista, Campus Flamboyant, 74845 090 Goiânia, GO, Brazil

Thiago Levi Silva Oliveira
Instituto de Ciências da Saúde, Universidade Paulista, Campus Flamboyant, 74845 090 Goiânia, GO, Brazil
Programa de Pós-Graduação em Ciências Farmacêuticas, Laboratório de Pesquisa de Produtos Naturais, Faculdade de Farmácia, Universidade Federal de Goi´as, 74605 220 Goiânia, GO, Brazil

Maria Teresa Freitas Bara, Edemilson Cardoso da Conceição and José Realino dePaula
Faculdade de Farmácia, Universidade Federal de Goiás, 74605 220 Goiânia, GO, Brazil

MariaHelena Rezende
Instituto de Ciências Biológicas, Universidade Federal de Goiás, Campus Samambaia, 74001 970 Goiânia, GO, Brazil

Pedro Henrique Ferri
Instituto de Química, Laboratório de Bioatividade Molecular, Universidade Federal de Goiás, Campus Samambaia, 74001 970 Goiânia, GO, Brazil

Masoud Negahdary, Gholamreza Mazaheri, Somyyeh Rad and Fariba Pishbin
Department of Biology, Payam-e-Noor University, Tehran 7371719578, Iran

Mohammadreza Hadi, Mojdeh Khosravian-hemami
Department of Biology, Shahrekord University, Shahrekord, Iran

Roya Malekzadeh
Department of Biology, Payam-e-Noor University, Taft, Iran

Mohammad Mahdi Saadatmand
Young Researchers Club, Islamic Azad University, Yazd Branch, Yazd, Iran

Saeed Rezaei-Zarchi
Department of Biology, Payam-e-Noor University, Taft, Iran

Sanda Rončević and Ivan Nemet
Laboratory of Analytical Chemistry, Department of Chemistry, Faculty of Science, University of Zagreb, Horvatovac 102 a, 10000 Zagreb, Croatia

Anica Benutić and Buga Gabelica
Croatian National Institute of Public Health, Enviromental Health Service Rockefellerova 7, 10000 Zagreb, Croatia

Kevin D. Quinn, Charmion I. Cruickshank and Troy D.Wood
Department of Chemistry, State University of New York at Buffalo, Buffalo, NY 14260, USA

Kristy L. Nowak-Lovato
Decision Applications Division, Los Alamos National Laboratory, Los Alamos, NM 87545, USA

Kirk D. Rector
Chemistry Division, Los Alamos National Laboratory, Los Alamos, NM 87545, USA

Leandro De Santis Ferreira, Norberto Peporine Lopes and Orghêda Luiza Araújo Domingues Zucchi
Departamento de Física e Química, Faculdade de Ciências Farmacêuticas de Ribeirão Preto, Universidade de S˜ao Paulo, 14040-903 Ribeirão Preto, SP, Brazil

Rosana Peporine Lopes and Mabel Norma Costas Ulbrich
Instituto de Geociências, Universidade de São Paulo, 05508-080 São Paulo, SP, Brazil

Thais Guaratini
Lychnoflora Pesquisa e Desenvolvimento em Produtos Naturais LTDA, Incubadora SUPERA, Campus da USP, 14040-900 Ribeirão Preto, SP, Brazil

Pio Colepicolo
Instituto de Química, Universidade de São Paulo, 05513-970 São Paulo, SP, Brazil

Ricardo Clapis Garla
Departamento de Botânica, Ecologia e Zoologia, Centro de Biociências, Universidade Federal do Rio Grande do Norte, 59072-970 Natal, RN, Brazil

Eurico Cabral Oliveira Filho
Instituto de Biociências, Universidade de São Paulo, 05508-090 São Paulo, SP, Brazil

Adrian Martin Pohlit
Coordenação de Pesquisa em Produtos Naturais, Instituto Nacional de Pesquisa da Amazônia, 69060-001 Manaus, AM, Brazil

Baosheng Liu, Chao Yang, Xiaona Yan, Jing Wang and Yunkai Lv
Key Laboratory of Medical Chemistry and Molecular Diagnosis, College of Chemistry & Environmental Science, Hebei University, Ministry of Education, Baoding 071002, China

Yixi Xie, Yu Lei, Shalini Shah, Hao Wu, Jian Wu, Elise Megehee and Enju Wang
Department of Chemistry, St. John's University, Queens, NY 11439, USA

Leandro De Santis Ferreira, Izabel Cristina Casanova Turatti and Norberto Peporine Lopes
Departamentos de Física e Química, Faculdade de Ciências Farmacêuticas de Ribeirão Preto, Universidade de São Paulo, 14040-903 Ribeirão Preto, SP, Brazil

Thais Guaratini
Lychnoflora Pesquisa e Desenvolvimento em Produtos Naturais LTDA, Incubadora SUPERA, Campus da USP, 14040-900 Ribeirão Preto, SP, Brazil

Pio Colepicolo
Instituto de Química, Universidade de São Paulo, 05513-970 São Paulo, SP, Brazil

Eurico Cabral Oliveira Filho
Instituto de Biociências, Universidade de São Paulo, P.O. Box 11461, 05508-090 São Paulo, SP, Brazil

Ricardo Clapis Garla
Departamento de Oceanografia e Limnologia, Centro de Biociências, Universidade Federal do Rio Grande do Norte, 59072-970 Natal, RN, Brazil

Najat K. Mohammed and Mohamed S. Mazunga
Department of Physics, University of Dar es Salaam, P.O. Box 35063, Dar es Salaam, Tanzania

Gisely C. Lopes, Renata Longhini, Marcos Luciano Bruschi, and João Carlos P. deMello
Programa de Pós-Graduação em Ciências Farmacêuticas, Departamento de Farmácia, Universidade Estadual de Maringá, Avenida Colombo, 5790, 87020-900 Maringá, PR, Brazil

Paulo Victor P. dos Santos
Student of Pharmacy, Universidade Estadual de Maringá, Avenida Colombo, 5790, 87020-900 Maringá, PR, Brazil

Adriano A. S. Araújo
Departamento de Fisiologia, Universidade Federal de Sergipe, Avenida Marechal Rondon, s/n, Cidade Universitária, 49100-000 São Cristóvão, SE, Brazil